# ALTERNATIVE FUELS

# Applied Energy Technology Series

## James G. Speight, Ph.D., *Editor*

# ALTERNATIVE FUELS

## Sunggyu Lee

*Robert Iredell Professor*
*Department of Chemical Engineering*
*The University of Akron*
*Akron, Ohio*

Taylor & Francis
*Publishers since 1798*

| USA | Publishing Office: | Taylor & Francis |
| --- | --- | --- |
| | | 1101 Vermont Avenue, N.W., Suite 200 |
| | | Washington, D.C. 20005-3521 |
| | | Tel: (202) 289-2174 |
| | | Fax: (202) 289-3665 |
| | Distribution Center: | Taylor & Francis |
| | | 1900 Frost Road, Suite 101 |
| | | Bristol, PA 19007-1598 |
| | | Tel: (215) 785-5800 |
| | | Fax: (215) 785-5515 |
| UK | | Taylor & Francis Ltd. |
| | | 1 Gunpowder Square |
| | | London EC4A 3DE |
| | | Tel: 0171 583 0490 |
| | | Fax: 0171 583 0581 |

*TP360*
*.L46*
*1996*

**ALTERNATIVE FUELS**

1 2 3 4 5 6 7 8 9 0   BRBR   9 8 7 6

This book was set in Times Roman by Pro Image Corporation. The editors were Christine Williams and Mary Prescott. Cover design by Michelle Fleitz. Printing and binding by Braun-Brumfield, Inc.

A CIP catalog record for this book is available from the British Library.
∞ The paper in this publication meets the requirements of the ANSI Standard Z39.48-1984 (Permanence of Paper)

**Library of Congress Cataloging-in-Publication Data**

Lee, Sunggyu.
  Alternative fuels / Sunggyu Lee.
    p.  cm. — (Applied energy technology series)

  1. Synthetic fuels.  I. Title.  II. Series.
TP360.L46   1996
662'.66—dc20

96-11367
CIP

ISBN 1-56032-361-2

# CONTENTS

v

# PREFACE

During the past several decades, there has been considerable increase in research and development in the area of environmentally acceptable alternative fuels. More recently, emphasis has been placed on more reliable sources of energy that are environmentally acceptable and satisfy current energy consumption patterns. This trend is going to be intensified for the next several decades due to the dwindling supply of petroleum reserves, emphasis on environmentally safe use of chemical technologies, and new discoveries of high-reserve gas wells all over the world. Many books, monographs, and research articles in these areas are referenced every day; however, these sources are widely scattered in the literature and lack consistency, which increases the significance of a more comprehensive textbook or a reference book. Students, researchers, and field engineers feel the need for one comprehensive textbook that covers this topic, with special emphasis on coal utilization, synthesis gas, coal liquid, slurry fuels, energy integration, oil shale, tar sands, geothermal energy, biomass conversion, and energy from solid wastes, from the viewpoints of synthesis and process chemistry, chemical engineering, environmental engineering, process technology, and engineering design. This book caters to such an audience.

This book aims to present comprehensive information regarding science and technology in the area of alternative fuels. Special emphasis is placed on various process treatments leading to clean use of coal and coal products, synthesis gas, coal liquids, methanol and higher alcohols, hydrocarbon synthesis, fermentation ethanol, shale oil crude, and slurry fuels. More attention is given to the energy resources that are directly tied to the alleviation of petroleum dependence and cleaner and better use of existing fuel resources than to energy sources that have only limited or specific applications. Process chemistry and technology and process economics are discussed in detail. Necessary thermodynamic and process data, as well as recent advances in technology, are also presented, which adds to the consistency of the book and enhances its value as a reference text.

Chapter 1 focuses on the current concerns in the area of energy, economics, and environment and highlights the importance of further development and the use of clean and alternative fuels. It also addresses the problems faced by the world in the current era in this regard.

Chapter 2 discusses in detail the use of coal, coal derivatives, and coal products as a source of fuel and feedstock. It discusses the characterization of coal, its petrochemistry, and structure. Such scientific information is crucially important for establishing the fundamental and scientific background needed for further development of alternative fuels as well as additional technological breakthroughs. Safety and environmental issues regarding the mining, transportation, and use of coal as a fuel are also discussed in detail. Not only will researchers in the field of biomass and coal conversion find the subject interesting, but also students in engineering and applied chemistry will be able to use this chapter to broaden their background. Special emphasis is placed on the fundamental coal chemistry and terminology, since these subjects serve as background for all fossil fuel conversion studies.

Chapter 3 deals with coal gasification to produce synthesis gas. Synthesis gas is a legitimate petrochemical feedstock for various chemical processes, including synthetic hydrocarbons, methanol, and oxygenated fuels. Very detailed coverage of various gasifiers and gasification procedures is presented in this chapter. Significant process features of various gasification processes are assessed and compared.

Chapter 4 discusses in detail coal liquefaction for alternative liquid fuels. Major subjects covered include coal pyrolysis, direct liquefaction, indirect liquefaction, and coal-oil coprocessing liquefaction. Recent advances in clean liquid fuels, including catalytic two-stage liquefaction, Fischer-Tropsch, methanol and dimethyl ether synthesis, and methanol-to-gasoline, are included in this chapter.

Chapter 5 is devoted to integrated gasification combined cycle (IGCC) technology because of its rapidly growing popularity in energy industries. An overview of alternative clean coal technology in addition to detailed discussion of IGCC technology is included. Various technological, environmental, energy efficiency–related, and economic issues of energy integration schemes are discussed. This chapter presents a state-of-the-art technology review of the IGCC process and provides a great many innovative ideas on energy integration.

Chapter 6 covers various topics related to coal slurry fuels, including transportation, internal combustion, handleability, and environmental issues. Special foci are kept on the use and development of coal water slurry and coal oil slurry.

Chapter 7 discusses in detail the use of oil shale and shale oil as a source of fuel. Petrochemistry, geology, and characterization of oil shale and shale oil are also included. Special emphasis is given to kinetics, extraction, and upgrading of shale oil. Detailed discussions of extraction processes both in situ and ex situ are presented.

Chapter 8 discusses in detail the extraction, characterization, and potential for future use of tar sands as a source for clean liquid fuel.

Chapter 9 deals with utilization of geothermal energy and provides scientific and technological updates in addition to statistical information. Various processes involving geothermal power plants and direct heat use are included in the discussion. Several application areas for geothermal energy are also detailed.

Chapter 10 presents the use of biomass as a source of fuel and discusses biomass conversion reactions resulting in derivatives that are again a source of fuel and chemical feedstocks. Fuels from agricultural sources are also included in this chapter. Various treatments, including direct combustion, gasification, liquefaction, pyrolysis, anaerobic digestion, and fermentation, are discussed. In-depth coverage is given to fuel ethanol fermentation.

Chapter 11 highlights recent advances in alternative sources of energy from solid wastes, which has rapidly become an issue of importance from both environmental and energy standpoints. Energy recovery processes involving municipal solid wastes (MSW), polymeric wastes, and spent tires are covered in detail. Some very innovative process ideas are also included.

This book will be a very useful handbook for engineers and scientists working in the area of fuel science and engineering, particularly in clean coal technology, $C_1$-chemicals, oxygenates, syncrude upgrading, fermentation ethanol, heat recovery from MSW, and cleaner use of fuels. This book compiles a large amount of information and data in a comprehensive but consistent manner, making it very convenient for researchers as well as students, and is also a good referral text for professors.

This book is the first of its kind, as most books in this area are edited books that merely cover preliminary or very specific issues such as recovery of methane and its use as a fuel and conversion.

Most other books in the energy field emphasize the use of coal, synthesis gas, oil shale, and biomass as sources of alternative fuel, with little focus on the technological, environmental, economical, and political implications. Furthermore, there is no single-authored book that covers the area of process chemistry, technology, and engineering for alternative fuels with special emphasis on coal, syngas, coal liquids, clean synthetic liquid fuels, slurry fuels, IGCC, oil shale crude, fermentation alcohols, energy from MSW, and geothermal energy with consistency and comprehensiveness.

This book can be used as a textbook for a three-credit-hour course entitled Alternative Fuels; as a cotextbook for a more general course on Fuel Engineering, Gas and Coal Technology, Petrochemicals, Alternative Energy and Feedstock, Energy Conversion, or Energy and Environment; or as a reference text for engineering design courses or advanced chemistry electives. The materials in this book have been offered to chemical engineering students under the course titles Fuel Engineering and Alternative Fuels, seven times for the past 12 years at The University of Akron. A number of problems appear at the ends of several chapters. Some of the problems are quite involved, whereas others are very straightforward. For energy and chemical industries, this book can serve not only as a high-circulation reference manual but also as an individual desk handbook. For R&D scientists and engineers, this book serves as a one-volume comprehensive reference that will provide necessary information regarding chemistry, technology, and alternative routes as well as scientific foundations for further enhancement and technological breakthroughs. Otherwise, such information would be available only from a comprehensive compilation of a great amount of literature from diverse sources.

Writing a book of this nature is not a solo endeavor. It requires the support and assistance of family, friends, and colleagues. In recognition of this support, I would like to thank the following people:

First and foremost are my wife, Kyung, and my children, Tracy, JoJo, and Leroy, whose patience and encouragement provided great moral support.

Dr. James Speight of the Western Research Institute was instrumental in encouraging me to write this book. Support and encouragement were also provided by my friends Conrad Kulik and Howard Lebowitz. Colleagues at the university and from the Fuels and Petrochemicals Division of the American Institute of Chemical Engineers provided many of the technical references necessary in an undertaking of this sort.

Finally, I would like to acknowledge Dr. Brian Kocher, Mr. Timothy Tartamella, Dr. Teresa Cutright, Mr. Sanjay Shah, Dr. Shahid Bashir, Dr. Medha Joshi, Mr. Mark Brundage, Mr. Ben Lopez, Mr. Abhay Sardesai, Mr. David Tucker, Mr. Theodore Thome, Mr. H. Bryan Lanterman, and Dr. Kathy Fullerton for their help in compiling scientific and technical information. I would also like to thank Ms. Jeanine Gray and Mrs. Maria Peters for editing and assembling the manuscript. The contributions of those mentioned made this book possible.

# GLOBAL ENERGY OVERVIEW

## 1.1 U.S. ENERGY SUPPLIES

The U.S. energy production during March 1995 totaled 5.9 quadrillion Btu, which is a 1.0% increase from the level during March 1994. Crude oil and natural gas plant liquids decreased 0.9%, coal production decreased 0.3%, and natural gas production remained about the same. However, all other forms of energy production combined were up 10.4 percent from the level of production during March 1994 [1]. Such statistical information is available from the Monthly Energy Review of the Energy Information Administration (EIA). Table 1 shows the energy summary for March 1995.

Energy consumption during March 1995 totaled 7.5 quadrillion Btu, 1.3% above the level of consumption during March 1994. For the month of March 1995, one can observe that the total energy production is substantially lower than the total energy consumption for the same month, indicating that the consumption/production pattern is highly seasonal, and the storage of energy and fuel is a very important issue. Consumption of natural gas increased 2.7%, petroleum products consumption increased 0.5%, and coal consumption decreased 2.5%. Consumption of all other forms of energy combined increased 8.1% from the level of March 1994. Other forms of energy include hydroelectric and nuclear electric power and electricity generated for distribution from wood, waste, geothermal, wind, photovoltaic, and solar thermal energy.

1

## Table 1 Energy summary for March 1995 (quadrillion Btu)

| | March | | | Cumulative January through March | | | | |
| --- | --- | --- | --- | --- | --- | --- | --- | --- |
| | 1995 | 1994 | Percent change[a] | 1995 | 1995 Daily rate | 1994 | 1994 Daily rate | Percent change[a] |
| Production[b] | 5.947 | 5.886 | 1.0 | 17.394 | 0.193 | 16.698 | 0.186 | 4.2 |
| Coal | 2.045 | 2.052 | −.3 | 5.835 | .065 | 5.433 | .060 | 7.4 |
| Natural gas (dry) | 1.659 | 1.658 | .0 | 4.905 | .054 | 4.826 | .054 | 1.6 |
| Crude oil[c] and natural gas plant liquids | 1.396 | 1.409 | −.9 | 4.068 | .045 | 4.098 | .046 | −.7 |
| Other[d] | .847 | .768 | 10.4 | 2.586 | .029 | 2.342 | .026 | 10.4 |
| Consumption[b] | 7.481 | 7.384 | 1.3 | 22.909 | .255 | 23.147 | .257 | −1.0 |
| Coal | 1.557 | 1.596 | −2.5 | 4.807 | .053 | 4.992 | .055 | −3.7 |
| Natural gas[e] | 2.146 | 2.091 | 2.7 | 6.905 | .077 | 7.041 | .078 | −1.9 |
| Petroleum products[f] | 2.898 | 2.883 | .5 | 8.516 | .095 | 8.649 | .096 | −1.5 |
| Other[g] | .881 | .815 | 8.1 | 2.681 | .030 | 2.464 | .027 | 8.8 |
| Net imports | 1.565 | 1.491 | 5.0 | 4.273 | .047 | 4.313 | .048 | −.9 |
| Coal[h] | −.166 | −.141 | 17.2 | −.455 | −.005 | −.346 | −.004 | 31.7 |
| Natural gas | .223 | .199 | 12.2 | .686 | .008 | .614 | .007 | 11.8 |
| Petroleum[i] | 1.474 | 1.386 | 6.3 | 3.947 | .044 | 3.922 | .044 | .6 |
| Other[j] | .033 | .047 | −28.8 | .094 | .001 | .122 | .001 | −23.0 |

[a] Based on daily rates prior to rounding.

[b] Due to a lack of consistent historical data, some renewable energy sources are not included. For example, in 1992, 3.0 quadrillion Btu of renewable energy consumed by U.S. electric utilities to generate electricity for distribution is included, but an estimated 3.0 quadrillion Btu of renewable energy used by other sectors is not included.

[c] Includes lease condensate.

[d] "Other" is hydroelectric and nuclear electric power and electricity generated for distribution from wood, waste, geothermal, wind, photovoltaic, and solar thermal energy.

[e] Includes supplemental gaseous fuels.

[f] Products obtained from the processing of crude oil (including lease condensate), natural gas, and other hydrocarbon compounds.

[g] "Other" is hydroelectric and nuclear electric power; electricity generated for distribution from wood, waste, geothermal, wind, photovoltaic, and solar thermal energy; and net imports of electricity and coal coke.

[h] Minus sign indicates exports are greater than imports.

[i] Crude oil, lease condensate, petroleum products, pentanes plus, unfinished oils, gasoline blending components, and imports of crude oil for the Strategic Petroleum Reserve.

[j] "Other" is net imports of electricity and coal coke.

Notes: Totals may not equal sum of components due to independent rounding. Geographic coverage is the 50 states and the District of Columbia.

Source: Reference 1.

Net energy imports by the United States during March 1995 totaled 1.6 quadrillion Btu, which is 5.0% above the level of March 1994. Net imports of natural gas were up by 12.2%, and net imports of petroleum increased 6.3%. Net exports of coal rose 17.2% from the level of the previous year.

Figure 1 shows the U.S. energy consumption, production, and imports for 1973–1994. During this period, the lowest production was recorded in 1975 at 59.86 quadrillion Btu, whereas the highest production was realized in 1990 at 67.85 quadrillion Btu. For the same period, the U.S. consumption ranged from 70.52 (in 1983) to 85.57 (in 1994) quadrillion Btu, whereas the lowest and

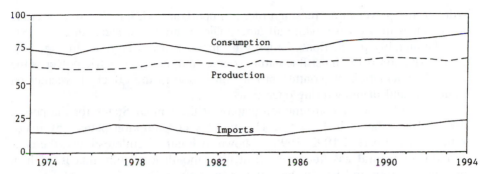

**Figure 1** Energy overview: consumption, production, and imports, 1973–1994 (units: quadrillion Btu). (Source: Reference 1.)

highest energy imports were recorded at 12.03 (in 1983) and 22.58 (in 1994) quadrillion Btu, respectively. From the statistics, one can also observe that the energy consumption is very strongly related to the economic strength of the country for a given year.

Figure 2 shows the U.S. energy production by major source for the period 1973–1994. As can be clearly seen, coal production is the most important, followed by the production of natural gas, crude oil, nuclear electric power, and

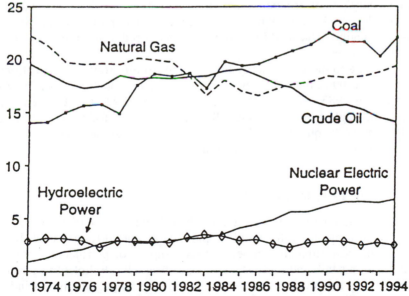

**Figure 2** U.S. energy production by major sources, 1973–1994 (units: quadrillion Btu). (Source: Reference 1.)

hydroelectric power in decreasing order of importance. U.S. production of crude oil continues to decrease, while all other major sources are increasing.

The monthly production of energy by major sources is given in Figure 3. Coal production is the most fluctuating source, and nuclear electric power is the next. The fluctuation in monthly production is due to the effect of weather on production and manufacturing operations.

The yearly energy consumption pattern of the United States for the period 1973–1994 is shown in Figure 4. The monthly consumption pattern by major sources for the period 1992–1995 is shown in Figure 5. Interestingly, the consumption of natural gas fluctuated most from month to month, that is, high in the winter season and low in the summer season.

Figures 6 and 7 show the yearly and monthly energy net imports by the United States, respectively. The net imports of petroleum products followed an up-and-down pattern, increasing (1974–1977), decreasing (1979–1982), and increasing (1985–1994). This fluctuating change may be attributed to energy policy changes, domestic crude oil price, international market price of crude oil, and international politics.

## 1.2 WORLD PRIMARY ENERGY PRODUCTION

In every phase of human life, consumption of energy in various forms is involved. Macroscopically, the energy consumption of a country is directly related to the country's total production activities as well as the living standards of the country. This is why energy resources are the most important resources for any country or region. The world has already experienced a major energy crisis and several minicrises.

According to the U.S. Energy Information Administration, International Energy Annual (1994), the total world primary energy production was 278.78 quadrillion Btu in 1983 and 343.10 quadrillion Btu in 1992. The U.S. energy production was highest at 63.79 quadrillion Btu in 1992, followed by the former USSR with 57.88 quadrillion Btu in 1992. Table 2 shows the world primary energy production by region and country [2].

## 1.3 CRUDE OIL RESERVES

Since the world energy crisis ignited by the oil embargo in 1973, we have been constantly hearing about the world's rapidly depleting oil reserves. We have also heard about the estimated number of years' supply of petroleum from various sources. All this talk and these statistics have been closely linked to government policy making, programs for alternative fuels, long-range planning of industries, and so forth. Table 3 shows the estimated proved world crude oil reserves in billions of barrels.

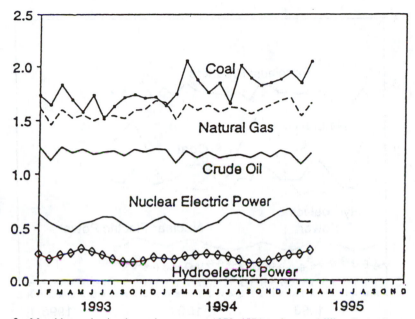

**Figure 3** Monthly production by major sources, 1973–1994 (units: quadrillion Btu). (Source: Reference 1.)

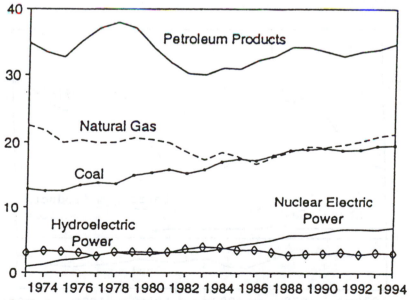

**Figure 4** Yearly energy consumption patterns in the United States (units: quadrillion Btu). (Source: Reference 1.)

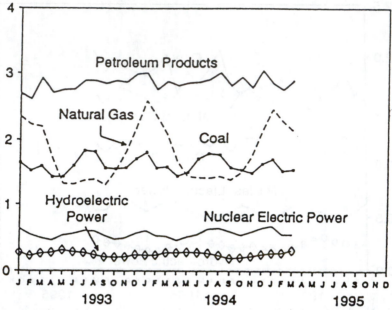

**Figure 5** Monthly energy consumption patterns in the United States (units: quadrillion Btu). (Source: Reference 1.)

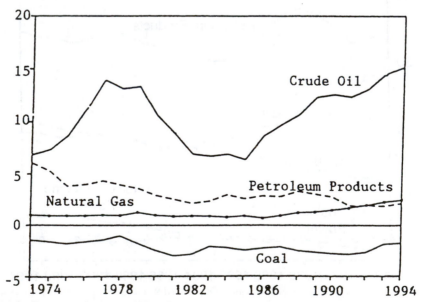

**Figure 6** Energy net imports, 1973–1994 (units: quadrillion Btu). (Source: Reference 1.)

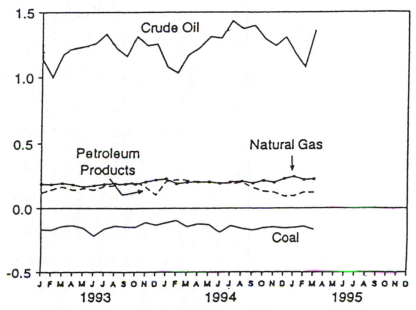

**Figure 7**  Monthly energy net imports (units: quadrillion Btu). (Source: Reference 1.)

Interestingly, Table 3 shows that the estimated proven world crude oil reserves have been increasing: from 338.67 billion barrels in 1965 to 999.76 billion barrels in (1995), which is a threefold increase. This is due to advances in recovery processes as well as increased activities in geological exploration.

Table 4 shows statistical data for world crude oil production by regions.

As can be seen from Table 4, worldwide crude oil production has remained fairly constant at a level of 21 to 22 billion barrels per year. Considering the current proven crude oil reserves of the world and the current rate of worldwide crude oil production, the currently proven reserve will last 45 years, if 100% recovery is assumed. This number inspires much optimism about the future petroleum supply. The same calculation would have given 30 (1965), 33 (1970), 37 (1975), 19 (1980), 35 (1985), 47 (1990), and 45 years (1995), respectively. The crude oil reserves and the estimated supply period of petroleum will undoubtedly affect the direction of alternative fuel development.

## 1.4 SUPPLIES OF NATURAL GAS IN THE UNITED STATES

The total dry natural gas production in the United States during April 1995 was an estimated 1.6 trillion cubic feet, which is 1% higher than the production during the previous April. However, consumption of natural and supplemental gas in April 1995 was 1.7 trillion cubic feet, 5% above the level in April 1994.

## Table 2 World primary energy production[a] (quadrillion Btu)

| Region and country | 1973 | 1974 | 1975 | 1976 | 1977 | 1978 | 1979 | 1980 | 1981 | 1982 | 1983 | 1984 | 1985 | 1986 | 1987 | 1988 | 1989 | 1990 | 1991 | 1992 |
|---|---|---|---|---|---|---|---|---|---|---|---|---|---|---|---|---|---|---|---|---|
| Canada | 9.50 | 9.38 | 8.89 | 8.68 | 9.03 | 9.14 | 9.99 | 10.04 | 9.75 | 9.66 | 10.14 | 11.01 | 11.80 | 11.71 | 12.32 | 13.17 | 13.10 | 13.07 | 13.96 | 14.36 |
| Mexico | 1.87 | 2.15 | 2.44 | 2.74 | 3.14 | 3.76 | 4.51 | 5.80 | 6.78 | 7.82 | 7.70 | 7.88 | 7.74 | 7.07 | 7.28 | 7.33 | 7.37 | 7.56 | 7.78 | 7.76 |
| United States | 61.95 | 60.72 | 59.73 | 59.81 | 60.14 | 61.03 | 63.71 | 64.64 | 64.30 | 63.79 | 61.08 | 65.71 | 64.58 | 64.05 | 64.62 | 65.79 | 65.85 | 67.66 | 67.36 | 66.68 |
| Total North America | 73.32 | 72.25 | 71.06 | 71.22 | 72.31 | 73.93 | 78.22 | 80.48 | 80.83 | 81.27 | 78.92 | 84.60 | 84.12 | 82.82 | 84.22 | 86.30 | 86.31 | 88.29 | 89.10 | 88.80 |
| Venezuela | 8.13 | 7.29 | 5.90 | 5.79 | 5.69 | 5.52 | 6.04 | 5.71 | 5.58 | 5.22 | 5.00 | 5.02 | 4.78 | 5.17 | 5.14 | 5.59 | 5.74 | 6.38 | 6.86 | 6.80 |
| Total Central and South America | 12.85 | 12.09 | 10.62 | 10.67 | 10.88 | 11.19 | 12.23 | 12.15 | 12.13 | 12.01 | 12.20 | 13.14 | 13.44 | 14.18 | 14.08 | 15.04 | 15.52 | 16.32 | 17.35 | 17.60 |
| France | 1.78 | 1.82 | 1.88 | 1.71 | 2.00 | 2.02 | 2.03 | 2.26 | 2.65 | 2.61 | 2.96 | 3.37 | 3.54 | 3.81 | 3.97 | 4.09 | 4.09 | 4.20 | 4.49 | 4.66 |
| Germany[b] | 4.92 | 4.95 | 4.87 | 5.14 | 5.12 | 5.12 | 5.44 | 5.44 | 5.57 | 7.47 | 7.41 | 7.78 | 8.27 | 8.02 | 7.87 | 8.11 | 7.90 | 7.29 | 6.31 | 6.14 |
| Italy | 1.06 | 1.07 | 1.08 | 1.11 | 1.15 | 1.13 | 1.10 | 1.07 | 1.11 | 1.16 | 1.11 | 1.17 | 1.18 | 1.26 | 1.19 | 1.27 | 1.20 | 1.19 | 1.28 | 1.32 |
| Netherlands | 2.52 | 2.86 | 3.16 | 3.44 | 2.90 | 2.49 | 2.69 | 3.32 | 3.10 | 2.67 | 2.62 | 2.71 | 2.82 | 2.71 | 2.78 | 2.55 | 2.63 | 2.63 | 2.94 | 2.91 |
| Norway | 0.83 | 0.88 | 1.22 | 1.47 | 1.47 | 2.07 | 2.68 | 3.02 | 3.06 | 3.12 | 3.42 | 3.66 | 3.83 | 3.98 | 4.48 | 4.74 | 5.77 | 5.96 | 6.22 | 6.80 |
| United Kingdom | 4.66 | 4.47 | 4.97 | 5.39 | 6.58 | 7.19 | 8.24 | 8.40 | 8.71 | 9.44 | 9.85 | 8.78 | 10.11 | 10.55 | 10.24 | 9.96 | 8.95 | 8.78 | 9.29 | 9.23 |
| Total Western Europe | 19.58 | 20.01 | 21.36 | 22.34 | 23.93 | 24.81 | 27.40 | 28.65 | 29.73 | 32.42 | 33.75 | 34.43 | 37.19 | 38.02 | 38.59 | 39.10 | 38.58 | 38.30 | 38.96 | 39.31 |
| Iran | 13.27 | 13.68 | 12.23 | 13.43 | 12.88 | 11.93 | 7.46 | 3.91 | 3.28 | 5.12 | 5.67 | 5.29 | 5.57 | 5.06 | 5.66 | 5.70 | 7.01 | 7.67 | 8.28 | 8.53 |
| Iraq | 4.35 | 4.25 | 4.89 | 5.25 | 5.08 | 5.53 | 7.49 | 5.45 | 2.16 | 2.19 | 2.17 | 2.61 | 3.09 | 3.66 | 4.58 | 5.97 | 6.47 | 4.54 | 0.69 | 1.01 |
| Kuwait | 6.75 | 5.67 | 4.68 | 4.84 | 4.46 | 4.90 | 5.78 | 3.88 | 2.70 | 1.98 | 2.51 | 2.76 | 2.44 | 3.36 | 3.77 | 6.34 | 4.32 | 2.83 | 0.43 | 2.48 |
| Saudi Arabia | 16.63 | 18.68 | 15.68 | 18.88 | 20.51 | 18.45 | 21.24 | 22.48 | 22.57 | 14.86 | 11.69 | 11.29 | 8.55 | 11.91 | 10.73 | 12.73 | 12.68 | 15.92 | 19.75 | 20.63 |
| United Arab Emirates | 3.27 | 3.56 | 3.56 | 4.20 | 4.45 | 4.09 | 4.12 | 3.89 | 3.45 | 3.00 | 2.91 | 3.00 | 3.29 | 3.68 | 4.21 | 4.25 | 4.99 | 5.51 | 6.27 | 6.23 |
| Total Middle East | 46.61 | 48.33 | 43.58 | 49.13 | 49.77 | 47.30 | 48.65 | 42.13 | 36.73 | 29.54 | 27.29 | 27.65 | 25.66 | 30.62 | 32.09 | 36.04 | 39.61 | 41.02 | 40.23 | 43.94 |

| | | | | | | | | | | | | | | | | | | | |
|---|---|---|---|---|---|---|---|---|---|---|---|---|---|---|---|---|---|---|---|
| Algeria | 2.47 | 2.35 | 2.34 | 2.65 | 2.63 | 3.28 | 3.16 | 2.75 | 2.95 | 3.11 | 3.46 | 3.71 | 3.77 | 3.55 | 4.01 | 4.02 | 4.28 | 4.52 | 4.81 | 4.83 |
| Libya | 4.72 | 3.32 | 3.27 | 4.26 | 4.53 | 4.40 | 4.63 | 4.03 | 2.57 | 2.61 | 2.52 | 2.53 | 2.46 | 2.43 | 2.29 | 2.73 | 2.69 | 3.18 | 3.44 | 3.46 |
| Nigeria | 4.45 | 4.89 | 3.88 | 4.51 | 4.50 | 4.06 | 4.95 | 4.50 | 3.18 | 2.86 | 2.77 | 3.12 | 3.35 | 3.30 | 3.04 | 3.29 | 3.88 | 4.05 | 4.26 | 4.49 |
| South Africa | 1.48 | 1.57 | 1.65 | 1.84 | 2.03 | 2.15 | 2.46 | 2.74 | 3.09 | 3.24 | 3.45 | 3.87 | 4.17 | 4.26 | 4.23 | 4.39 | 4.28 | 4.19 | 4.19 | 5.06 |
| Total Africa | 14.81 | 13.93 | 13.19 | 15.49 | 16.44 | 16.70 | 18.30 | 17.35 | 15.09 | 15.43 | 16.12 | 17.52 | 18.43 | 18.14 | 18.53 | 19.54 | 20.47 | 21.46 | 23.39 | 23.70 |
| Australia | 3.25 | 3.44 | 3.64 | 3.35 | 3.43 | 3.51 | 3.66 | 3.53 | 3.89 | 4.04 | 4.24 | 4.41 | 5.33 | 5.49 | 6.02 | 5.81 | 6.15 | 6.71 | 6.30 | 6.66 |
| China | 13.14 | 14.25 | 15.11 | 15.35 | 16.16 | 18.05 | 18.53 | 18.32 | 18.10 | 19.14 | 20.47 | 22.39 | 21.60 | 25.34 | 26.25 | 27.50 | 29.15 | 29.29 | 29.89 | 30.18 |
| India | 2.22 | 2.36 | 2.68 | 2.88 | 2.95 | 3.14 | 3.30 | 3.35 | 4.09 | 4.43 | 4.88 | 5.22 | 5.53 | 5.97 | 5.71 | 5.97 | 6.53 | 6.96 | 6.71 | 6.94 |
| Indonesia | 2.90 | 2.99 | 2.90 | 3.38 | 3.85 | 3.71 | 3.83 | 4.16 | 4.27 | 3.68 | 3.84 | 4.27 | 4.24 | 4.33 | 4.36 | 4.49 | 4.86 | 5.14 | 5.70 | 6.09 |
| Total Far East and Oceania | 26.17 | 28.04 | 29.71 | 30.59 | 32.09 | 34.53 | 35.99 | 36.53 | 37.47 | 38.88 | 41.57 | 45.18 | 49.50 | 51.56 | 53.24 | 54.94 | 58.26 | 59.93 | 60.94 | 62.75 |
| Poland | 4.94 | 5.09 | 5.37 | 5.00 | 5.18 | 5.36 | 5.51 | 5.28 | 4.54 | 5.16 | 5.25 | 5.37 | 5.54 | 5.72 | 5.79 | 5.87 | 5.49 | 4.72 | 3.78 | 3.74 |
| Former USSR | 39.28 | 41.77 | 43.62 | 47.46 | 49.79 | 52.11 | 53.88 | 55.67 | 56.71 | 57.88 | 59.27 | 61.30 | 63.17 | 65.73 | 67.61 | 69.82 | 69.94 | 68.86 | 62.51 | NA |
| Total Eastern Europe and former USSR | 51.44 | 54.17 | 56.51 | 60.23 | 62.95 | 65.40 | 67.52 | 69.20 | 69.67 | 69.23 | 70.80 | 73.02 | 75.04 | 77.92 | 79.86 | 81.99 | 81.76 | 79.08 | 70.80 | 67.01 |
| Total World | 244.78 | 248.81 | 246.03 | 259.67 | 268.37 | 273.87 | 288.31 | 286.50 | 281.66 | 278.78 | 280.65 | 295.54 | 303.37 | 313.26 | 320.61 | 332.97 | 340.51 | 344.41 | 340.76 | 343.10 |

[a] Includes only crude oil, lease condensate, natural gas plant liquids, dry natural gas, coal net hydroelectric power, and net nuclear power.
[b] Beginning with 1982, Germany includes both the former East Germany and the former West Germany.

*Source:* U.S. Energy Information Administration. International Energy Annual, 1994.

**Table 3 Estimated proven world crude oil reserves annually as of January 1 (billions of barrels)**

| Region or country | 1965 | 1970 | 1975 | 1980 | 1985 | 1990 | 1995 |
|---|---|---|---|---|---|---|---|
| United States | 30.99 | 29.63 | 34.25 | 29.81 | 28.45 | 26.50 | 22.96 |
| Canada | 6.18 | 8.62 | 7.17 | 6.80 | 7.08 | 6.13 | 5.04 |
| Latin America | 25.53 | 29.18 | 40.58 | 56.47 | 83.32 | 125.03 | 129.07 |
| Middle East | 212.18 | 333.51 | 403.86 | 361.95 | 398.38 | 660.25 | 660.29 |
| Africa | 19.40 | 54.68 | 68.30 | 57.07 | 55.54 | 58.84 | 62.18 |
| Australia and Asia | 11.61 | 13.14 | 21.05 | 19.36 | 18.53 | 22.55 | 44.45 |
| Western Europe | 2.04 | 1.78 | 25.81 | 23.48 | 24.43 | 18.82 | 16.57 |
| Communist nations | 30.75 | 60.00 | 111.40 | 90.00 | 84.10 | 84.10 | NA |
| USSR and Eastern Europe | NA | NA | NA | NA | NA | NA | 59.20 |
| Total | 333.67 | 530.53 | 712.42 | 644.93 | 699.81 | 1002.21 | 999.76 |

*Sources:* 1965–1975. US. American Petroleum Institute; 1980–1990. US Energy Information Administration; 1995. *Oil and Gas Journal.*

Imports of natural gas in April 1995 were 225 billion cubic feet, 10% higher than imports in April 1994. Stocks of working gas (gas available for withdrawal) in underground natural gas storage reservoirs at the end of April 1995 totaled 1.4 trillion cubic feet, 17% above the level of stocks available.

Tables 5 and 6 show U.S. natural gas production data for the period 1973 through 1994. Several important facts can be pointed out:

1. For the 20-year period, the yearly gross withdrawals changed from 24,067 (1973) to 18,659 (1983) billion cubic feet.
2. Viewing the 1994 statistics, the monthly gross drawings of natural gas stayed at a relatively constant level, about 2000 billion ft$^3$ per month.

In Tables 5 and 6, the following terms are used [3]:

Total dry gas production = total wet gas production − extraction loss
Total wet gas production = total marketed production
= gross withdrawals − repressuring
− nonhydrocarbon gases removed
− vented and flared

In general, the total dry gas production is approximately 80% of the gross withdrawals of natural gas.

Figure 8 shows the U.S. consumption, dry production, and imports of natural gas for the period 1973–1994. Figure 9 shows the monthly data for consumption,

**Table 4 World crude oil production by area
(billion barrels)**

| Region or country | 1965 | 1970 | 1975 | 1980 | 1985 | 1990 | 1993 |
|---|---|---|---|---|---|---|---|
| United States | 2.85 | 3.52 | 3.05 | 3.15 | 3.27 | 2.68 | 2.52 |
| Canada | 0.30 | 0.46 | 0.52 | 0.52 | 0.53 | 0.57 | 0.61 |
| Latin America | 1.68 | 1.92 | 1.60 | 2.04 | 2.28 | 2.49 | 2.67 |
| Western Europe | 0.15 | 0.14 | 0.20 | 0.89 | 1.37 | 1.53 | 1.70 |
| Middle East | 3.05 | 5.09 | 7.16 | 6.74 | 3.76 | 6.00 | 6.69 |
| Africa | 0.81 | 2.22 | 1.83 | 2.23 | 1.77 | 2.17 | 2.25 |
| Asia | 0.25 | 0.50 | 0.81 | 1.00 | 1.14 | 2.26 | 2.37 |
| Communist nations | 1.98 | 2.87 | 4.33 | 5.20 | 5.35 | NA | NA |
| USSR | NA | NA | NA | NA | NA | 3.96 | 2.86 |
| Total | 11.06 | 16.72 | 19.50 | 21.76 | 19.49 | 21.66 | 21.67 |

*Sources*: 1965–1970. International Petroleum Annual; 1975. USEIA. World Crude Oil Production Annual; 1980–1990. USEIA. Petroleum Supply Annual; 1993. *Oil and Gas Journal*. Worldwide Production Report Issue.

dry production, and imports for 1993–1995. It is interesting to note that the dry production and imports do not fluctuate monthly, whereas consumption varies seasonally. Figures 10 and 11 show the breakdown of natural gas consumption by different sectors. Whereas the industrial consumption fluctuates substantially from year to year, the residential consumption stays nearly constant. Figures 12 and 13 show the yearly and monthly data for underground storage. The total underground storage by the end of 1994 reached a level of 7 trillion cubic feet.

Table 7 shows the data for the U.S. natural gas trade by country. The data indicate that U.S. imports of natural gas depend mainly on Canada.

## 1.5 COAL OVERVIEW

Figure 14 shows the U.S. production, consumption, and net exports of coal in million short tons for the period 1973–1994 [1]. Figure 15 shows the pattern of U.S. production, consumption, and net exports of coal on a monthly basis for 1992–1995. Both production and consumption on a monthly basis fluctuate noticeably, even though the peaks for the two do not match. Figure 16 shows the breakdown for coal consumption by sector for 1973–1994. Remarkably, coal consumption by electric utilities has grown consistently at a steep rate. For the past 20 years, the coal consumption by electric utilities has doubled, whereas all other uses have practically remained at very low levels.

The U.S. coal production and consumption for 1993 were 945 and 926 million short tons, respectively. Table 8 presents the world coal production data for the period 1980–1992 [2]. The U.S. production of coal accounted for 25% of the world production in 1992. In the United States, coal is becoming more

### Table 5 U.S. national gas production (billion cubic feet)

| Year | Gross withdrawals | Total dry gas production |
|------|------|------|
| 1973 | 24,067 | 21,731 |
| 1975 | 21,104 | 19,236 |
| 1977 | 21,097 | 19,163 |
| 1979 | 21,883 | 19,663 |
| 1981 | 21,587 | 19,181 |
| 1983 | 18,659 | 16,094 |
| 1985 | 19,607 | 16,454 |
| 1987 | 20,140 | 16,621 |
| 1989 | 21,074 | 17,311 |
| 1991 | 21,750 | 17,698 |
| 1992 | 22,132 | 17,840 |
| 1993 | 22,729 | 18,244 |
| 1994 | 23,679 | 18,244 |

*Source*: Energy Information Administration *Natural Gas Monthly*, June 1995.

### Table 6 1994 U.S. monthly natural gas production (billion cubic feet)

| Month | Gross withdrawals | Total dry gas production |
|------|------|------|
| January | 2,045 | 1,623 |
| February | 1,843 | 1,462 |
| March | 2,037 | 1,614 |
| April | 1,943 | 1,552 |
| May | 2,003 | 1,597 |
| June | 1,906 | 1,533 |
| July | 1,965 | 1,579 |
| August | 1,951 | 1,568 |
| September | 1,890 | 1,516 |
| October | 1,987 | 1,562 |
| November | 2,014 | 1,599 |
| December | 2,096 | 1,640 |
| Total | 23,679 | 18,845 |

*Source*: Energy Information Administration *Natural Gas Monthly*, June 1995.

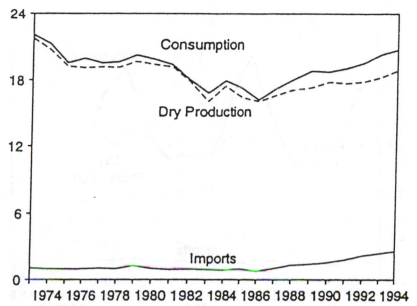

**Figure 8** Overview of natural gas consumption, dry production, and imports, 1973–1994 (units: trillion cubic feet). (Source: Reference 1.)

and more important as fuel for electricity generation. The abundance of coal in the United States makes coal strategically most important. The issues have been environment-friendly utilization of coal and alternative use of coal in an unconventional fashion. With dwindling supplies of petroleum, coal and natural gas are likely to become more important as petrochemical feedstocks and starting materials for transportation fuels, especially via synthesis gas technology.

## 1.6 NUCLEAR ENERGY

On March 31, 1995, there were 109 operable nuclear generating units in the United States, with a collective summer capacity of 99.0 million kilowatts of electricity. In March 1995, U.S. nuclear generating units produced a total of 52 net tetrawatt-hours (billion kilowatt-hours) of electricity, that is, 6% more than in March 1994. Nuclear units generated at an average capacity factor of 70.4%, which is four percentage points higher than in March 1994. Nuclear power supplied 22.2% of the total utility-generated electricity in March 1995, compared with 21.1% in March 1994. As of March 31, 1995, there were 116 domestic nuclear generating units in all stages of construction and operation. Figure 17 shows the net generation of electricity by sources for the period 1973–1994.

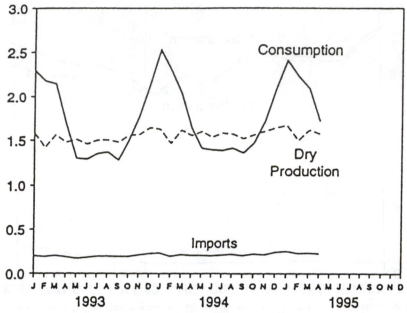

**Figure 9** Monthly overview of natural gas consumption, dry production, and imports, 1973–1994 (units: trillion cubic feet). (Source: Reference 1.)

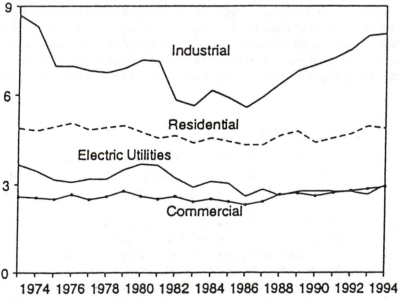

**Figure 10** Natural gas consumption by major sectors, 1973–1994 (units: trillion cubic feet). (Source: Reference 1.)

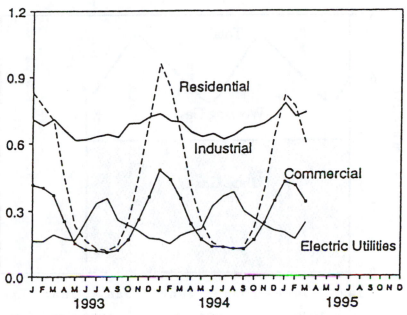

**Figure 11**  Monthly consumption of natural gas by major sectors (units: trillion cubic feet). (Source: Reference 1.)

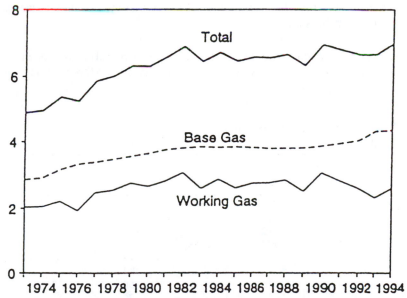

**Figure 12**  Year-end underground storage of natural gas, 1973–1994 (units: trillion cubic feet). (Source: Reference 1.)

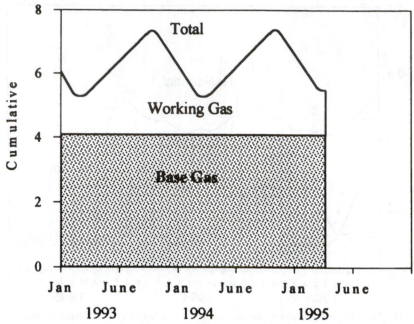

**Figure 13** Month-end underground storage of natural gas, 1973–1994 (units: trillion cubic feet). (Source: Reference 1.)

Figure 18 shows the monthly variation of capacity factor. Based on these figures, the following can be said:

1. The nuclear portion of domestic electricity net generation has been slightly increasing, from 19.8% (March 1993), to 21.1% (March 1994), and to 22.2% (March 1995).

**Table 7 The U.S. natural gas trade (billion cubic feet)**

| Country | Imports 1974 | Imports 1984 | Imports 1994 | Exports 1974 | Exports 1984 | Exports 1994 |
|---|---|---|---|---|---|---|
| Canada | 959 | 755 | 2500 | | | |
| Algeria | 0 | 36 | 51 | | | |
| Other | 0 | 52 | 7 | | | |
| Total | 959 | 843 | 2558 | | | |
| Canada | | | | 13 | | 48 |
| Mexico | | | | 13 | 2 | 47 |
| Japan | | | | 50 | 53 | 63 |
| Total | | | | 76 | 55 | 157 |

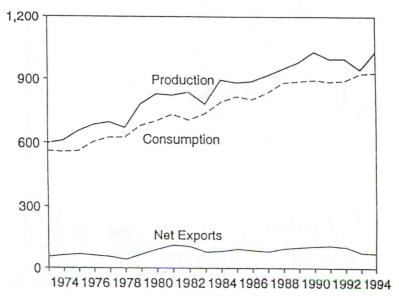

**Figure 14** U.S. production, consumption, and net imports of coal (million short tons). (Source: Reference 1.)

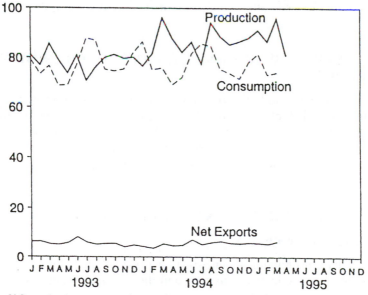

**Figure 15** U.S. production, consumption, and net exports of coal on a monthly basis for 1992–1995 (million short tons). (Source: Reference 1.)

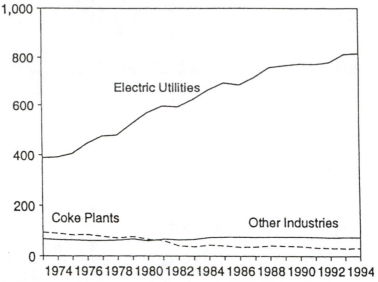

**Figure 16** Breakdown of coal consumption by sector for 1973–1994 (million short tons). (Source: Reference 1.)

2. The capacity factor fluctuates quite significantly from month to month. The average capacity factor is approximately 70%.
3. The number of generating units has not increased since 1989.

## 1.7 ELECTRIC POWER GENERATION FROM RENEWABLE ENERGY SOURCES

Table 9 shows the world electric power generation from renewable energy sources in 1990 [5]. Table 10 shows the data for direct use of geothermal power worldwide. Table 11 shows the statistics for destination of solar thermal collector shipments by state or territory in 1989. Tables 12 and 13 show the wind power generation data for the world and for California, respectively.

## 1.8 ENERGY, ENVIRONMENT, AND ECONOMY

### 1.8.1 Energy Appetite

The per capita consumption of energy is a good indicator of the standard of living, the level of productive activities, and the general strength of economy of a country.

**Table 8 World coal production (million short tons)**

| Region and country | 1980 | 1981 | 1982 | 1983 | 1984 | 1985 | 1986 | 1987 | 1988 | 1989 | 1990 | 1991 | 1992(p) |
|---|---|---|---|---|---|---|---|---|---|---|---|---|---|
| Canada | 40 | 44 | 47 | 50 | 63 | 67 | 64 | 67 | 78 | 78 | 75 | 78 | 72 |
| Mexico | 8 | 8 | 8 | 10 | 10 | 9 | 9 | 12 | 12 | 11 | 12 | 5 | 6 |
| United States | 830 | 824 | 838 | 782 | 896 | 884 | 890 | 919 | 950 | 981 | 1029 | 996 | 998 |
| Total North America | 878 | 876 | 894 | 841 | 969 | 960 | 964 | 998 | 1040 | 1069 | 1116 | 1080 | 1076 |
| Total Central and South America | 16 | 14 | 16 | 16 | 18 | 20 | 23 | 26 | 28 | 33 | 33 | 38 | 38 |
| Germany | NA | NA | 551 | 545 | 560 | 575 | 565 | 541 | 549 | 538 | 504 | 388 | 346 |
| West Germany | 239 | 241 | NA | NA | NA | NA | NA | NA | NA | NA | NA | NA | NA |
| United Kingdom | 141 | 138 | 137 | 127 | 55 | 104 | 119 | 115 | 117 | 111 | 98 | 106 | 94 |
| Yugoslavia | 52 | 58 | 60 | 65 | 72 | 75 | 77 | 78 | 78 | 82 | 84 | 74 | 74 |
| Total Western Europe | 543 | 561 | 881 | 880 | 836 | 914 | 927 | 904 | 906 | 903 | 863 | 738 | 682 |
| South Africa | 127 | 144 | 151 | 161 | 179 | 192 | 195 | 195 | 200 | 194 | 191 | 197 | 192 |
| Total Africa | 133 | 149 | 157 | 167 | 184 | 198 | 201 | 202 | 208 | 202 | 200 | 205 | 201 |
| Australia | 116 | 130 | 140 | 146 | 153 | 186 | 187 | 209 | 198 | 216 | 231 | 236 | 249 |
| China | 684 | 683 | 734 | 787 | 870 | 961 | 985 | 1023 | 1080 | 1162 | 1162 | 1213 | 1226 |
| India | 125 | 142 | 148 | 158 | 168 | 173 | 188 | 207 | 208 | 230 | 250 | 249 | 263 |
| Total Middle East, Far East, and Oceania | 1036 | 1071 | 1137 | 1205 | 1307 | 1447 | 1489 | 1571 | 1616 | 1753 | 1778 | 1839 | 1886 |
| Czechoslovakia | 136 | 137 | 139 | 140 | 143 | 140 | 139 | 137 | 137 | 130 | 119 | 111 | 102 |
| East Germany | 285 | 294 | NA | NA | NA | NA | NA | NA | NA | NA | NA | NA | NA |
| Poland | 254 | 219 | 250 | 258 | 267 | 275 | 286 | 290 | 294 | 275 | 237 | 231 | 227 |
| Former USSR | 790 | 703 | 713 | 707 | 700 | 714 | 741 | 750 | 761 | 761 | 775 | 624 | NA |
| Total Eastern Europe and Former USSR | 1566 | 1456 | 1211 | 1209 | 1224 | 1243 | 1282 | 1291 | 1312 | 1296 | 1229 | 1053 | 1092 |
| Total world | 4173 | 4128 | 4296 | 4319 | 4538 | 4783 | 4887 | 4994 | 5112 | 5258 | 5220 | 4952 | 4975 |

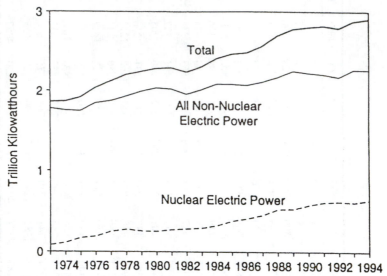

**Figure 17**   Net generation of electricity by source for 1973–1994. (Source: Reference 1.)

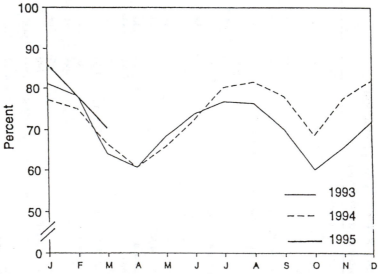

**Figure 18**   Monthly variation of capacity factor. (Source: Reference 1.)

**Table 9 World electric power generation from renewable energy sources in 1990**

| Energy source | Production (megawatt-hours) | | | Percent of renewable energy generation | Percent change 1989 to 1990 |
|---|---|---|---|---|---|
| | Nonutility producers | Utility producers | Total | | |
| Biomass (total) | 43,500,387 | 2,067,270 | 45,567,657 | 12.9 | 6.1 |
| Agricultural waste | 2,124,853 | —[a] | — | — | — |
| Landfill gas | 1,918,527 | — | — | — | — |
| Municipal solid waste | 9,470,190 | — | — | — | — |
| Subtotal (agricultural waste, landfill gas, and municipal solid waste) | 13,513,570 | 1,062,351 | 14,575,921 | 4.1 | 19.7 |
| Wood | 29,986,817 | 1,004,919 | 30,991,736 | 8.8 | 0.7 |
| Geothermal | 6,927,193 | 8,581,228 | 15,508,421 | 4.4 | 5.1 |
| Hydroelectric | 9,795,720 | 279,893,000 | 289,688,720 | 81.8 | 5.3 |
| Solar | 681,261 | 2,448 | 683,709 | 0.2 | 40.7 |
| Wind | 2,529,920 | 398 | 2,530,318 | 0.7 | 14.4 |
| Total | 63,434,481 | 290,544,344 | 353,978,825 | 100.0 | 5.5 |

[a]A dash indicates that the data are not available.

*Source*: Adapted by World Information Systems from Edison Electric Institute, *1990 Capacity and Generation of Non-Utility Sources of Energy*. Washington, DC: Edison Electric Institute, 1991, and data from the U.S. Energy Information Administration.

Human appetites for energy have constantly developed in a direction of better quality and greater amount. The trend of human preferences in energy can be summarized as follows:

1. Humans try to reduce individual human efforts by using generated energy.
2. The convenience of handling and transportation is considered more important than the conversion efficiency.
3. Governments are emphasizing environmentally friendly utilization of energy.
4. Humans are reluctant to switch from one fuel to another if they are quite different in phases, colors, odors, chemical structures, and so on.
5. Public awareness of environmental problems has been constantly improving.
6. Public acceptance of nuclear energy is still not very positive, even though it has been improving [4].

## 1.8.2 Environmental Issues

All phases of energy production, distribution, and consumption involve environmental issues. The issues include: (1) mining-related problems; (2) gener-

**Table 10 Direct use of geothermal power worldwide (megawatts of heat)**

| Country | 1985 | 1990 |
|---|---|---|
| Africa | | |
|   Algeria | — | 13 |
|   Ethiopia | — | 38 |
|   Tunisia | — | 90 |
| Asia | | |
|   China | 393 | 2,143 |
|   Japan | 2686 | 3,321 |
|   Philippines | 1 | — |
|   Taiwan | 11 | 11 |
|   Thailand | — | 0.4 |
| Central and South America | | |
|   Colombia | 12 | 12 |
|   Guatemala | — | 10 |
|   Mexico | 28 | — |
| Europe | | |
|   Austria | 4 | 4 |
|   Belgium | — | 93 |
|   Bulgaria | — | 293 |
|   Czechoslovakia | 24 | 105 |
|   Denmark | 1 | 1 |
|   France | 300 | 337 |
|   Germany | 6 | 8 |
|   Greece | — | 18 |
|   Hungary | 1001 | 1,276 |
|   Iceland | 889 | 774 |
|   Italy | 288 | 329 |
|   Poland | 9 | 9 |
|   Romania | 251 | 251 |
|   Switzerland | 23 | 23 |
|   Turkey | 166 | 246 |
|   Yugoslavia | 10 | 112.7 |
| North America | | |
|   Canada | 2 | 2 |
|   United States | 339 | 463 |
| Oceania | | |
|   Australia | 11 | 11 |
|   New Zealand | 215 | 258 |
| USSR | 402 | 1,133 |
| Total | 7072 | 11,385.1 |

**Table 11  Solar thermal collector shipments by state in 1989**

| Destination | Total shipments[a] (square feet) | Destination | Total Shipments (square feet) |
|---|---|---|---|
| Alaska | 0 | Nevada | 27,216 |
| Alabama | 4,372 | New Hampshire | 387 |
| Arizona | 199,016 | New Jersey | 47,653 |
| Arkansas | 0 | New Mexico | 41,076 |
| California | 6,907,148 | New York | 49,366 |
| Colorado | 12,891 | North Carolina | 1,465 |
| Connecticut | 58 | North Dakota | 0 |
| Delaware | 0 | Ohio | 7,286 |
| District of Columbia | 0 | Oklahoma | 256 |
| Florida | 3,117,225 | Oregon | 87,525 |
| Georgia | 2,270 | Pennsylvania | 10,418 |
| Hawaii | 82,192 | Puerto Rico | 231,463 |
| Idaho | 0 | Rhode Island | 35 |
| Illinois | 25,541 | South Carolina | 310 |
| Indiana | 0 | South Dakota | 0 |
| Iowa | 192 | Tennessee | 7,286 |
| Kansas | 0 | Texas | 38,195 |
| Kentucky | 0 | Utah | 699 |
| Louisiana | 0 | Vermont | 67 |
| Maine | 24,598 | Virgin Islands | 0 |
| Maryland | 37 | Virginia | 26,937 |
| Massachusetts | 8,333 | Washington | 309 |
| Michigan | 19,306 | West Virginia | 310 |
| Minnesota | 0 | Wisconsin | 9,780 |
| Mississippi | 0 | Wyoming | 4,372 |
| Missouri | 25,320 | | |
| Montana | 0 | Total: United States | 11,020,910 |
| Nebraska | 0 | Total: Exports | 461,017 |
| | | Total: United States and Exports | 11,481,981 |

[a] The totals may not equal the sum of the components because of independent rounding.

Source: U.S. Energy Information. Administration. *Solar Collector Manufacturing Activity: 1989* (DOE/EIA-0174). Washington, DC: U.S. Energy Information Administration. 1991. p. 7.

ation site contamination; (3) accidental spills; (4) emissions of environmentally hazardous substances; (5) safety hazards in mining, transportation, and in end uses; (6) underground waterway contamination; (7) particulates and trace minerals; (8) treatment of process wastes and wastewater.

Because of much stricter environmental regulations, environmentally acceptable, benign, or friendly utilization of energy has become more important than ever. The Clean Air Act Amendment of 1990 is a good example of such regulations, which initiated several programs related to the preservation of the environment. Examples are control of $SO_x$ and $NO_x$ emissions and the oxygenated fuel program.

### Table 12 World wind power generation

| Country | Energy output (million kilowatt-hours) | Percent of total |
|---|---|---|
| Denmark | 730 | 19.4 |
| Germany | 90 | 2.4 |
| India | 40 | 1.1 |
| Netherlands | 90 | 2.4 |
| United States | 2777 | 73.9 |
| California | 2742 | 73.0 |
| Altamont Pass | 1100 | 29.3 |
| Palm Springs | 580 | 15.4 |
| Solano | 100 | 2.7 |
| Tehachapi Pass | 962 | 25.6 |
| Hawaii | 35 | 0.9 |
| Other | 30 | 0.8 |
| Total | 3757 | 100.0 |

*Source*: Reference 5.

### Table 13 Wind power generation in California

| Year | Energy output (million kilowatt-hours) |
|---|---|
| 1982 | 6 |
| 1983 | 49 |
| 1984 | 195 |
| 1985 | 671 |
| 1986 | 1218 |
| 1987 | 1727 |
| 1988 | 1818 |
| 1989 | 2079 |
| 1990 | 2500[a] |
| 1991 | 2700[a] |
| 1992 | 2850[a] |

[a] These figures are estimates.
*Source*: Reference 5.

## 1.8.3 Development of Alternative Fuels

Since the energy crisis in 1973, interest in alternative fuels has been quite strong. Both long-term and near-term solutions and technologies have been developed. A great deal of process and scientific information has been gathered in the literature. The world energy development program has taken a few turns for various reasons, some of which are listed below:

1. Because of the recessions in the early 1980s and subsequent spending cuts, long-term research and development became lower in priority.
2. Because of the relatively steady price of oil, clean liquid fuels via alternative routes have become noncompetitive without government subsidies.
3. Because of the Clean Air Act Amendments of 1990, additional restrictions on the use of conventional fuels have been implemented. These include stack emission control, reformulated gasoline, and oxygenated fuel programs. New products such as MTBE, ETBE, and TAME have been introduced into the marketplace.
4. Energy conversion technologies now have additional socioeconomic burdens from the environmental sector. Not only efficient energy conversion but also environmental protection and preservation have become critically important.
5. Energy integration concepts have been very popular, because of easy adaptation to existing facilities as well as enhanced efficiencies.
6. Corporate reengineering moves in the United States in the 1990s have substantially reduced the R&D workforce in the field. This deemphasized fundamental and scientific research at the expense of immediate goal-oriented technical services.

Nonetheless, the field of alternative fuels is strategically important to the long-term future of human life as well as better utilization of global resources. The twenty-first century is destined to become a "3-E" (energy-environment-economy) era.

## REFERENCES

1. Energy Information Administration (EIA) of U.S. DOE, *Monthly Energy Review*, June 1995.
2. American Petroleum Institute. *Basic Petroleum Data Book*, vol. 15, no. 1, January 1995.
3. Speight, J. G. *Fuel Science & Technology Handbook*, p. 31. New York: Marcel Dekker, 1990.
4. Kraushaar, J. J., and R. A. Ristinen. *Energy and Problems of a Technical Society*, rev. ed. New York: Wiley, 1988.
5. *1990 Capacity and Generation of Non-Utility Sources of Energy*. Washington, DC: Edison Electric Institute, 1991.

## PROBLEMS

1. The estimated number of years for petroleum reserves to meet the current level of demand has been changing from year to year. Also, the number of years estimated was not necessarily decreasing with time. How can you explain this seemingly paradoxical estimation?
2. Define the unit tetrakilowatt-hours.
3. Define renewable energy and nonrenewable energy. Provide lists of renewable energies and nonrenewable energies.
4. Which of the following are seasonally fluctuating?
   (a) Coal production in the United States
   (b) Coal imports by the United States
   (c) Natural gas production in the United States
   (d) Natural gas consumption in the United States
   (e) Electricity generation by nuclear energy
   (f) Petroleum crude production in the United States
5. Discuss the factors that affect the directions of alternative fuel development.
6. What do you think are important factors in choosing energy sources in general? If you had to make a choice for process heating, what would be the factors you would consider with high priority?
7. As basic forecasting techniques, the following methods can be considered:
   1. Delphi method
   2. Market research
   3. Panel discussions
   4. Visionary forecast
   5. Historical analogy
   6. Moving average
   7. Exponential smoothing
   8. Box-Jenkins method
   9. X-11
   10. Trend projections
   11. Regression model (simple, multiple)
   12. Econometric model
   13. Anticipation survey
   14. Input-output analysis
   15. Diffusion index
   16. Leading indicator
   17. Life cycle analysis

   What methods do you think are suitable for prediction of the following? What are the reasons for your choice?

(a) The total coal production of the world in 2020
(b) The U.S. natural gas import from Canada in 2005
(c) The world electric power generation by geothermal energy in 2010
(d) The U.S. natural gas production in 2005 based on the statistics for 1985–1994
(e) Monthly natural gas production in 2000

# COAL, COAL DERIVATIVES, AND COAL PRODUCTS

Coal is a typical conventional solid fuel that has long history of human exploitation. It has become a main source of energy generation ever since the Industrial Revolution. Despite the many environmental concerns and misgivings in recent years, the relative importance of coal as fuel has not lost any ground. The energy crisis in 1973 made coal a politically reliable source of fuel. Electric power generation in the United States is heavily dependent on coal combustion, 57% in 1994. The trend is not going to be much different in the twenty-first century, except that more responsible and environmentally friendly use of coal will be even more strongly supported. A much stricter emission control policy was implemented by the Clean Air Act Amendments of 1990.

It is not my intention to discuss conventional coal chemistry and technology in this chapter. Instead, environmentally friendly use of coal as a clean solid fuel as well as alternative uses of coal and its products via emerging process routes will be discussed. In particular, conversion of coal into alternative clean liquid and solid fuels will be considered. In this regard, coal is an excellent alternative source, replacing depleting petroleum as a petrochemical feedstock.

## 2.1 COAL AS A SOURCE OF CONVENTIONAL FUEL

### 2.1.1 Coal Petrography

Petrography is an alternative classification of coal; instead of using ranks, it emphasizes the compositional description of coal as a rock material. This classification is based on visual observation of light reflected or transmitted through thin coal sections. The relationship between petrographic or rank classification and chemical reactivity remains obscure in most applications.

Coal is a complex sedimentary rock derived from a variety of plant materials that have undergone various physical and chemical transformations and are mainly composed of macerals. Modern petrographic classification is concerned with the microscopic separation of coal macerals according to color and consistency. Macerals are organic substances derived from plant tissues and exudates that have been incorporated in sedimentary strata, variably subjected to decay, and then compacted, hardened, and chemically altered by natural processes [1].

The macerals are identified microscopically by their form, reflectance, color, shape, and relief or polishing hardness. Based on their appearance and physical characteristics, macerals are divided into three main groups:

**Vitrinite**: Mainly originates from plant structural elements composed of lignin and cellulose. Vitrinite is the main constituent of coal (50–90% by weight) and is characterized by a graphitic banded structure with a highly vitreous luster. It appears translucent dark or light orange in transmitted light and dark to light gray in reflected light.

**Exinite**: Consists of macerals derived from the resinous and waxy parts of plants such as spores, cuticles, resins, fungal bodies, and algae. These macerals are the most aliphatic and hydrogen-rich group and consist mainly of sporinite, suberinite, sesinite, and alginite macerals. Exinites are characterized on the basis of their origin rather than their chemical and physical properties.

**Inertinite**: This group of macerals is derived from degraded woody tissues. They are opaque to transmitted light and bright in incident light. They have the highest reflectance and carbon content in coal. Based on their shape, size, degree of preservation of cellular structure, and intensity of charring, they are divided into six groups: fusinite, semifusinite, macrinite, micrinite, sclerotinite, and inertoderinite.

The maceral carbon content increases and the atomic H/C ratio decreases in the order of exinite, vitrinite, inertinite. Differences in maceral compositions are, however, much less noticeable for high-rank coals. Maceral behavior during devolatilization indicates the strong influence of petrography. Vitrinite is the plasticizing, coke-forming portion of the coal structure. Exinite fluidizes and decomposes to tars and gases. Inertinite neither plasticizes nor devolatilizes. Selected properties associated with coal ranks are shown for various coals in Table 1.

## 2.1.2 Structure and Chemistry of Coal

Coal is physically and chemically a heterogeneous rock that consists mainly of organic material (macerals) with inorganic materials (minerals) interspersed as shown in Figure 1 [2].

**Table 1 Typical variations of coal properties with rank**

| Rank (classification) | C (%) (dmmf)[a] | VM[b] (%) | H₂O (%) | Agglomerating | Coalification age | Rank |
|---|---|---|---|---|---|---|
| Lignite (A,B) | 65–72 | 40–50 | >15 | No | Young | Low |
| Subbituminous (A,B,C) | 72–76 | 35–50 | 10–15 | No | | |
| HVC bituminous | 76–78 | 35–45 | 5–10 | Highly | | |
| HVB bituminous | 78–80 | 31–45 | 3–5 | Commonly | | |
| HVA bituminous | 80–87 | 31–40 | 1–2 | Commonly | | |
| MV bituminous | 89 | 22–31 | <1 | Commonly | | |
| LV bituminous | 90 | 14–22 | <1 | Commonly | | |
| Anthracite (meta-, semi-) | 93 | <14 | <1 | No | Old | High |

[a]dmmf, dry, mineral matter–free basis.
[b]VM, volatile matter

As discussed before, the macerals consist mainly of three groups: vitrinite, exinite (or liptinite), and inertinite. Vitrinite is the main group and is derived from woody plant material (mainly lignin). Exinite is derived from lipids and waxy plant substances, and inertinite is derived from degraded woody tissues.

The structural determination of coals is a very difficult and important problem facing chemists and engineers. Some factors that affect the determination of coal structure are [3, 4]:

- Absence of individual compounds in quantities large enough to permit isolation
- Complex nature of molecules varying in size, origin, and structure
- Mix of highly reactive functional groups
- Insolubility of most coals

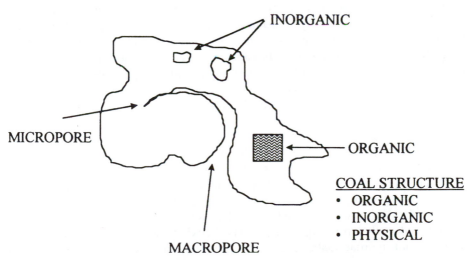

**Figure 1** Coal is a sedimentary rock.

- Nonvolatility of most coals
- Noncrystallinity of most coals
- Presence and variety of inorganic minerals
- Variation between coals and coals within the same screen
- Lack of communication between structural chemists and processing engineers

For all of these reasons, coal structure research is still a challenging task. In general, there are two approaches to the determination of coal structure: (1) degrade the coal macromolecules into representative fragments and derive the original structure from the structures of such fragments and (2) characterize coal directly without the destructive investigation of solid materials. The latter method typically utilizes infrared (IR) and solid-state nuclear magnetic reasonance (NMR) spectroscopy, X-ray diffraction and small-angle scattering, X-ray absorption spectroscopy (XAS, XANES, EXAFS), X-ray photoelectron spectroscopy (XPS), transmission electron microscopy (TEM), and Mossbauer spectroscopy [4].

Because of the heterogeneity of coals, the concept of a unique structure or simple repetitive structure cannot be justified. On the other hand, that does not mean there cannot be a concept to outline the macromolecular structure that would help understanding of concepts of coal processing, such as combustion, liquefaction, gasification, beneficiation, and desulfuriztion.

The aromatic-hydroaromatic model and the aliphatic-polyamantane model are the two principal models that claim to represent the organic structure of coals. The aromatic-hydroaromatic model of coal as described by Davidson is widely accepted and presents a picture of a skeletal structure consisting of clusters of condensed aromatic nuclei [5]. The clusters range in size from one to several rings, with an average of three rings for coals with 80% to 90% C and the number increasing rapidly above 90% as the anthracites are approached. The clusters are connected by hydroaromatic linkages with a small number of heteroatom (mainly oxygen) linkages. The coal models are constructed by considering the molecular weight, size, linkages between aromatic units, aromatic-to-aliphatic carbon ratio, elemental composition, and experimental results for chemical and thermal reactions.

One of the first coal models was proposed by Given in 1960 and is shown in Figure 2 [6]. It is a three-dimensional model and is based on X-ray diffraction, NMR, and other analytical information. However, this structure has a lower aliphatic-to-aromatic ratio than is expected today, with the hydroaromatic moieties being the major linkages. A two-dimensional model (although the presence of folded rings such as 9,10-dihydrophenanthrene means the structures are not flat) was proposed by Gibson and is shown in Figure 3 [5, 6]. A buckled-sheet structure for coal as derived from X-ray data was proposed by Cartz and Hirsch in 1960 and is shown in Figure 4 [6]. The buckled sheet consists of condensed aromatic and hydroaromatic rings bearing only short alkyl substituents (methyl

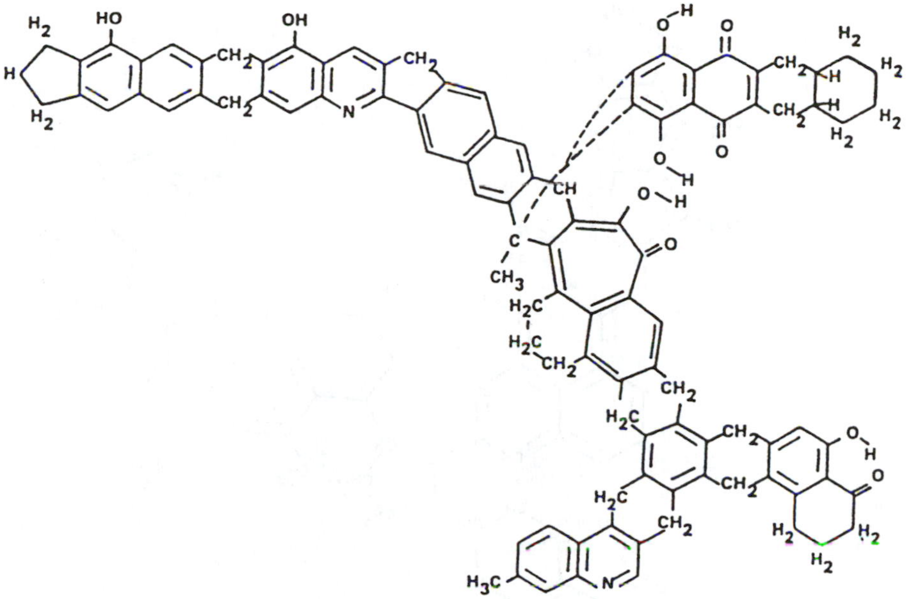

**Figure 2** An early structural model by Given (1960) of goal with the concept of a third dimension. (Source: Reference 6.)

and ethyl) with bridging and nonbridging methylenes adjoining the aromatic rings. A model proposed by Hill and Lyon in 1962 utilizing a complex network of condensed aromatic and heteroaromatic nuclei invoked the concept that tetrahedral three-dimensional carbon-carbon bonds are present in coal molecules, as shown in Figure 5 [6].

A model for a bituminous coal proposed by Wiser in 1975 is shown in Figure 6 [7]. It is illustrated by more flexible linkages of ether sulfide and carbon-carbon bridges with numerous functional groups. A model based on the 9,10-dihydrophenanthrene structure in which molecules exist in a tangled state was proposed by Pill in 1979 as shown in Figure 7 [5, 6].

A model for a high-volatile bituminous Pittsburgh seam coal was proposed by Solomon in 1981 and is shown in Figure 8 [7]. This model describes the pyrolytic reactivity of coal with linkages that produce chemically reactive gases and tar under pyrolysis conditions. Space-filling models for low-, intermediate-, and high-rank coal molecules were proposed by Spiro and Kosky in 1982 and are shown in Figure 9 [7]. The molecules are designed to conform to experimentally determined parameters such as chemical composition, aromaticity, and ring index. The low-rank coal appears fluffy, porous, and random with most interior atoms exposed as surface. The intermediate-rank coal model is more flat and oriented, with closed pores and occasional noncoplanar protrusions because

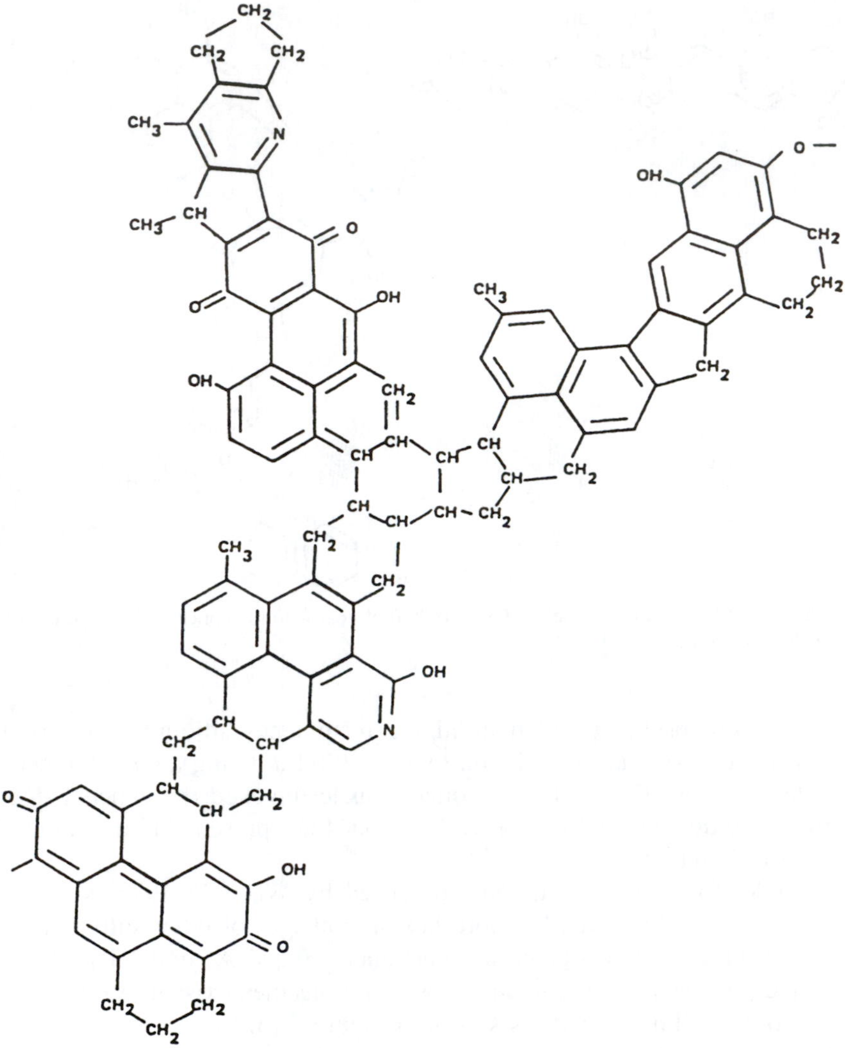

**Figure 3** A Proposed Molecular Model for a Vitrinite (82%C) by Gibson (1978). (Source: References 5 and 6.)

of aliphatic, alicyclic, and hydroaromatic moieties. The high-rank coal model is of a higher order with graphitic domains.

A new reactive model of coal structure is based on detailed chemical analyses of both coal and products from various liquefaction schemes. These products are characterized in terms of elemental distribution, aromaticity, functional groups, chemistry, and reactivity for a vitrinite-rich high-volatile bituminous coal with molecular weight of 10,000 was proposed by Shinn in 1994 and is shown

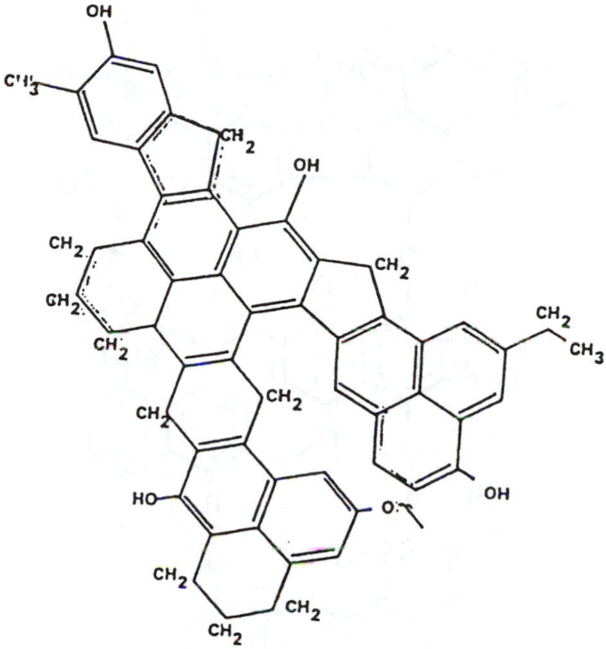

**Figure 4** A buckled-sheet structure for coal as derived from X-ray data by Cartz and Hirsch (1960). (Source: Reference 6.)

in Figure 10 [7, 9]. A model explaining the pyrolysis and hydropyrolysis behavior of a brown coal was proposed by Huttinger and Michenfelder in 1987 and is shown in Figure 11 [7, 10]. This model was developed from results of elemental analysis, pyrolysis experiments, titration studies, and extrapolation of literature data for similar coals. A hypothetical polyamantane configuration simulating coals from 76 to 90% C was proposed by Chakrabarthy and Berkowitz in 1974 and is shown in Figure 12 [5, 6]. This model requires that the polyamantane units increase in size with increase in rank and that benzenoyl carbons occupy the periphery of units. These units would have to aromatize as bituminous coals pass to the anthracite stage. Also, the coalification process would be interrupted to produce a stable aliphatic carbon skeleton from the original plant matter and then to produce individual benzene rings.

## 2.1.3 Characterization of Coal

Coal is characterized by a wide variety of properties and compositions, because it is a complex and heterogeneous sedimentary rock. The coal properties are characterized by proximate analysis, ultimate analysis, and heating value.

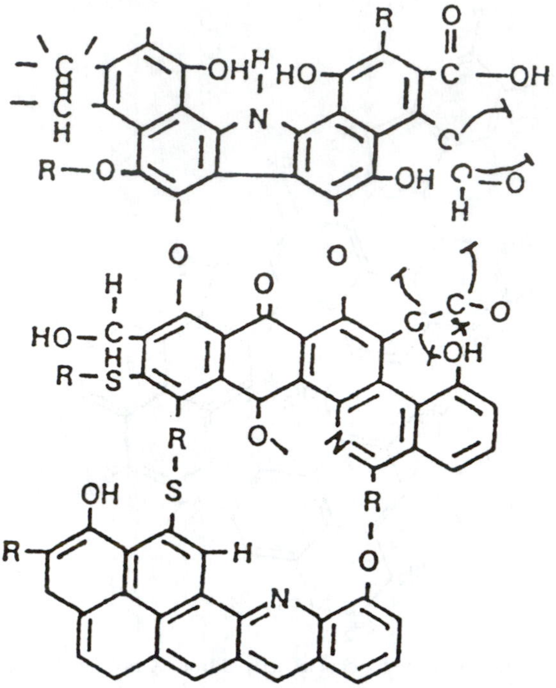

**Figure 5** Representative section of a hypothetical coal molecule using the concept of a network of condensed aromatic and heteroaromatic nuclei by Hill and Lyon (1962). (Source: Reference 6.)

**Proximate analysis.** Proximate analysis (ASTM D 3172) serves as a simple means for analyzing the thermal behavior of coals by determining moisture content, volatile matter content, ash content, and by difference the fixed carbon content. The moisture content of coal (ASTM D 3173) is determined by measuring the percentage weight loss of approximately a 1-g sample at a temperature of 105–107°C in an inert atmosphere for 1 hour.

The volatile matter (VM) content of coal is the percentage of weight loss corrected for moisture that occurs when coal is heated in a specified apparatus under standardized conditions. It consists mainly of combustible gases such as hydrogen, carbon monoxide, methane, and other hydrocarbons; tar vapors; and incombustible gases such as carbon dioxide and steam. The determination of the volatile matter content of coal (ASTM D 3175; ISO 562) is important because the data are used in coal classification systems. It is also an evaluation of the stability of coal for combustion and carbonization.

Ash content is the residue derived from the mineral matter during complete incineration of coal. Ash content is usually very variable even in the same type of coal and is typically determined by burning 1 to 2 g of coal at a temperature

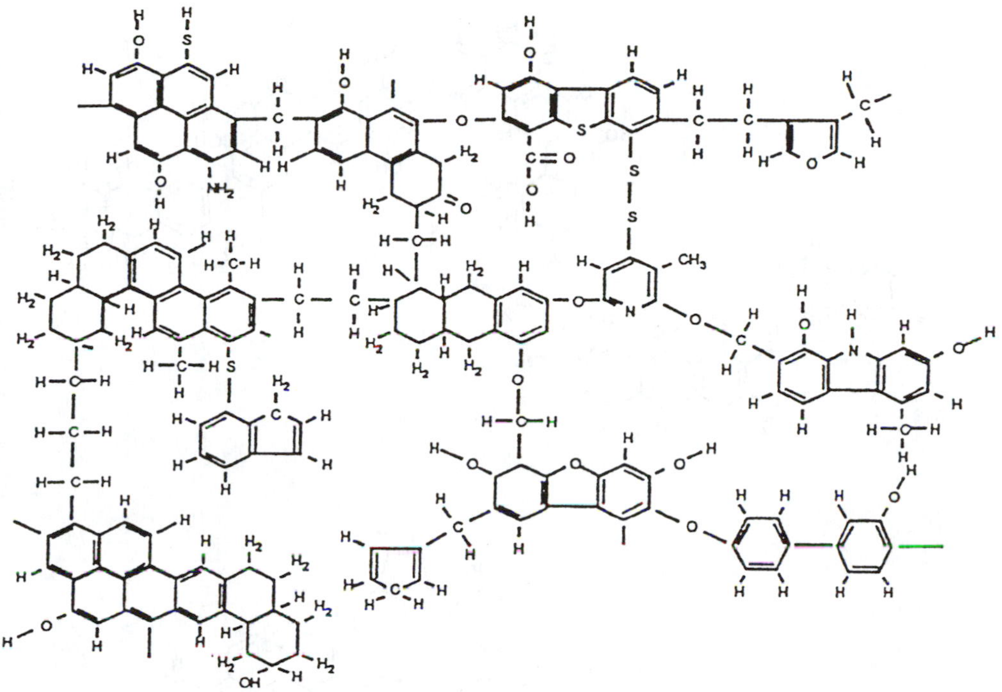

**Figure 6** Model of a high-volatile bituminous coal by Wiser (1975). (Source: Reference 7.)

range from 700 to 950°C in a ventilated muffle furnace. Ash is normally of sandy, bright yellowish color.

The material remaining after the determination of moisture, volatile matter, and ash is called fixed carbon. It is calculated by subtracting the percentages of moisture, volatile matter, and ash contents from 100%. Fixed carbon plus ash represents the approximate yield of coke from coal. Proximate analysis is very frequently handled by an instrument called a proximate analyzer.

**Ultimate analysis.** Ultimate analysis (ASTM D 3176) determines the composition of carbon, hydrogen, nitrogen, sulfur, ash, and oxygen content by difference, regardless of their origin. Carbon and hydrogen are often considered the most important constituents of coal and account for 70–95% and 2–6% of the organic substance of coal, respectively. Analytical procedures (ASTM D 3178; ISO 609 and 625) are used to determine carbon and hydrogen by burning an exact amount of coal in a closed system with the products of the combustion ($CO_2/H_2O$) being determined by absorption. Nitrogen also occurs in the organic matter of coal and accounts for about 1–2%. It is measured mainly by use of the Kjeldahl method (ASTM D 3179; ISO 332 and 333) or the standard method

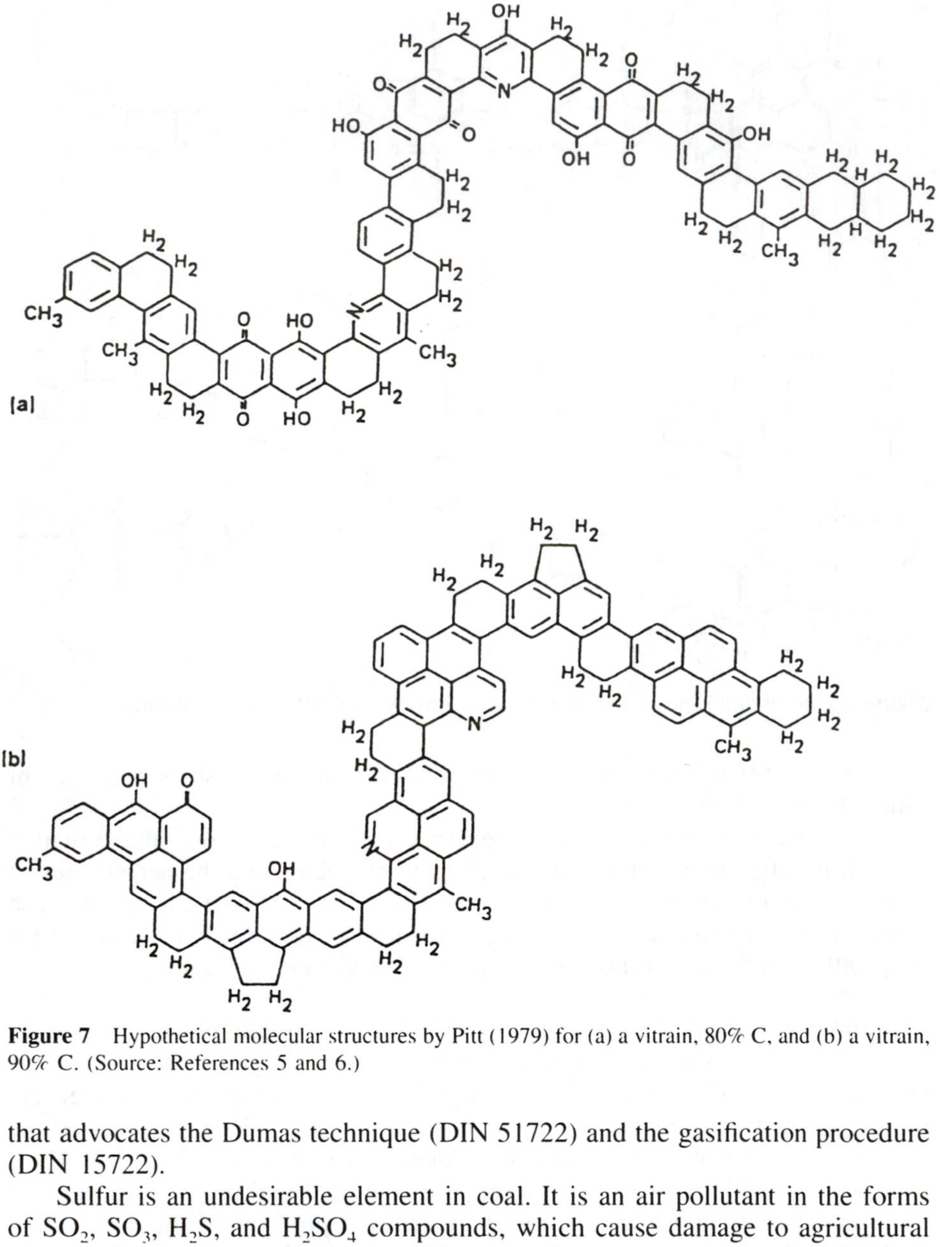

**Figure 7** Hypothetical molecular structures by Pitt (1979) for (a) a vitrain, 80% C, and (b) a vitrain, 90% C. (Source: References 5 and 6.)

that advocates the Dumas technique (DIN 51722) and the gasification procedure (DIN 15722).

Sulfur is an undesirable element in coal. It is an air pollutant in the forms of $SO_2$, $SO_3$, $H_2S$, and $H_2SO_4$ compounds, which cause damage to agricultural crops and corrosion to almost everything (metals, stones, concrete, and clothing) and contribute to acid rain. The sulfur content of coal approximately ranges from 0.2 to 8% and sulfur species are present in coal in the forms of inorganic

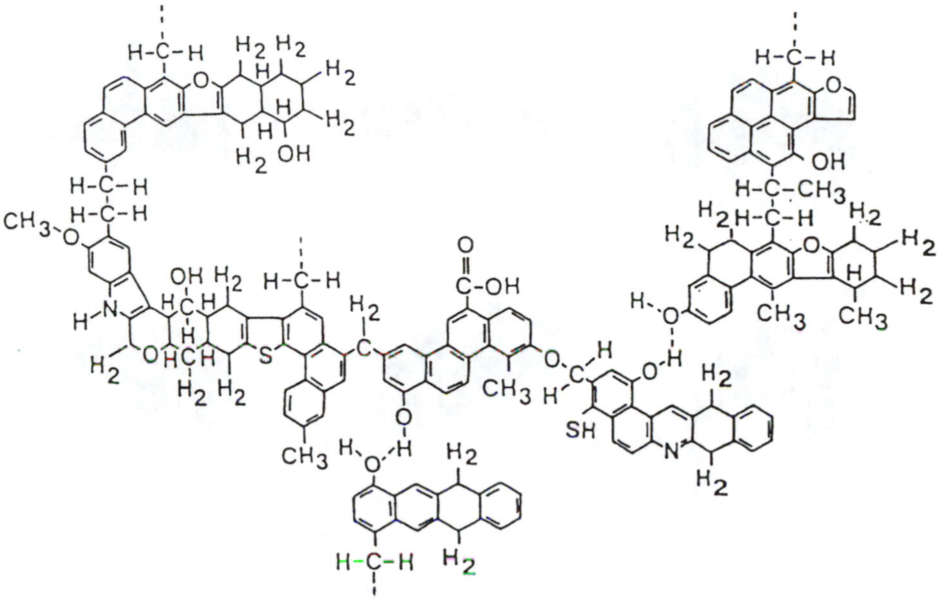

**Figure 8** Hypothetical structural model (Solomon, 1981) of a Pittsburgh high-volatile bituminous goal. (Source: Reference 7.)

sulfur (pyrite and marcasite, $FeS_2$, and sulfate, $FeSO_4$) and organically bound sulfur.

Ultimate analysis measures the total sulfur content of coal without distinguishing among the forms in which it occurs. Organic sulfur is usually determined by difference between the total sulfur and the inorganic (pyritic and sulfatic) sulfur measured by use of ASTM methods (ASTM D 2492; ISO 157) or by the conventional Powell and Parr method.

The amount of oxygen in coal is determined by an indirect (BS 1016) method (subtracting the percentages of the other four elements from 100%). The disadvantage of this procedure is that all the experimental errors in the carbon, hydrogen, nitrogen, and sulfur measurements are accumulated into the calculated value of oxygen.

**Heating value.** The heating value or calorific value (CV) of coal is a direct measurement of the chemical energy stored in it. Therefore, it is an important parameter for determining the value of coal as a fuel. The heating values of coals vary widely from less than 6000 Btu/lb to more than 14,000 Btu/lb. This parameter is used in ASTM rank classification of major coal types. The direct and indirect methods used in determining the calorific value of coals are described in the following subsections.

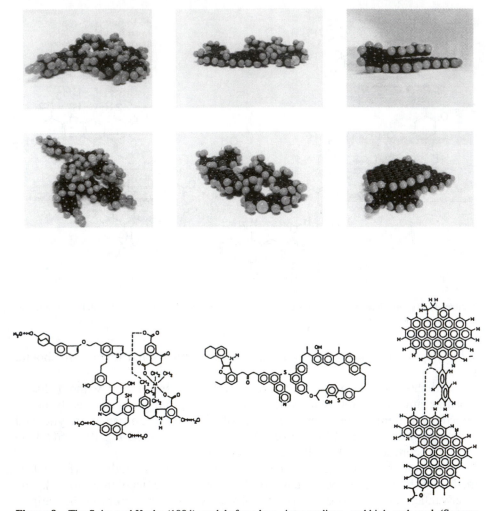

**Figure 9** The Spiro and Kosky (1984) models for a low-, intermediate-, and high-rank coal. (Source: References 7 and 8.)

*Calorimetric determination.* The use of calorimeters provides a direct measure of the heat of combustion. This determination involves measuring the amount of heat released on burning a known amount of coal in a closed vessel under pressurized oxygen. An oxygen bomb calorimeter can be used for this measurement.

*Dulong's formula.* The calorific value of coal can be estimated from the ultimate analysis by the use of Dulong's equation [11], which states:

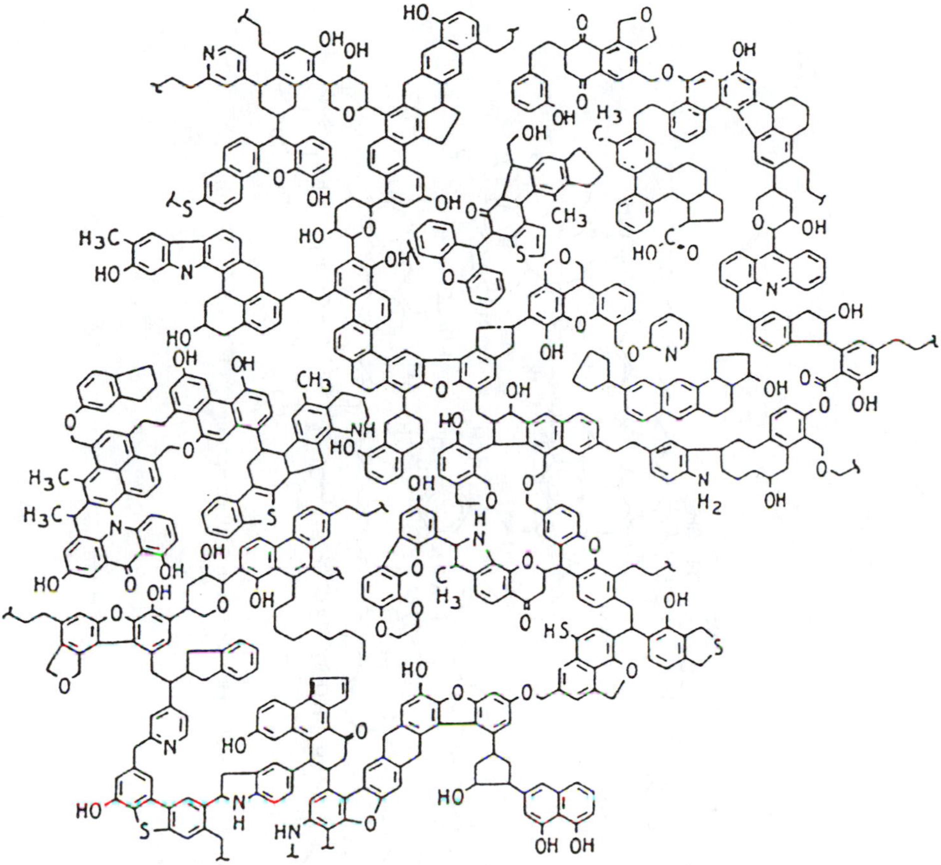

**Figure 10** The Shinn (1984) model of a bituminous coal [7, 9].

$$CV = 145.44X_C + 620.28(X_H - X_O/8.0) + 40.5Xs \qquad (1)$$

where CV is the calorific value (Btu/lb) on a dry, ash-free (daf) basis and the $X$'s denote the weight percentages of the elements carbon, hydrogen, oxygen, and sulfur as designated by the subscripts. The coefficients of the $X$'s are the calorific values of carbon, hydrogen, and sulfur, respectively, from left to right.

*Goutal's formula.* The calorific value of coal can be estimated from the proximate analysis by the use of Goutal's formula [11]

$$CV = 14760 + a(VM) \qquad (2)$$

where VM is the percentage of volatile matter and $a$ is a parameter that is a function of VM on a dry, ash-free basis.

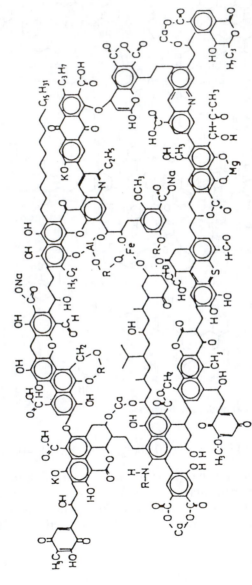

**Figure 11** The Huttinger and Michenfelder (1987) model of brown coal, comparing $C_{270}H_{240}N_3S_1O_{90}$ [7, 10].

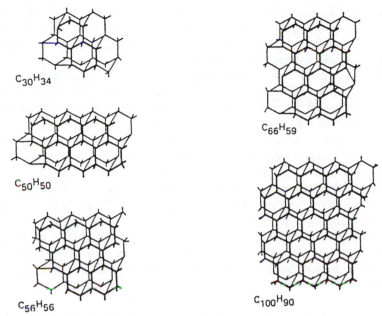

**Figure 12** The Berkowitz and Chakrabartty hypothetical polyamantane configurations simulating coals from 76 to 90% C [5, 6].

***Ergun's formula.*** The calorific value of coal can be estimated from the moisture content by the use of Ergun's formula [11]

$$CV = 19000 - 1000\sqrt{12.0 + 3.1M} \tag{3}$$

It is based on a moist, mineral matter–free basis. $M$ is the moisture content of coal.

***Classification of coal.*** The widespread occurrence and the diversity of coal uses have resulted in a number of coal classification systems. Most of these systems use the calorific value, fixed carbon or volatile matter, and the agglomerating properties of coal as a basis for their classifications. The most widely used classifying systems worldwide are the ASTM, the British National Coal Board (United Kindom), and the International Organization for Standardization (ISO) classifications.

***ASTM classification of coals by rank.*** Coal classification by rank is the most common classification. It is a measure of the degree of coalification that the organic plant sediment has reached in the metamorphosis from peat to graphite. Rank gives a qualitative indication of the amount of carbon in the coal, which increases as it matures. However, rank is not necessarily directly equated to the carbon content of coal or any specific coal structure.

The ASTM designation (D 388-92a) as shown in Table 1 is the standard specification for the classification of coal by rank in the United States. This system uses volatile matter and fixed carbon results from proximate analysis and heating values as indicators of rank, which results in the classification of all coals into a series of classes and groups. According to this system, coals with higher rank (having less than 31% volatile matter) are classified according to their percentages of fixed carbon from the proximate analysis based on a dry, ash-free basis. On the other hand, coals with lower rank (having more than 31% volatile matter) are classified by their calorific values on a moist basis. Also, in this system, the agglomerating characteristics of the coals are used to distinguish between adjacent categories.

The variations in energy content and composition of coals on a moist, ash-free basis with coal rank are shown in Figure 13 [1, 7]. From this figure it can be seen that the energy content of coal increases as the moisture content decreases from the lignite stages to the low-volatile bituminous coal stages. The calorific value of coal also decreases because of a decrease in the volatile matter especially in the anthracitic stages. Brief descriptions of the ranks of coals are as follows:

**Lignite**: The lowest in rank among the coals, with a high moisture content and a high volatile matter content, which makes it easy to ignite. However, it burns with a smoky flame and slags when it dries on exposure to air and can experience spontaneous combustion in storage or shipment. Lignite is a low-quality coal because it has a low heating value (about one-half that of bituminous coal). The coal is relatively soft and ranges in color from brown to black. In the coal market, this type of coal is very inexpensive; however, dewatering and demineralization become an issue.

**Subbituminous coal**: It is an intermediate rank between lignite and bituminous coal; the coal has matured to the point that the woody texture is no longer apparent. It is black in color and has the same tendency toward slagging and spontaneous combustion as lignite. However, it contains enough volatile matter to be easily ignitable. Subbituminous coal is a better quality than lignite because it has a relatively higher heating value, lower sulfur content, and normally burns cleaner.

**Bituminous coal**: It is the most known and widely used rank of coal, having lower moisture and volatile values and higher heating values than subbituminous coal. The coal is black with a banded appearance and is the only coal that may exhibit agglomerating or caking behavior, making it useful for producing coke to be used in the steel industry. It burns with a clean hot flame.

**Anthracite**: The highest rank of coal with a very low volatile matter content. It burns with a hot, clean flame with no smoke or soot and has a low moisture content (about 3%) and low sulfur content, is jet black, and has a high luster. It is the hardest and most dense coal and does not form coke when heated.

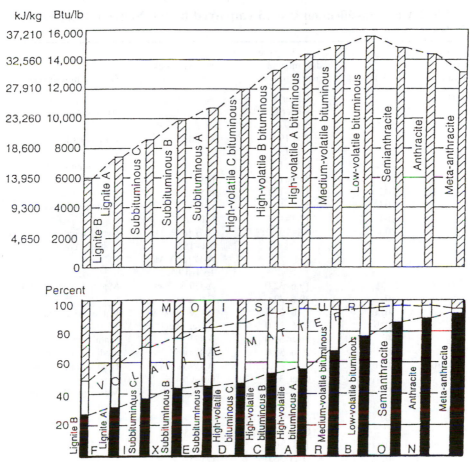

**Figure 13** Schematic of variations of energy content and composition on a moist ash-free basis with coal rank [1, 7].

***The British National Coal Board (BNCB) classification system.*** The BNCB coal classification system, shown in Table 2 [6], is based on the coke-forming characteristics of the various coals as well as the Gray-King coke type test on the dry mineral-free volatile matter. In this system of classification, a three-digit code number is used to describe each particular coal, with lower numbers assigned to higher rank coals. The groups of coals at 100 and above are classified by use of volatile matter alone.

***International Organization for Standardization (ISO) coal classification.*** This system is based on the dry, ash-free volatile matter and moist, ash-free calorific value, which is similar to the basis of the ASTM coal classification system. In this system, coals are divided into two major types: hard coals, coals with a

## Table 2 Coal classification system employed by the National Coal Board (UK)

| Group | Class | Volatile matter, dry, mineral matter free % | Gray-King coke type[a] | General description |
|---|---|---|---|---|
| 100 | | Under 9.1 | A | |
| | 101[b] | Under 6.1 | | Anthracites |
| | 102[b] | 6.1–9.0 | A | |
| 200 | | 9.1–19.5 | A-G8 | Low-volatile steam coals |
| | 201 | 9.1–13.5 | A-G | |
| | 201a | 9.1–11.5 | A-B | Dry steam coals |
| | 201b | 11.6–13.5 | B-C | |
| | 202 | 13.6–15.0 | B-G | |
| | 203 | 15.1–17.0 | B-G4 | Coking steam coals |
| | 204 | 17.1–19.5 | G1-G8 | |
| | 206 | 9.1–19.5 | A-B for V.M 9.1–15.0 A-D for V.M 15.1–19.5 | Heat-altered low-volatile steam coals |
| 300 | | 19.6–32.0 | A-G9 & over | Medium-volatile coals |
| | 301 | 19.6–32.0 | G4 & over | |
| | 301a | 19.6–27.5 | G4 & over | Prime coking coals |
| | 301b | 27.6–32.0 | G4 & over | |
| | 305 | Over 32.0 | A-G3 | Mainly heat altered |
| | 306 | 19.6–32.0 | A-B | medium-volatile coals |
| 400–900 | Over 32.0 | | A-G9 & over | High-volatile coals |
| 400 | | Over 32.0 | G9 & over | |
| | 401 | 32.1–36.0 | G9 & over | Very strongly caking coals |
| | 402 | Over 36.0 | | |
| 500 | | Over 32.0 | G5-G8 | |
| | 501 | 32.1–36.0 | | Strongly caking coals |
| | 502 | Over 36.0 | G5-G8 | |
| 600 | | Over 32.0 | G5-G4 | |
| | 601 | 32.1–36.0 | | Medium caking coals |
| | 602 | Over 36.0 | G1-G4 | |
| 700 | | Over 32.0 | B-G | |
| | 701 | 32.1–36.0 | | Weakly caking coal |
| | 702 | Over 36.0 | B-G | |
| 800 | | Over 32.0 | C-D | |
| | 801 | 32.1–36.0 | | Very weak caking coals |
| | 802 | Over 36.0 | C-D | |
| 900 | | Over 32.0 | A-B | |
| | 901 | 32.1–36.0 | | Noncaking coals |
| | 902 | Over 36.0 | A-B | |

[a]Coals of groups 100 and 200 are classified by using the parameter of volatile matter alone. The Gray-King coke types quoted for these coals indicate the ranges found in practice and not criteria for classification.

[b]To divide anthracites into two classes, it is sometimes convenient to use a hydrogen content of 3.35% (dmmf) instead of volatile matter of 6.0% as the limiting criterion. In the original Coal Survey rank coding system the anthracites were divided into four classes then designated 101, 102, 103, and 104. Although the present division into two classes satisfies most requirements, it may sometimes be necessary to recognize four or five classes.

calorific value larger than 10,260 Btu/lb; and brown coals, coals with calorific values less than 10,260 Btu/lb. The groups and subgroups of hard coals are separated by their caking and coking properties. Three-digit numbers are used to describe coals in this classification and are defined as follows: the first figure indicates the class of coal, the second figure indicates the group into which the coal falls, and the third figure is an indication of the subgroups. The classification parameters and the numerical symbols representing the coal groups and sub-groups of hard coals are listed in Table 3 [6, 11]. The international system for the classification of brown coals shows a value either as a fuel or as a chemical raw material and is based on the total moisture (ash-free basis) and the yield of tar as shown in Table 4 [6].

A comparison of class numbers for the international system of coal classi-fication and the ASTM system of coal classification is shown in Figure 14 [6]. Another schematic comparison of the various systems of coal classifications is shown in Table 5 [6].

Other classifications of coals such as geological age classification, banded structure classification, Alpern coal classification, and a classification profile chart developed by Hensel in 1981 from published coal data on a moisture- and ash-free basis as shown in Table 6 [7] are not discussed here.

## 2.1.4 Power Generation by Coal Combustion

Coal is the most abundant fossil fuel in both the world and the United States, accounting for an approximately 75% of the total world and U.S. resources of fossil fuels. It is the most important fuel for power generation, with more than 50% of the coal produced in the world used in electric power generation [1, 12]. Electric power demand is expected to continue to rise at a relatively high rate at least in the next 20 years [13] in spite of various energy measures to conserve scarce petroleum and natural gas resources. Based on the necessity to change to more abundant forms of primary energy consumption and the fact that any pri-mary energy source can be enhanced by conversion to electrical energy, the natural trend to an energy form of greater convenience and ease of use, the trend since 1975 was to develop nuclear power for electricity generation. Growing public concerns in Western countries over nuclear energy development directed many of these countries to start looking at coal and coal products as a fuel for the world's new power stations, reducing their dependence on petroleum, fuel gas, and nuclear energy in the immediate term. Therefore, the most conventional solid fuel is still going to be the major source of energy in the next century. The issues are going to be clean and alternative uses of coals.

Most of the power plants in the world use steam turbines in which high-pressure steam, generated by the use of hot gases from combustion, spin the turbines that drive the generators. The most common power generation plants are summarized as follows [14].

# Table 3  International system for the classification of hard coals

Code numbers:

The first figure of the code number indicates the class of the coal, determined by volatile-matter content up to 33% volatile matter and by calorific parameter above 33% volatile matter.
The second figure indicates the group of coal, determined by caking properties.
The third figure indicates the subgroup, determined by coking properties.

| Groups (determined by coking properties) | | | Code numbers — Class | | | | | | | | | | Subgroups (determined by coking properties) | | |
|---|---|---|---|---|---|---|---|---|---|---|---|---|---|---|---|
| **Group number** | **Free-swelling index (crucible swelling number)** | **Roga index** | 0 | 1 (A / B) | 2 | 3 | 4 | 5 | 6 | 7 | 8 | 9 | **Subgroup number** | **Audibert-Amu dilatometer** | **Gray-King** |
| 3 | >4 | >45 | | | | | 435 | 535 | 635 | | | | 5 | >140 | >$G_K$ |
| | | | | | | 334 | 434 | 534 | 634 | 734 | | | 4 | >50–140 | $G_1$–$G_2$ |
| | | | | | | 333 | 433 | 533 | 633 | 733 | | | 3 | >0–50 | $G_1$–$G_4$ |
| | | | | | | 332 a  b | 432 | 532 | 632 | 732 | 832 | | 2 | ≡0 | E–G |
| 2 | 2½–4 | >20–45 | | | | 323 | 423 | 523 | 623 | 723 | 823 | | 3 | >0–50 | $G_1$–$G_4$ |
| | | | | | | 322 | 422 | 522 | 622 | 722 | 822 | | 2 | ≡0 | E–G |
| | | | | | | 321 | 421 | 521 | 621 | 721 | 821 | | 1 | Contraction only | B–D |
| 1 | 1–2 | >5–20 | | | 212 | 312 | 412 | 512 | 612 | 712 | 812 | | 2 | ≡0 | E–G |
| | | | | | 211 | 311 | 411 | 511 | 611 | 711 | 811 | | 1 | Contraction only | B–D |
| 0 | 0–½ | >0–5 | 000 | 100 (A B) | 200 | 300 | 400 | 500 | 600 | 700 | 800 | 900 | 0 | Nonsoftening | A |
| **Class number →** | | | 0 | 1 | 2 | 3 | 4 | 5 | 6 | 7 | 8 | 9 | | | |

Class parameters:

| | 0 | 1 | 2 | 3 | 4 | 5 | 6 | 7 | 8 | 9 |
|---|---|---|---|---|---|---|---|---|---|---|
| **Volatile matter (dry, ash-free) →** | 0–3 | >3–6.5 (A)  >6.5–10 (B)  [>3–10] | >10–14 | >14–20 | >20–28 | >28–33 | >33 | >33 | >33 | >33 |
| **Calorific parameter[a] →** | — | — | — | — | — | — | >13,950 | >12,960–13,950 | >10,980–12,960 | >10,260–10,980 |

As an indication, the following classes have an approximate volatile-matter content of:
Class 6: 33–41% volatile matter
Class 7: 33–44% volatile matter
Class 8: 35–50% volatile matter
Class 9: 42–50% volatile matter

Classes: Determined by volatile matter up to 33% volatile matter and by calorific parameter above 33% volatile matter.

[a] Gross calorific value on moist ash-free basis (86°F/96% relative humidity) Btu per pound.

*Notes:* Where the ash content of coal is too high to allow classification according to the present systems, it must be reduced by laboratory float-and-sink method (or any other appropriate means). The specific gravity …

## Table 4 International system for the classification of brown coals

| Group | Group parameter tar yield (dry, ash-free %)[a] | Code number | | | | | |
|---|---|---|---|---|---|---|---|
| 40 | >25 | 1040 | 1140 | 1240 | 1340 | 1440 | 1540 |
| 30 | 20–25 | 1030 | 1130 | 1230 | 1330 | 1430 | 1530 |
| 20 | 15–20 | 1020 | 1120 | 1220 | 1320 | 1420 | 1520 |
| 10 | 10–15 | 1010 | 1110 | 1210 | 1310 | 1410 | 1510 |
| 00 | 10 or less | 1000 | 1100 | 1200 | 1300 | 1400 | 1500 |
| Class no. | | 10 | 11 | 12 | 13 | 14 | 15 |
| Class parameter (i.e., total moisture, ash-free, %)[b] | | 20 and less | 20–30 | 30–40 | 40–50 | 50–60 | 60–70 |

[a]Gross calorific value below 10,260 Btu/lb. Moist, ash-free basis (86°F, 96% relative humidity).
[b]The total moisture content refers to freshly mined coal. For internal purposes, coals with a gross calorific value over 10,260 Btu/lb (moist, ash-free basis), considered in the country of origin as brown coals, but classified under this sytem to ascertain, in particular, their suitability for processing. When the total moisture content is over 30% the gross calorific value is always below 10,260 Btu/lb.

1. **Pulverized coal-fired (PCF) plant.** Most of the old coal-fired power plants were built as PCF plants. The efficiency of electricity generation for a PCF plant ranges from 35–40%; in other words, for every 100 units of energy supplied in the coal delivered to the plant, 35-40 units reached the consumer as electricity. PCF power plants are used to generate power from about 100 MW to more than 1000 MW.

2. **Fluidized bed combustion (FBC) system.** The advantages of the FBC system are the formality of the temperature brought about by the intimate mixing of gas and coal particles and efficient heat transfer to immersed surfaces in the fluidized bed. A circulating fluidized bed combustion (CFBC) system in which high gas velocities are used to transport bed material from one vessel to another was developed because of the relatively high $SO_2$, and

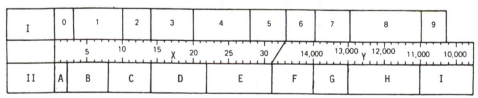

**Figure 14** Comparison of I. class numbers (International System) with II. coal rank (ASTM system): A = Metaanthracite; B = Anthracite; C = Semianthracite; D = Low-volatile bituminous coal; E = Medium-volatile bituminous coal; F = High-volatile A bituminous coal; G = High-volatile B bituminous coal; H = High-volatile C bituminous coal and subbituminous A coal; I = Subbituminous B coal; X = Volatile-matter parameter (Parameters in the International System are on an ash-free basis; in ASTM system, they are on mineral-matter free basis.); and Y = Calorific-value parameter (No upper limit of calorific value for class 6 and high volatile A bituminous coals) [28].

# Table 5 Comparison of various coal classification systems

| Classes of the International System | Parameters | | Classes of national systems | | | | | | | |
| --- | --- | --- | --- | --- | --- | --- | --- | --- | --- | --- |
| | Volatile matter content | Calorific value (calculated to standard moisture content) | Belgium | Germany | France | Italy | Netherlands | Poland | United Kingdom | United States |
| 0 | 0–3 | | | | | Antraciti speciali | | Meta-antracyt | | Meta-anthracite |
| 1A | 3–6.5 | | Maigre | Anthrazit | Anthracite | Antraciti comuni | Anthraciet | Antracyt | Anthracite | Anthracite |
| 1B | 6.5–10 | | | | | | | Polantracyt | | |
| 2 | 10–14 | | ¼ gras | Magerkohle | Maigre | Carboni magri | Mager | Chudy | Dry steam | Semianthracite |
| 3 | 14–20 | | ½ gras ¾ gras | Esskohle | Demi gras | Carboni semi-grassi | Esskool | Polkoksowy Metakoksowy | Coking steam | Low volatile bituminous |
| 4 | 20–28 | | | Fettkhole | Gras a courte flamme | Carboni grassi corta flamma | Vetkool | | Medium volatile coking | Medium volatile bituminous |
| 5 | 28–33 | | | Gaskohle | Gras proprement dit | Carboni grassi media flamma | | Ortokoksowy | | High volatile bituminous A |
| 6 | >33 (33–40) | 8450–7750 | | | | Carboni da gas | Gaskool | Gazowo koksowy | High volatile | High volatile bituminous B |
| 7 | >33 (32–44) | 7750–7200 | | | Flambant gras | Carboni grassi da vapore | Gasvla-mkool | Gazowy | | High volatile bituminous C |
| 8 | >33 (34–46) | 7200–6100 | | Gas Flammkohle | Flambant sec | Carboni secchi | Vlamkool | Gazowoplomienny | | Subbituminous |
| 9 | >33 | 66100 | | | | | | Plomienny | | |

**Table 6 Classification profile chart**

| Type | Average analysis—moisture and ash free | | | | | | |
|---|---|---|---|---|---|---|---|
| | Volatile matter (%) | Hydrogen (wt. %) | Carbon (wt. %) | Oxygen (wt. %) | Heating value (kJ/kg) | $\frac{C}{H}$ | $\frac{C+H}{O}$ |
| Anthracite | | | | | | | |
| Meta- | 1.8 | 2.0 | 94.4 | 2.0 | 34425 | 46.0 | 50.8 |
| Anthracite | 5.2 | 2.9 | 91.0 | 2.3 | 35000 | 33.6 | 42.4 |
| Semi- | 9.9 | 3.9 | 91.0 | 2.8 | 35725 | 23.4 | 31.3 |
| Bituminous | | | | | | | |
| Low-volatile | 19.1 | 4.7 | 89.9 | 2.6 | 36260 | 19.2 | 37.5 |
| Med-volatile | 26.9 | 5.2 | 88.4 | 4.2 | 35925 | 16.9 | 25.1 |
| High-volatile A | 38.8 | 5.5 | 83.0 | 7.3 | 34655 | 15.0 | 13.8 |
| High-volatile B | 43.6 | 5.6 | 80.7 | 10.8 | 33330 | 14.4 | 8.1 |
| High-volatile C | 44.6 | 4.4 | 77.7 | 13.5 | 31910 | 14.2 | 6.2 |
| Subbituminous | | | | | | | |
| Subbit. A | 44.7 | 5.3 | 76.0 | 16.4 | 30680 | 14.3 | 5.0 |
| Subbit. B | 42.7 | 5.2 | 76.1 | 16.6 | 30400 | 14.7 | 5.0 |
| Subbit. C | 44.2 | 5.1 | 73.9 | 19.2 | 29050 | 14.6 | 4.2 |
| Litnite | | | | | | | |
| Litnite A | 46.7 | 4.9 | 71.2 | 21.9 | 28305 | 14.5 | 3.6 |

particulates emission and low power output relative to combustor cross-sectional area. The CFBC is used to generate electric power in the range of 20 to 150 MW.

3. **Fully fired combined cycle (FFCC).** The FFCC system is a conventional boiler topped by a gas turbine (GT) for the flue gas that is used for the burners of the coal-fired boiler as combustion air. This combined cycle has almost 5% higher efficiency than a normal PCF plant.

4. **Pressurized fluidized bed combustion combined cycle (PFBC).** It is an FBC topped with a gas turbine (GT) in which the coal is burned at about 850°C in a supercharged boiler that is integrated into the GT combustor. A net efficiency up to 43% can be reached, and emissions of $SO_2$ and $NO_x$ can be reduced more efficiently than those of atmospheric FBC plants due to this combination. The PFBC is used to generate electric power of about 150 MW.

5. **Integrated coal gasification combined cycle (IGCC).** IGCC has outstanding merits in energy efficiency and operation economics via innovative energy integration strategies. Details can be found in Chapter 5.

6. **Pressurized pulverized coal-fired combined cycle (PPCC).** PPCC is based on the use of pulverized coal as a fuel in the combustor of a gas turbine. An efficiency of over 50% could be reached using PPCC power plants if the problem of removing adherent ash particles and corrosion caused mainly by alkaline constituents at about 1300°C could be solved.

The efficiency of a power station can be doubled by use of a combined heat and power (CHP) scheme that converts the primary fuel into energy and utilizes the waste heat for industrial processes [13]. The efficiencies for the different power generation systems are shown in Figure 15 [15].

The main disadvantage of the use of coal as a fuel for power generation is that it can cause severe environmental problems due to particulate, sulfur dioxide, and nitrogen oxides emissions if environmental protection measures are not taken to reduce these pollutants. However, these measures result in a reduction of plant efficiency in addition to additional cost. To conclude, more work should be done in order to produce more useful energy from coal efficiently and to minimize the environmental problems of using coal for power generation.

## 2.1.5 Environmental Regulations

As discussed, the worldwide demand for electricity is expected to rise dramatically over the next two decades. This means that coal production and use will also increase, and so will global emissions of sulfur dioxide ($SO_2$), nitrogen oxides ($NO_x$), and carbon dioxide ($CO_2$) in addition to the solid wastes.

The growing public awareness of and sensitivity to pollution of the environment have put increased pressure on environmental regulators to protect the environment. This increased public awareness has had a major impact on all industries that generate pollution of one form to another. Environmental issues must be considered as part of the overall energy picture. Most countries have taken steps to reduce emissions from pollutant-generating plants including coal-fired plants. The ranges of national emission standards (as guidelines for particulates, $SO_2$, and $NO_x$, mg/m$^3$) for new power plants in different countries are shown in Table 7 [16].

The 1988 Council of European Communities Directive on Large Combustion Plants has set reduction targets for $SO_2$ and $NO_x$ emissions at plants over 50 MW. By the year 2003 the overall emission of $SO_2$ in the European Community countries is to decline by 58% and $NO_x$ emissions are to decline by 40% [17]. In the United States the 1990 Clean Air Act Amendments establish emission reduction targets for $SO_2$ and $NO_x$, the major contributors to acid rain, as summarized in Table 8 [18].

The 1990 Clean Air Act Amendments do not include the global trend to reduce $CO_2$ emissions and thus reduce the threat of global warming. Also, solid wastes generated by coal-fired power plants are exempted from their regulation under section 3001 of the Resource Conservation and Recovery Act (RCRA).

The 1990 Clean Coal Air Act Amendments and other environmental regulations can be met by the development and use of clean coal technologies or by the modification, replacement, or repowering technologies such as atmospheric fluidized bed combustion (AFBC), pressurized fluidized bed combustion (PFBC), integrated gassification combined cycle (IGCC), integrated gasification

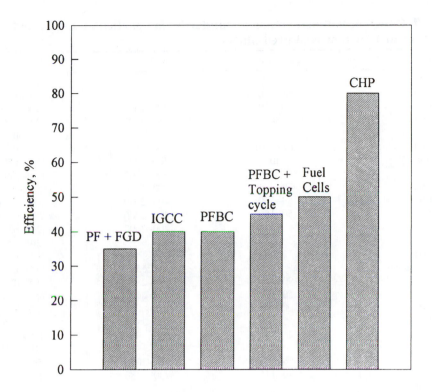

PF + FGD = pulverized fuel combustion with flue gas desulphurisation
IGCC = integrated gasification combined cycle
PFBC = pressurised fluidised bed combustion
CHP = combined heat and power (cogeneration)

**Figure 15** Efficiencies for the different power generation systems [15].

fuel cell system, and magnetohydrodynamic (MHD) systems. These systems can achieve $SO_2$ and $NO_x$ emission reductions as shown in Table 9 [19]. They can also increase efficiency, reduce wastes, and give more operational flexibility and greater cost effectiveness.

The utilization of clean coal technologies would reduce current and future $CO_2$ emission levels below those achievable by the use of conventional technologies as shown in Table 10 [19].

## 2.1.6 $SO_x/NO_x$ Problem

$SO_x$ emission is the most serious pollution problem resulting from the use of coal. Both organic and inorganic (pyrites) sulfur are oxidized to sulfur dioxide

**Table 7 Range of national emission standards$^a$ for particulates, SO$_2$ and NO$_x$ (mg/m$^3$) for new coal-fired plants**

| Country | Particulates | SO$_2$ | NO$_x$ |
|---|---|---|---|
| **Current** | | | |
| Australia | 65–210 | — | 535–860 |
| Austria | 50–150 | 200–1620 | 200–500 |
| Belgium | 50 | 250–2000 | 200–800 |
| Canada | 125 | 715 | 740 |
| Czechoslovakia | 100–250 | 500–2500 | 650 |
| Denmark | 40 | 400–2000 | 200–650 |
| EEC | 50–100 | 400–2000 | 650–1300 |
| Finland | 430–570 | 380–1270 | 135–405 |
| France | 50–100 | 400–2000 | 650–1300 |
| FRG | 50–150 | 400–2000 | 200–1500 |
| Italy | 50 | 400–2000 | 200–650 |
| Japan | 50–300 | $^b$ | 205–980 |
| Netherlands | 50 | 200–700 | 100–650 |
| New Zealand | 105 | 1255 | 2005 |
| Poland | 190–3700 | 540–2890 | 95–705 |
| Spain | 50–100 | 400–2000 | 650–1300 |
| Sweden | 35–50 | 160–270 | 80–540 |
| Switzerland | 55–160 | 430–2145 | 215–535 |
| Taiwan | 25–500 | 2145–4000 | 720–2050 |
| Turkey | 140–235 | 430–1875 | 750–1690 |
| UK | 50–300 | 400–3000 | 500–650 |
| USA | 40–125 | 740–1480 | 615–980 |
| **Proposed** | | | |
| Finland | 60 | — | — |
| FRG | 150$^c$ | — | — |
| Netherlands | 20 | — | — |
| Switzerland | — | 430 | — |
| Taiwan | — | 1430 | — |
| USA | 60 | — | — |

$^a$Emission standards are at 6% O$_2$, STP (0°C. 101.3 kPa) on dry flue gas.
$^b$Set on a plant-by-plant basis according to nationally defined formulas.
$^c$Only for plants <1 MW.

(SO$_2$) during the combustion process. Depending on the combustion conditions, a small amount of sulfur trioxide (SO$_3$) may be formed by direct oxidation of SO$_2$.

Coal combustion contributes about one-fourth the sulfur oxides present in the atmosphere. Some of this sulfur dioxide is converted to sulfur trioxide and then to sulfuric acid or sulfate salts. About 15% of the sulfur in coal is retained in the ash, and 40% of the sulfur is deposited as sulfur dioxide on surfaces (buildings, plants, cars, etc.) or by reacting with them. Another 20% of the sulfur is washed from the air as sulfurous acid in rain or snow, and the remaining 25% is washed from the air as sulfuric acid or sulfate salts in rain or snow [12].

## Table 8 Summary of the Clean Air Act Amendments of 1990

| Affected emission sources | Steam electric plants of 75 MW or greater, 1985 emission rates of 1.2 lb/MBtu $SO_x$ or greater | |
|---|---|---|
| Total $SO_x$ emission | Phase I | Phase II |
| Deadlines | December 31, 1994 | December 31, 1999 |
| $SO_x$ emission limits | 2.5 lb/Mbtu | 1.2 lb/Mbtu |
| Emission cap | Existing capacity capped at 1985–1987 average emissions rate and emissions from new units/ capacity require offsets | |
| Total $NO_x$ reduction | $2.0 \times 10^6$ ton/yr by phase II | |
| Emission trading | Federal allocation | |
| | Priority reserve at $1500/ton | |
| | Federal auction | |
| | Special allowances | |
| | Open market | |
| $SO_2/NO_x$ substitution | Under study before formulating regulations (1-1-94) | |
| New sources performance | Present NSPS would be reinforced by new standards (NSPS) | |
| | Revised emission cap on new emissions sources of 8.9 million tons/yr of $SO_2$ | |
| Clean coal technology | Deadline December 2003. Streamline permitting | |
| Incentives | FERC rate-making incentives | |

Sulfur gases may have long-term effects on human, animal, and plant life and have bad odors. The deposition of sulfurous and sulfuric acids can cause corrosion or decomposition of many of the materials on which these acids fall.

$NO_x$ emission from coal is another significant pollution problem. During combustion nitrogen in coal is oxidized to a mixture of nitrogen oxides ($NO_x$). The main products are nitric oxide (95%), nitrogen dioxide, and trace amounts of nitrous oxide. Unlike the case of $SO_x$, only a small amount of $NO_x$ formed during combustion comes from coal; the majority comes from the reaction of nitrogen in the air with oxygen at the high temperatures of combustion. Nitrogen oxides from coal combustion represent about 5% of the total nitrogen oxides in the air.

The primary methods for reducing emissions of $SO_2$ and $NO_x$ are to increase the efficiency of coal use, the use of precombustion measures, the use of con-

## Table 9 $SO_2$ and $NO_x$ reduction through repowering technologies

| Technology | $SO_2$ removal (%) | $NO_x$ removal (%) |
|---|---|---|
| AFBC | 90 | 60–80 |
| PFBC | 90–95 | 60–85 |
| IGCC | 99 | 80–95 |
| MHD | 99 | 80–95 |
| Fuel cells | 99 | 80–95 |

**Table 10 Comparison of conversion efficiencies and CO$_2$ emissions of coal utilization technologies**

| Technology | Efficiency (%) | CO$_2$ (gr./million tons) |
|---|---|---|
| Conventional | 33 | 3.1 |
| AFBC | 36 | 2.8 |
| PFBC | 40 | 2.5 |
| IGCC | 42 | 2.4 |
| Fuel cells | 50 | 2.0 |
| MHD | 55 | 1.8 |

current treatment, and the use of postcombustion measures or flue gas treatment [16, 20]. The efficiency of coal use can be increased by improving combustion, reducing heat loss, using more auxiliary equipment, and reducing heat.

Two technologies are considered in precombustion coal cleaning measures. The first is advanced physical coal preparation (beneficiation), which removes higher levels of ash and sulfur than is possible with the currently available processes; an example of this process is the two-stage fourth flotation process. The second is the coal-water mixture (CWM) technology, with which it is possible to convert existing oil-fired capacity to a coal-based fuel while minimizing modifications to the combustor and the ancillary equipment. The CWM can be made from raw coal or from coal that has been beneficiated. Details can be found in Chapter 6.

Four technologies are considered in the combustion category. The first employs the slagging combustor, where coal is burned at moderate temperatures, minimizing NO$_x$ production, and sorbents are used to control SO$_2$ emissions. The second is the integrated coal gasification combined cycle (IGCC) technology, which has the potential for very efficient use of coal while providing high levels of control for atmospheric emission associated with coal utilization. The third technology is the atmospheric fluidized bed combustor (AFBC), which can use lower quality coals while maintaining environmental standards at or near the levels required in the United States. The fourth technology is the pressurized fluidized bed combustor (PFBC), which gives better efficiency than the AFBC while maintaining high environmental standards and is used mostly in new electric power generation plants with a capacity of 200–300 MW or more.

Several posttreatment technologies are designed to control SO$_2$ and NO$_x$ emissions when coal is burned. Two of these technologies are capable of reducing both SO$_2$ and NO$_x$ emissions by 50–60%. They are the limestone injection low NO$_x$ burner (LIMB) and gas reburning–sorbent injection (GRSI). Two other technologies capable of meeting U.S. standards for SO$_2$ emissions are the dual-alkali and spray-dryer processes. By use of an appropriate additive, these technologies can achieve some NO$_x$ control. Another technology is the NO$_x$SO$_x$

process, in which high levels (up to 90%) of $SO_2$ and $NO_x$ control can be achieved simultaneously. In this process, it is possible to regenerate the sorbent and produce a marketable by-product such as elemental sulfur or fertilizer. The last of the posttreatment technologies is the selective catalytic reduction (SCR) process, which achieves high efficiencies of removal of $NO_x$ from oil- and coal-fired facilities.

## 2.1.7 Ash Disposal Problem

The ash content of coal ranges from about 2 to 30% by weight. Ash in both of its forms, fly ash and bottom ash, is the main solid residue produced by conventional combustion processes. Fly ash typically accounts for 10–85% of the total ash content, depending on the type of facility and coal used. Fly ash particles range between 0.5 and 200 $\mu$m in diameter and consist mainly of mineral (inorganic) matter that has been altered by the high processing temperatures in the combustion process. Fly ash is removed from the gaseous combustion effluent by one or more of the following control devices: mechanical cyclones, electrostatic precipitators, fabric filters, and wet scrubbers. Bottom ash is a coarser, larger, and heavier material that is usually removed from the bottom of the boiler by sluicing [1].

The increase in use of coal for power generation and other industries, along with the addition of more sophisticated air pollution control devices, has led to a larger volume of coal ashes and other wastes. Their disposal is a major problem for the coal industry as well as the environment. An example showing the large amount of ash produced by a power plant is the White Bluff power plant of the Arkansas Power and Light Company (AP&L), which burns comparatively low-ash-content subbituminous coal (contains 5–7% ash by weight). If the plant is operated at full capacity 22–32 tph of coal ash is produced (80% is fly ash and about 20% is bottom ash) by each of the two boiler units [21].

This example indicates the difficulty of eliminating or reducing the large amount of ash produced by the use of coals in power generation plants and other industries. Typically, coal ashes are disposed of in mines as fillers, dumped into waste ponds, and dumped into landfills on the site where coal is burned or treated. These methods of ash disposal have a potential impact on the environment through the contamination of ground and surface water by leaching of trace contaminants, disruption of land and contamination of land surfaces, or deterioration of the air quality by resuspension of collected fine fly ashes in the atmosphere during the disposal process.

These problems have led to increased research efforts to find commercial and economical ways to take advantage of the large amounts of coal ashes produced. One way to reduce the amount of ash to be disposed of is to market the ash commercially. An example is fly ash, which can be used in concrete and concrete products, resulting in a lower water requirement, higher ultimate strength, improved workability, lower ultimate permeability, and less heat generation during curing. Fly ash can also be used in the cement industry as an inexpensive source of silica and alumna. Other uses are in soil stabilization, asphalt paving mixes, ceramic products, and as an agricultural limiting agent. Bottom ash has little commercial use and is usually disposed of in landfills.

## 2.1.8 Particulate Emission Control

Particulate matter evolving during coal combustion is produced from the mineral matter. Particulate emissions from coal-fired power plants are a function of the fly ash/bottom ash ratio, which in turn is a function of the type of boiler, the nature of the coal and ash, and the control devices employed.

Cyclones, electrostatic precipitators (ESPs), wet scrubbers, and fabric filters are the devices most frequently used for particulate matter control. The design of particulate matter control processes depends on the properties of fly ash, such as its particle size and composition, the composition of the coal used, and present and future environmental regulations. The particle size of ash depends on the type of boiler, boiler temperature, ash melting point, slag viscosity, and use of additives to change the fusion point.

Cyclone separators use centrifugal force to separate particulates from flue gas. They are the simplest and most economical particulate control devices. Cyclones are most efficient when used to preclean large particles ahead of other control devices, with efficiencies as high as 85–90% being achievable. However, their collection efficiency can drop to about 30% when used to remove submicrometer particulates.

Electrostatic precipitators (ESPs) are the most common control devices used in the electric power industry. ESPs consist of two electrodes between which the effluent gases pass, acquire an electric charge, and are attracted to one of the electrodes. ESPs remove particles of all sizes with an efficiency ranging between 97 and 99.95%, but these efficiencies are considerably lower for particle sizes between 0.1 and 1.0 $\mu$m than for either larger or smaller sizes [22]. The collection efficiency of ESPs can be influenced by variations in chemical composition, temperature, water vapor content, sulfur trioxide concentration, particulate matter loading, and particle size distribution.

Wet scrubbers are used to remove particulates where an $SO_2$ removal system must be installed. The scrubber can be used to remove both $SO_2$ and particulates. This eliminates the need to install two scrubbers, one for particulate removal and the other for $SO_2$ removal. The use of wet scrubbers as control devices has many disadvantages, such as the high pressure required to remove small particles, corrosion of the wetted surfaces, and reduction in the gas buoyancy resulting from cooling.

Fabric filters or baghouses are particulate collectors with efficiencies higher than 99.5%. Their removal ability is affected less by changes in particle sizes, ash analysis, gas composition, and temperature than that of ESPs. However, they require a large amount of space, have a high pressure drop, have the possibility of fire or explosion, and have low collection efficiencies for some coal ashes.

## 2.1.9 Acid Rain

Emissions of nitrogen oxides ($NO_x$) and sulfur oxides ($SO_x$) into the atmosphere can cause the formation of acid rain. If these oxides are in the atmosphere, they can be further oxidized over time. In this form they combine with water vapor in the air, producing nitric and sulfuric acids, which fall as acid rain. Acid rain formation schemes are summarized in Table 11 [23].

Acid rain has a lower pH than normal rain (pH of about 5.6). Normal rain is, typically, slightly acidic from atmospheric $CO_2$ [24]. Acid rain has devastating effects on lakes, soils, and vegetation. Some lakes have become so acidic that fish, plants, and other aquatic life can no longer survive. The number of plant nutrients in the ground is reduced, leading to problems of tree growth and eventually the destruction of forests. Human health problems and property damage (e.g., eroding coated film surface) have also resulted.

Acid rain can spread over a large area; it is not limited to the emitting source but can spread over hundreds of miles. The fact that these oxides (mainly sulfur dioxide) can be transported through the atmosphere over hundreds of miles has made acid rain, once a local issue, an international issue.

The combustion of coal produces approximately 25% of the total $SO_x$ and 5% of the $NO_x$ gases in the atmosphere [12]. However, options are available to reduce the contribution of coal to the total $SO_x$ and $NO_x$ gases in the atmosphere

**Table 11  Acid rain formation processes**

| | |
|---|---|
| **Formation of sulfuric acid** | |
| Step I | The oxidation of sulfur when coal is burned |
| | $S + O_2 = SO_2$ |
| Step II | Further oxidation of sulfur in the atmosphere |
| | $SO_2 + 0.5O_2 = SO_3$ |
| Step III | Combination with water vapor to form acid |
| | $SO_3 + H_2O = H_2SO_4$ |
| **Formation of nitric acid** | |
| Step I | Nitrogen fixation in hot engines and furnaces |
| | $N_2 + O_2 = 2NO$ |
| Step II | Further oxidation of nitrogen in the atmosphere |
| | $NO + 0.5O_2 = NO_2$ |
| Step III | Combination with water vapor to form nitric acid (and more NO) |
| | $3NO_2 + H_2O = 2HNO_3 + NO$ |

and to meet the standards set by the new environmental regulations. They are the use of low-sulfur coals, beneficiation of high-sulfur coals, and installation of equipment to increase flue gas desulfurization efficiency or reduce $NO_x$ formation, as discussed previously.

### 2.1.10 Greenhouse Effect

The greenhouse effect is caused by accumulation of greenhouse gases such as carbon dioxide ($CO_2$), methane ($CH_4$), chlorofluorocarbons (CFCs) such as the refrigerant Freon gases, nitrous oxide ($N_2O$), and ozone ($O_3$). The presence of these infrared-absorbing gases in large quantities in the atmosphere contributes to the greenhouse effect (global warming) by allowing incoming solar radiant energy to penetrate to the earth's surface while absorbing infrared radiation emanating from the earth. Figure 16 [25] demonstrates the greenhouse effect (radiation flows expressed as a percentage of total incoming or outgoing energy).

An increase in the planet's temperature due to the continuous emission of greenhouse gases may result in extensive melting of the polar icecaps. A large amount of water would be released, raising sea level to a point that could result in flooding of many of the world's major cities and destruction of productive agricultural land.

The relative importance of greenhouse gases in terms of their effect, estimated lifetime, and percent contribution is summarized in Table 12 [25–27].

Coal utilization worldwide contributed to ~32% of the $CO_2$ emitted in 1990 and was responsible for ~20% of the enhanced greenhouse effect, of which 19%

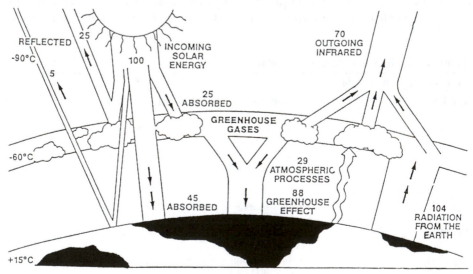

**Figure 16** Demonstration of the greenhouse effect [25].

**Table 12 The greenhouse gases**

| | Carbon dioxide ($CO_2$) | Methane ($CH_4$) | CFCs and halons | Nitrous oxide ($N_2O$) |
|---|---|---|---|---|
| Global increase annual basis, % | 0.5 | 1 | 6 | 0.4 |
| Life, yr | 120 | 10.5 | 110 | 132 |
| Effect relative to $CO_2$, times | 1 | 30 | 20,000 | 150 |
| Contribution to the greenhouse effect, % | 55 | 1 | 2 | 6 |
| Sources | a | b | c | d |

a Fossil fuel burning, deforestation and land use, and burning of biological matter.
b Product of bacterial breakdown of organic matter in swamps, paddy fields, waste dumps, animals, natural gas, and bituminous coal mining leaks.
c Coolants in refrigerators, air-conditioners, aerosols, production of plastic foams, solvents, and fire extinguisher testing.
d Fertilizers, fossil fuel, and biomass burning; changes in land use; and vehicle emissions.

was due to the emission of $CO_2$ and the other 1% due to the emission of $CH_4$ and $N_2O$ gases [26]. About half of the 20% contribution of coal to the greenhouse gases came from coal-fired power generation plants, where more than 50% of the coal produced worldwide is utilized.

Because $CO_2$ is the single largest contributor to the greenhouse gases from the use of coal for power generation plants, most studies concentrate on how to reduce the $CO_2$ emission to the atmosphere. Some options for reducing $CO_2$ emissions from coal in power generation, other than conservation of energy by reducing overall demand, are as follows: the use of supercritical steam cycles may result in a $CO_2$ reduction of up to 20%; the use of PFBC and coal combined cycles may reduce $CO_2$ emissions by 25% or as much as 37% with pulverized coal–natural gas combined cycles; the use of coal-fired MHD and fuel cells may reduce $CO_2$ emissions by about by 17–44%; and cogeneration using all these technologies could result in $CO_2$ reduction of 60% [26].

Postcombustion control through recovery and disposal of $CO_2$ consumes energy and reduces power plant efficiency. Gas-liquid separators (absorption), gas-solid separators (adsorption), and gas fractionation systems that depend on cryogenic or membrane separation techniques are three different methods available for the separation and recovery of $CO_2$ from flue gases. These gases can be disposed of in solution-mined salt domes or in the deep ocean or can be utilized by natural or industrial processes [28].

## 2.1.11 Safety Issues in Coal Mining

Mining has been and will continue to be a dangerous and risky business; but new regulations and measures, increased awareness and training of the miners, and the use of advanced technologies and safety measures in coal production operations have resulted in a reduction of the mine accidents. In the United States, 3242 miners died in 1907 while producing 800 million tons of coal, but in 1991 59 miners died while producing 1 billion tons of coal [29]. In Great Britain, 1453 miners died in 1909 and this number dropped to 28 miners in 1986 [30]. The most important safety issues to be dealt with in the coal mining industry are summarized as follows:

**Mine ventilation.** Continuous ventilation is required to provide miners with fresh air and to remove hazardous and harmful gases. In Europe, for example, the average weight of air passing daily through a coal mine is about six times the total weight of coal produced daily; large mines can require fresh air at a rate of more than 0.5 million cubic ft$^3$ per minute [6].

**Mine gases.** The most harmful and hazardous gases produced during mining operations are known as damps. They are summarized as follows:

1. Firedamp (methane) is the most common gas and is a by-product of coali-fication. The amount of methane varies from one coal mine to another, and up to 5000 cubic feet of methane can be liberated for each ton of coal produced. Methane is colorless, tasteless, and odorless, making its presence not easily detectable. It is a lighter gas than air, highly flammable gas that forms explosive mixtures if it has a concentration in air of 5–15%. In order to keep the amount of methane in mines under critical levels (~1%) to avoid the formation of explosive methane-air mixtures, large volumes of air are circulated. Methane is usually liberated gradually in coal mines; however, sometimes an "outburst," a sudden large violent production of methane, is liberated, typically accompanied by large quantities of broken rocks and dust.
2. Whitedamp (carbon monoxide) is produced as a by-product of explosions and mine fires as a result of incomplete combustion of carbon, slow oxidation of the coal, or the use of diesel engines.
3. Blackdamp (chokedamp, carbon dioxide) is found mainly in old working or badly ventilated mines. It is formed in mines as a result of combustion or oxidation processes such as coal fires, explosions, or lamp flames. It is not poisonous but can cause asphyxiation by displacing breathable oxygen. The danger arises mainly because $CO_2$ is heavier than air, so it tends to accumulate and cause a lack of oxygen in lower levels of the mines where miners work.
4. Stinkdamp (hydrogen sulfide), evident from an odor of rotten eggs, is a highly poisonous gas that can be fatal in a few minutes at a concentration of about 0.1% in air. It is produced in the mines when coal is heated out of contact with air or catches fire or when an acidic mine water reacts with pyrites in the coal. $H_2S$ can also be formed in small quantities with methane resulting from natural process or outbursts.
5. Afterdamp (mixed gases) is a mixture of gases that can be found in a mine after an explosion or fire. The composition of these mixed gases varies and

depends on the nature of the explosion or fire, the amount of material burned or exploded, and the conditions. Typically, it has a high carbon monoxide content and is deficient in oxygen, which significantly increases the dangers for miners and rescuers.

**Mine fires.** Mine fires usually occur as a result of spontaneous combustion of coal, coal fines, and mixed gases by electrical sparks, electrical malfunction, welding, cutting, and smoking. Mine fires are dangerous because they can initiate an explosion and the coal itself may be a lasting supply of fuel. These fires are very difficult to control and may result in a large number of bodily injures and casualties.

**Rock or roof falls.** Rock and roof falls are some of the most serious causes of death and injuries to miners and rescuers. The risk of rock or roof falls can be reduced by enhanced safety measures, such as compliance with general roof support requirements and local conditions, safety training of the miners, and modification of equipment design and operating procedures in areas of unsupported or dangerous roofs.

**Dust explosions.** Dust explosions, especially caused by methane ignition, are more dangerous than simple firedamp explosions, because of their violent behavior and widespread effects. The danger of coal dust explosions can be reduced by mixing limestone dust with coal dust to convert it into a noncombustible mixture or by deposition of a highly hygroscopic salt to absorb moisture from the air and convert the coal dust to mud, which hardens into a subsolid substance.

Coal dust can cause black lung disease in humans, which results from the accumulation of coal dust in the small sacs in the lungs. The danger of dust can be reduced by spraying water at the point of origin to keep dust levels down, infusing water at coal faces, improving ventilation or dust collection, and using dust masks.

Other safety issues in coal mining are the proper use and functioning of the operational equipment, the availability and readiness of a well-trained rescue team, and proper mine planning control for design of improved safety features. Some of the features of mine planing are safety pillars, escape ways, sufficient air supplies, roof support, and equipment selection.

## 2.1.12 Safety and Economic Issues in Coal Transportation

The transportation cost of coal is critical to the coal industry because it can affect the delivered price of coal more than the production cost. More intense market competition has forced the coal industry to study and compare the different methods of coal transportation in terms of their effectiveness, economics, practicality, and safety. The most significant modes of coal transportation are summarized as follows:

**Railroads.** Railroad transportation is the most economical and important means of long-distance transportation of coal. It handled about 59% of the coal transported in the United States in 1983 [31] and about 75% of the coal transported worldwide [12]. The development and use of the unit train, which consists of coal cars loaded at one mine and unloaded at one destination on each trip and moves in each direction on a prefixed schedule, allows shippers to take advantage of the lowest rates possible and to have faster and more effective service.

Disadvantages and concerns about using railroad transportation in general and the unit train in particular are related to disruptions on local streets and crossings that may delay essential hospital, fire, and police services and increased air pollution from both locomotives and auto traffic idling at grade crossings and coal dust particles escaping from open hopper cars.

**Trucks.** Truck transportation of coal is expensive. It is economical only for moving coal short distances from the mine to preparation plants or moving it to or from long-distance transportation by trains and barges. Trucks in the United States handled 14% of the coal moved in 1983 [31].

**Water transportation.** This is the cheapest way of moving coal other than railroad transportation. It includes barge and self-propelled vessels. It is the second most important mode of coal transportation in the United States, with about a 16% share in 1983 [31]. The use of water transportation depends on the location of coal mines and the delivery points: whether they are near a waterway where barges can be used or far from a waterway, where other transportation modes should be used. Water transportation usually involves another mode of transportation, such as rail, truck, pipeline, and converters to feed it or to move its loads to the delivery site.

**Pipeline transportation.** Coal slurry pipelines are another way to transport coal from mines to consumers. Coal is ground and mixed into a slurry, which is pumped to the consumer end, where coal is recovered. The economics of coal slurry pipelines depend on the distance the coal will be transported (the longer the better), the volumes (high or low), whether there is an existing rail service, and long-term contracts (should be for at least 20 years). The slurry pipeline fixed cost is very high, but once the pipeline is constructed the operating cost is low compared with other modes of coal transportation. A disadvantage of the use of pipelines for coal transportation is that they require a large amount of water or other fluids, such as liquefied petroleum gas, fuel oil, crude oil, methanol, and carbon dioxide, to be able to transport the coal.

The most successful pipeline in the United States is the 273-mile-long, 18-in.-diameter Black Mesa coal pipeline, which transports about 5 million tons of coal each year from the Black Mesa mine in Arizona to two 750-MW power-generating units in southern Nevada [12, 31, 32]. Despite environmental problems such as those arising during the construction of the pipelines and the opposition of local groups (farmers) and railroads, which fear competition, slurry pipelines offer a very good method of coal transportation for the future.

**Conveyors.** Conveyors are used to transport coal short distances, usually up to 10 miles from the coal mines, to power plants or transporting terminals. They have a high operating cost, and for conveyors to be economical a large and constant volume of coal must be a available to transport. Conveyors and slurry pipelines accounted for about 11% of coal transportation in the United States in 1983 [31].

Building power plants (mine-mouth plants) near the coal mines in order to eliminate long-distance transportation of coal has been studied and practiced in the past few years. In this case conveyor belts or short-haul trucks are used to move the coal from the mine to the plant for use in the production of electricity, which is transported to consumers by extrahigh-voltage (EHV) transmission lines.

## 2.1.13 Future and Trend Analysis

Coal is the most abundant fossil fuel, constituting approximately 75% of the total world resources of fossil fuels. It is estimated that at current consumption levels, coal reserves will last about 250 years, while oil and natural gas reserves will last about 43 and 67 years, respectively [33, 34].

The fast increase in the world's population and growth in developing countries will result in increased demand for energy and the services it provides, such as heating, cooling, cooking, mobility, and motive power. To meet this increase in demand, coal will be the most logical choice as an energy alternative because of its abundance compared with other sources such as oil and natural gas, high level of mining industry technology, and practical world experience in using coal. Most important is the geographic diversity of coal reserves, which it can be found in more than 100 countries [34], making coal one of the most secure sources of the energy essential for economic growth and social stability.

The main use of coal is as a fuel for electric power plants, where more than 50% of the coal produced in the world is used [1, 12]. In 1990 coal was the basis for about 39% of the world's electricity consumption, compared with about 14% for gas, 12% for oil, 17% for nuclear, and 19% for hydroelectric. The consumption of electricity is expected to grow at 2–2.5% up to 2020, which means that the amount of coal required to produce electricity will increase steadily in the near future [35].

The second greatest use of coal is for production of coke. Coke is one of the most important raw materials in the iron and steel industry and is used to liberate iron from its ores. Whether the use of coal in the steel industry will increase in the future depends on the total requirements for steel, the development and use of new technologies such as direct reduction of iron ore to steel, and the development of small steel plants that reuse steel scrap, resulting in decreased use of coal.

Another use of coal is for the production of chemical products such as solvents, fertilizers, drugs, dyes, perfumes, and artificial flavors. Most of these are produced as by-products of coke production for the steel industry. The contribution of coal in the production of chemicals could be increased in the future by the construction and use of new coal gasification and liquefaction plants.

Other uses of coal that may be increasingly important in the future are in the production of liquid fuels by direct or indirect processes to replace fuels made from petroleum; production of methanol, a possible substitute for gasoline; and production of synthetic gases (high- or low-heating-value gases).

Increased use of coal in the future depends on several factors. One is the development of more economical and efficient process technologies based on the use of coal that are able to compete with other sources of energy including oil, natural gas, and nuclear. A second factor is the development of new technologies to meet strict environmental regulations imposed by governments worldwide and to solve environmental problems related to the use of coal, such as emission of sulfur oxides, nitrogen oxides, and carbon monoxide gases, resulting in the greenhouse effect and acid rain. The last factor is how long oil and natural gas resources will last.

## 2.2 COAL AS A SOURCE OF CLEAN FUEL

Coal has broad industrial importance. Largely, though, it is used as a feedstock in the production of petrochemicals from syngas and as a raw material in the production of electricity at combustion facilities. Its continued use worldwide is dependent on many factors, some related directly to the economic climate of the chemical industry, others related to the current level of technology and its ability to produce an environmentally acceptable effluent stream.

Generally speaking, coal has competed with natural gas as a feedstock in many areas. The use of coal over natural gas has many advantages:

• Coal is more economical than natural gas.
• Coal is present in much larger reserves than natural gas.
• Coals is ideal for the production of CO-rich syngas by steam reforming, which is desired for many industrial processes.

Disadvantages of coal use:

- Coal requires processing after, during, or before combustion to minimize $SO_x$ and $NO_x$ emissions. Elaborate control devices are necessary to monitor flue gas compositions to ensure that contaminant levels are within acceptable limits.
- Coal often has a high water content.
- High ash levels in some coals cause fouling, slagging, corrosion, and erosion problems in boilers.

Overall, the future use of coal will depend on many factors, not least of which is the availability of dwindling natural gas reserves and the cost effectiveness of using coal. Some of the factors in the use of coal as a clean fuel are discussed in the following sections.

## 2.2.1 Environmental Acceptability

The future use of coal, in relation to natural gas, will depend on several factors, many of which have already been outlined. An important factor in the use of coal over the next 10 years will be the economics of developing or retrofitting combustion facilities with the necessary equipment to reduce undesirable emissions to acceptable levels and its relation to switching to natural gas.

Legislation has and will continue to dictate the "acceptable" level of emissions in the United States. European countries such as the United Kingdom, France, Sweden, and Germany are also adopting control schemes to reduce harmful products of combustion.

The problem has been defined by the legislators in several parts. They are all directly related to the chemical composition of coal and the reactions it undergoes during combustion and subsequently with the atmosphere after it is released into the environment.

Acid rain has been a problem for almost two decades, especially in Germany and France, where strict regulations on emission controls are not present on a national scale. Acid rain arises from sulfur and nitrogen present in coal. In the combustors, sulfur is converted to $SO_2$ and $SO_3$ (i.e., $SO_x$) and nitrogen is converted to NO and $N_2O$ ($NO_x$). $SO_x$ and $NO_x$ emissions create a problem when they come into contact with moisture in the atmosphere and form sulfuric and nitric acids. The acid, which precipitates as acid rain, attacks trees, destroys vegetation, alters the pH of ponds and streams, and in general creates a great imbalance in the ecosystem.

Acid rain also attacks and breaks down the cement of brick buildings and destroys landmarks and statues. In general, anything that is exposed to the environment is subject to some level of destruction by acid rain.

Acid rain is not the only problem associated with coal burning. Particulate emissions also create problems, especially when the particulates contain the high levels of trace metals commonly found in coal. Greenhouse gases like CO, $CO_2$, and $CH_4$ must also be monitored and controlled.

Overall, the use of coal in combustion facilities poses serious environmental concerns that must be addressed for continued operation. To get a better understanding of the requirements placed on combustion sites and the regulations they must abide by, it is necessary to examine the current set of regulations imposed by legislators on industrial facilities.

**Clean Air Act.** The main legislative work that dictates the acceptable level of emissions is the Clean Air Act Amendments (CAAA) of 1990, also known as the acid rain laws. The main feature of the amendments is Title IV, which relates specifically to acid rain emissions. The approach taken by the Department of Energy is unique in that it is a market-based approach to cleanup rather than a "command and control" attack. The CAAA have three essential elements [36].

1. A 10-million-ton reduction in annual $SO_2$ emissions split into phase I and phase II and a 2-million-ton reduction in $NO_x$ emissions from 1980 levels.
2. A nationwide cap on $SO_2$ emissions of 8.9 million tons a year.
3. An elaborate $SO_2$ allowance accounting and trading scheme.

This system of allowances gives the amendments their market-based approach. Each allowance is equivalent to 1 ton of $SO_2$ removed. Allowances representing $SO_2$ emissions reduced beyond what is required in the legislation can be earned, banked, leased, bought, or sold like money, stocks, and other financial instruments. Their value fluctuates with the market. They are, in a sense, the currency of clean air. Although California has used credits for some industrial emissions, no free-market approach to environmental protection on this scale has been attempted anywhere in the world. "It is a grand experiment with the world's eyes upon it" [36].

**Emission restrictions**

*Details of the legislation.* Title IV of the CAAA calls for a two-step program to reduce $SO_2$ emissions nationwide by 10 million tons from 1980 levels by the year 2000. Afterward, the cap of 8.9 million tons a year takes effect. The reduction deadlines involve two phases. The first phase is focused on cutting the emissions of 110 of the largest $SO_2$-emitting central stations to an annual average rate of 2.5 lb/million Btu multiplied by an average of their fuel use between 1984 and 1987. This phase I reduction accounts for half of the total $SO_2$ reductions that will occur. The deadline for phase I went into effect on January 1, 1995. Table 13 provides a listing of the power plants affected by phase I [37, 38].

Phase II begins January 1, 2000, and affects 785 plants (more than 2100 units) in 27 states. Phase II regulations tighten the annual emissions of phase I plants and set restrictions on all plants with generating capacities greater than

## Table 13  Power plants affected by phase I

| State | Number of targeted plants | Generation capacity MW |
|---|---|---|
| Alabama | 2 | 3,363 |
| Florida | 2 | 2,771 |
| Georgia | 5 | 8,444 |
| Iowa | 6 | 977 |
| Illinois | 8 | 6,002 |
| Indiana | 15 | 11,275 |
| Kansas | 1 | 158 |
| Kentucky | 10 | 4,649 |
| Maryland | 3 | 2,390 |
| Michigan | 1 | 650 |
| Minnesota | 1 | 163 |
| Missouri | 8 | 6,547 |
| Mississippi | 1 | 750 |
| New Hampshire | 1 | 459 |
| New Jersey | 1 | 299 |
| New York | 5 | 2,408 |
| Ohio | 15 | 14,131 |
| Pennsylvania | 9 | 7,675 |
| Tennessee | 4 | 6,331 |
| Wisconsin | 6 | 2,740 |
| West Virginia | 6 | 7,363 |
| Total targeted capacity | | 89,545 |

25 MW and all new utility units after that. Average emission rates for phase II are set at 1.2 lb/MMBtu. Units using a clean coal technology will receive a 4-year extension of the deadline.

$NO_x$ emissions are also regulated under Title IV. Here, however, the market-based approach of allowances is not present. Rather, strict limits on emissions are applied depending on the type of boiler unit used and in a command-and-control manner [38]. Specifically, dry-bottom and wall- or tangentially fired units must reduce $NO_x$ emissions by 2 million tons in phase I. Wall-fired unit emissions are set at 0.50 lb/million Btu, tangentially fired units at 0.45 lb/million Btu. The limits are based on the use of low-$NO_x$ burners (LNB) and on an annual average. $NO_x$ limits for other types of boilers, also based on LNBs, are to be specified by January 1, 1997. In addition, the Environmental Protection Agency must set revised New Source Performance Standards (NSPS) for $NO_x$ emissions from fossil-fired units. Although the CAAA does not permit allowance trading between $SO_2$ and $NO_x$, it does allow studying the concept [36].

Based on the analysis of Sargent & Lundy engineers Peter W. Guletsky, Jerry A. Sadlowski, and Rajendra P. Gaikwad, the relationship between Title I and Title IV has not been clearly defined by the EPA. Their analysis of the situation concluded that the more stringent emission limit applicable to any particular plant will prevail and that CAAA seems to establish low-NO$_x$ burners as the model technology for current plants under Title I, Reasonably Available Control Technology (RACT), and Title IV. For new plants, selective catalytic reduction (SCR) and selective noncatalytic reduction (SNCR) technology may be required [38].

For phase I, group II boilers (i.e., cyclones, cell burners, wall-fired wet-bottom boilers, and other utility-grade boilers), EPA must establish requirements by 1997. The CAAA does not yet specify the date on which these units must comply, and Desai and Coffey pointed out that the EPA can set the date at its own discretion. It is likely that compliance will be required by January 1, 2000 [38].

Figure 17 provides a schedule for the DOE Clean Coal Technology Program, the deadlines corresponding to DOE clean coal technology, and also those dictated by the Clean Air Act Amendments.

The CAAA is set up with a series of exceptions that dictate the ways in which allowances may be earned and obtained from various sources. It is this aspect of the legislation that gives rise to confusion, not the existence of the allowances themselves. These exceptions, among other things, determine deadline extensions, allocation of allowances due to overcompliance and special situations, and early compliance features [36].

1. Affected sources that meet the SO$_2$ emissions limits earn only the number of allowances that permit continued operation of the unit. For sources that

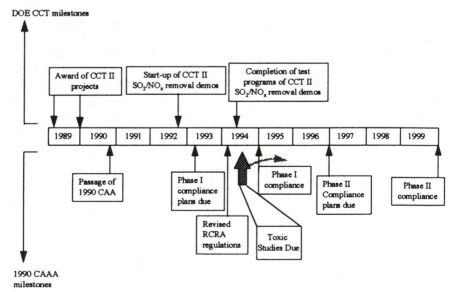

**Figure 17** Milestone schedule for the DOE Clean Coal Technology Program and 1990 Clean Air Act Amendments compliance [38, 50].

overcomply, early additional allowances can be used to cover other sources in the owner's system or can be used on the open market. This applies in phases I and II.

2. Sources that employ a technology in phase I that removes at least 90% of the $SO_2$ are entitled to a 2-year extension on the compliance deadline and bonus allowances from a 3.5-million set-aside reserve. In phase I these allowances amount to the actual emissions rate subtracted from 2.5 lb/million Btu, in phase II (through 1999) to 1.2 lb/million Btu minus the actual emissions rate. Units that comply early using scrubbers in phase I receive two-for-one bonus allowances from this reserve.

3. Another set-aside of 200,000 allowances has been specifically assigned to units in Ohio, Indiana, and Illinois—the states hit hardest economically by phase I. Yet another 50,000 bonus allowances per year are assigned to units in 10 other midwestern states that make reductions in phase I.

4. Units that repower in phase II with qualifying control technologies may extend their compliance deadlines by up to 4 years and circumvent NSPS rules if they do not increase emissions of any regulated pollutant. During the extension period, the unit gains nontransferable allowances. After the unit is repowered, the allowances become transferable. Intent to repower must be expressed by the end of 1997, with preliminary design and engineering reported by January 1, 2000.

5. To ensure that allowances are available to those that do not automatically obtain them, EPA controls 2.8% of all the allowances allocated in phase I and phase II. These total 150,000 in phase I and 250,000 in phase II and will be disbursed by special auctions. Supplementing this pool are 50,000 allowances available for direct sale beginning in the year 2000.

6. A reserve of 300,000 allowances have been created for qualifying conservation programs and renewable energy sources. They are allocated based on $SO_2$ emissions avoided.

7. Miscellaneous allowances are allocated for a variety of special situations. These involve so-called dirty units in clean states and clean units in dirty states and even apply to specific utilities hit hard by CAAA compliance. For example, coal-fired units that emit $SO_2$ at a rate below 0.6 lb/million Btu in dirty states are entitled to increase emissions by 20% of their baseline rates. In clean states, such plants would receive a share of the bonus allowances through year 2010.

8. Nonaffected sources can "opt in" to the allowance trading program. If this route is selected, the source must submit a compliance plan and permit application and generally subject itself to all of the Title IV provisions.

**Other titles.** CAAA consist of 11 titles. The most important one for electric power generation facilities is Title IV, which relates directly to $SO_x$ and $NO_x$ emissions. A few of the other titles also affect power generation directly. These include Title I on nonattainment, Title III concerning air toxic emissions, Title

V involving permits, and Title VII on enforcement for criminality associated with failure to comply [36].

- Title I is focused on urban area air quality and addresses the issue of ozone formation and particulate matter smaller than 10 μm, PM-10. This could mean problems for sources that do not reach compliance standards for $NO_x$, which could be considered a precursor to ozone formation [36].
- Title III regulates the emission of hazardous compounds. Specifically, 189 compounds have been identified as known or suspected to threaten public health. Regulation of air toxics will be based on scientific and engineering studies conducted over the next 3 years. Mercury, which has been separated out, will be studied over the next 4 years. Sources that are found to be significant emitters of these compounds will be expected to apply maximum achievable control technology. The emission problem will be reevaluated after the controls have been applied. Of specific concern to power plants are vapor-phase compounds and compounds attached to particulate matter, which are both subject to Title III provisions [36].
- The EPA's enforcement power is strengthened by Title V, which sets forth the drafting of an operating permit containing all of a given source's pollution obligations. Sources must submit periodic reports regarding the extent to which they have complied with their emission obligations [36].
- Finally, Title VII also strengthens EPA's power by giving it the authority to issue criminal penalties of up to $2 million and field citations of up to $5000. Criminal penalties for knowingly violating the CAAA are now categorized as felonies. As some have interpreted Title VII, chief executive officers of companies cited for violations could now face felony charges and imprisonment for negligent releases of hazardous pollutants, record-keeping infractions, and so on [36].

**Strategies for compliance.** With the phase I deadline already past and phase II deadline quickly approaching, power utilities are searching for quick answers to emission control problems. Quick solutions are desired not only to meet EPA deadlines but also to minimize phase I expenses and have time to consider options for phase II. Also, they give the $SO_2$ allowance market time to develop. By choosing a low-cost, quick solution to phase I requirements, industries may concentrate on the most economical, technically accommodating approach for phase II compliance. This approach minimizes a quick, expensive solution to a problem that fails in the long run [40]. Most industry sources agree that utilities have six basic strategies for CAAA compliance [38].

1. Fuel switching
2. Retrofitting control technology
3. Emissions-based dispatching

4. Allowance transactions
5. Unit repowering
6. Plant retirement

Although literature sources disagree on which strategy will be the dominant one, most agree that the two most popular options among existing power plant operations are **fuel switching and blending** and installation of **scrubbers** with an existing flue gas desulfurization (FGD) system to obtain compliance ratings. Figure 18 illustrates the options taken for phase I compliance and the capacity each option is expected to handle, as of February 1994. Trends reported here were predicted by industry experts [40].

From Figure 18, the majority of phase I strategies will be concentrated on complete or partial switching of fuel from a high- to a low-sulfur level. Increased confidence in the availability of low-sulfur coal has significantly reduced the estimates of phase I scrubbers that would be installed. Less than 14,000 MW of capacity is now expected to scrub in phase I [40].

*Fuel switching.* Switching from a high- to a low-sulfur coal will be a dominant CAAA compliance strategy. "It involves relatively low capital investment, imposes less onerous time constraints, it takes advantage of very favorable contracts for coal currently available from both the West and from central Appalachia, it can generate excess allowances within a very short time, and it allows a utility to buy some time to observe how the allowance market develops or how select clean-coal technologies evolve" [36].

Fuel switching is not as straightforward as, perhaps, it sounds. A large amount of retrofitting is necessary in switching from a high-sulfur coal to a low-sulfur coal. This is largely due to the design of furnaces, which are often based on a specific type or rank of coal as shown in Figure 19. In fact, changing from one coal to another affects virtually every aspect of a power plant's operation. Performance problems brought about by fuel switching are due largely to the new fuel's volatile manner, ash content and properties, moisture and sulfur content, and heating value [41, 42]. As shown in Table 14, the properties of moisture level, percent ash, and percent sulfur vary widely with rank of coal.

Once a specific low-sulfur coal is deemed acceptable, based on proximate and ultimate analysis, operating parameters like gas flows, temperatures, and volumes must be calculated.

Ash characteristics are an important parameter when selecting a desired coal feedstock. This is due largely to the impact that ash has on the combustion equipment, emissions control systems, and ash handling and disposal systems. Factors to be examined include ash fusability temperatures; dolomite and ferric concentrations; and base-to-acid, iron-to-calcium, silica-to-aluminum, and iron-to-dolomite ratios. These parameters determine an ash's slagging, fouling, corrosion, erosion, abrasion, and carbon burnout characteristics. All of these are

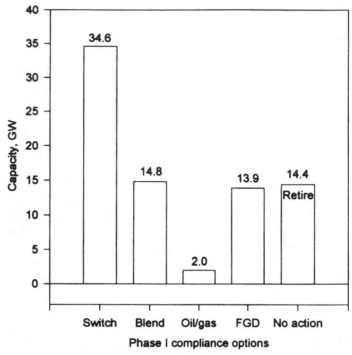

**Figure 18** Expected phase I compliance strategies.

important features to consider when selecting an alternative feedstock for use in an existing unit [41].

- **Slagging**: Slag refers to the fused or resolidified molten materials that form primarily on furnace walls or other surfaces. Slag usually forms at locations where temperatures are high. Changes to slagging potential affect furnace sizing, arrangement of radiant and convective hearing surfaces, and the number of soot blowers required [41].
- **Fouling**: Fouling is a buildup that occurs in the lower temperature regions of the combustion system. It affects the design and maintenance of superheaters, reheaters, and furnace waterwalls. Fouling has been found to be directly related to the alkaline content of ash, although the total quantity of alkali is not necessarily an accurate index of fouling potential. Sodium in coal ash has been found to be a key link in the fouling formation. Specifically, the vapor-phase sodium in the combustion gases promotes contact and subsequent bonding between ash particles and metal surfaces as gas cools and the sodium condenses. The high-alkali ash found in many western coals has been linked to a rapid buildup of deposits. This, in turn, has influenced the design of tube spacing in convective passes [41].

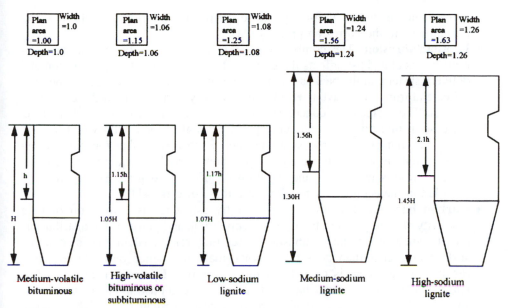

**Figure 19** Furnace size dimensions for different coal feeds [41].

• **Corrosion**: Corrosion of metal components takes place during the normal operation of a combustor. This breakdown of metal usually occurs primarily on waterwalls near the firing zone, in high-temperature superheater and reheater sections, and in low-temperature gas passes and air heaters. Corrosion has bene found to be related to a number of ash-containing components. Specifically, alkali and alkaline earth elements, iron, and sulfur are the most significant

**Table 14 Typical analyses, U.S. coals [41]**

| Property | Med-vol.[a] bituminous | High-vol. bituminous | Subbituminous | Low-sodium lignite | Medium-sodium lignite |
|---|---|---|---|---|---|
| $H_2O$, % | 5.0 | 15.4 | 30.0 | 31.0 | 30.0 |
| Ash, % | 10.3 | 15.0 | 5.8 | 10.4 | 28.4 |
| Sulfur, % | 1.8 | 3.2 | 0.3 | 0.6 | 1.7 |
| Fusion temp, F | | | | | |
| initial deform | 2170 | 1990 | 2200 | 2080 | 2120 |
| softening | 2250 | 2120 | 2250 | 2200 | 2380 |
| fluid | 2440 | 2290 | 2290 | 2310 | 2700 |
| Btu/lb, as fired | 13,240 | 10,500 | 8125 | 7590 | 5000 |
| ash, lb/MM Btu | 7.8 | 14.3 | 7.1 | 13.7 | 56.8 |
| Fuel feed,[b] 1000 lb/hr | 405 | 520 | 705 | 750 | 1175 |
| Ash out,[b] 1000 lb/hr | 42 | 78 | 41 | 78 | 334 |

[a]vol., volatility.
[b]Based on constant heat output, nominal 600-MW unit, adjuted for moisture content.

constituents related to high-temperature corrosion. Chloride also plays an important role in the corrosion process [41].

- **Erosion/abrasion**: The combination of increased velocity and hardness of both raw coal and ash affects the degree of erosion in the fuel preparation and delivery system, in the backpass convective section, and in the steam generator. Silica and iron oxide have been identified as key participants in the erosion of fly ash. The increase in concentration of these components will result in an increase in fuel flow rate and subsequently a higher force with which fly ash comes in contact with process equipment. High postcombustion particle velocity results in a large tendency for particles to stick to equipment wall surfaces or results in erosion and/or abrasion of some degree. Shields may be installed in the economizer to overcome these effects [41].

- **Carbon burnout**: Unburned carbon in fly ash leads is a function of char reactivity, the temperature field, $O_2$ concentration, and particle size distribution, all of which affect the combustion rate. Low-rank coal produces a higher surface area char than does bituminous coals and hence burns faster. As a result, low-rank coal is considered more reactive. This is not unreasonable, since low-rank coals often have a tendency to combust spontaneously upon sitting and are generally considered more reactive than their higher ranked counterparts [41].

*Flue gas desulfurization and scrubbing.* Flue gas desulfurization (FGD) has proved to be an efficient alternative for many eastern power generation facilities. A primary feature of FGD is that its technology has dominated the systems bought by electric utilities. It represents a known quantity with an experience base in the United States and offers relatively low marginal $SO_2$ removal costs, especially when at least 90% removal is desired for an extended period [40]. An example is the power facility of Henderson Municipal Power & Light. The utility is located in Kentucky and is erecting scrubbers for its 350-MW station. Two engineering consultants for the company concluded that it would cost the company about 10% less to scrub than to switch to a nonlocal low-sulfur coal. However, the saving is spread over a 20-year period and anticipates increases in rail transportation, operation, and maintenance costs but not allowance credit trades or sale [40].

Although scrubber capital costs are among the highest of all commercial options for CAAA compliance, their use affords a high degree of flexibility; scrubbers are the easiest to implement in existing operations, are relatively cost effective, and may be used quickly. Also, high removal efficiencies accommodate a wide variety of coals for combustion, and scrubbers have virtually no impact on boiler or electrostatic precipitator operation or performance [40].

In addition, downtime encountered with installation of a scrubbing unit is limited only to ductwork tie-ins and FGD system start-up [40]. This can be a distinct advantage for facilities facing compliance deadlines.

**Table 15 Comparison of capital costs for compliance [41]**

| Technique used | Cost, $/MM Btu (1988 value) |
|---|---|
| Scrubber retrofit | 160–290 |
| Plant upgrades needed in order to switch: | |
|     Boiler upgrades | 20–80 |
|     Coal handling | 0–25 |
|     Total capital cost with switch | 20–105 |

*Economics of fuel switching versus FGD.* The economics of fuel switching versus the FGD system are summarized in Tables 15 and 16. Some key features of the comparison are the high capital costs for the scrubber retrofit in comparison with the modest capital cost for coal switching; capital cost of coal switching ranges from 10 to 66% of FGD system cost. Operating and maintenance costs for fuel switching are much higher than for the FGD system, however.

Switching to a poorer quality, lower rank coal fuel as a means of achieving compliance may cause derating of the boiler by 3–20%. Lower sulfur coal, which has a comparable heating value and also burns with minimal slagging or fouling, can be expensive. Inexpensive, low-sulfur coal with undesirable ash properties may require extensive modification to the steam generator and balance-of-plant equipment or may result in a derating and reduction in generating capacity. Prices are currently about $5/ton, although they are expected to in-

**Table 16 Impact on operating and maintenance costs [41]**

| Technique used | Cost, $/MM Btu |
|---|---|
| Scrubbing | |
|     Operation and maintenance | 0.32 |
|     Capacity and energy penalty | 0.11 |
|     Total O&M cost, scrubbing | 0.43 |
| Switching | |
|     Nonfuel operating cost | 0.02–0.07 |
|     Boiler and related costs | 0.00–0.18 |
|     Coal receiving and handling | 0.02–0.25 |
|     Fuel-related costs | 0.00–0.60 |
|     Low-sulfur coal price premium | 0.00–0.35 |
|     Increased transportation cost | 0.45–0.95 |
|     Total O&M cost, switching | 0.02–1.20 |

crease to as high as $20/ton as the demand for this compliance fuel escalates [41, 43].

*Advanced coal cleaning.* Advanced physical coal cleaning is aimed at removing pyritic sulfur from coal by taking advantage of the physical differences between the mineral matter and the clean coal fractions of a coal particle. By incorporating an efficient, low-cost method for sulfur removal, utilities are given a larger selection of coal from which they can draw, thereby effectively increasing the supply of low-sulfur coal. Also, utilities are given more options for complying with federal emission regulations [44, 45].

Areas producing coal with a high pyritic sulfur content, between 55 and 80% pyritic sulfur, will be excellent candidates for an advanced physical coal cleaning process. These areas containing coal of high pyritic content are found in regions of western Pennsylvania, eastern and southern Ohio, northern West Virginia, western Indiana, southern Illinois, and northwestern Kentucky, which are known for their high-sulfur coal. The U.S. DOE has been supporting the investigation of three promising physical coal cleaning processes: selective agglomeration, advanced froth flotation, and advanced cycloning [44, 45].

The objective in their investigation was twofold:

1. Provide electric utility industries with additional options for maintaining CAAA compliance that is cost effective.
2. Examine the possibility of successfully removing trace elements from the coal matrix. These trace elements have been identified as precursors for the formation of hazardous air pollutants.

Several criteria were used to determine successful removal of sulfur; the process must first be effective, from an efficiency standpoint, in the removal of pyritic sulfur from the raw coal. Specifically, guidelines were set up to test whether the process could effectively reject 85%+ of the coal's original pyritic sulfur content. Second, the clean coal should maintain a substantial part of its initial heating value; once again, 85%+ was set as the acceptable limit for Btu recovery of the cleaned coal. Finally, the ash content of selected samples was examined. Six percent or less ash was found in the cleaned coal product for the moderate- and high-sulfur coals found in the eastern and midwestern United States [45].

*Selected agglomeration.* Agglomeration is a surface-based coal cleaning process in which a liquid immiscible in water is used to agglomerate coal particles under high-shear conditions. The hydrophilic mineral matter, including pyrite, remain dispersed in the aqueous medium. The coal agglomerates are physically separated from the aqueous phase, often by screening. The final product shape and size may be modified through a reconstitution step, such as pelletization [45].

Selective agglomeration tests were run on coal samples including Upper Freeport, Pittsburgh No. 8, and Kentucky No. 9 with a size range from 28 to 200 mesh. Two agglomeration solvents were used for all coals, specifically, diesel fuel and heptane, which were identified as the two most promising liquids based on bench-scale tests. The results showed that, for all particle sizes, the minimum value of 85% Btu recovery was achieved for both agglomerating liquids. However, the pyritic sulfur rejection (PSR) ranged between 67% and 81%, which was below the target value of 85% set by DOE [45].

Based on the bench-scale test results, a proof of concept (POC) scale (1.8 to 2.7 t/h or 2 to 3 stph) was constructed. The diesel oil agglomeration-pelletization configuration was selected from the bench-scale runs as the most technically mature and the most promising for commercial application and was therefore used in the POC scale runs [45].

Results for the POC scale were similar to those obtained in the bench scale. The Btu recovery for all three coals was at or above the 85% coal. However, as with the bench-scale results, the PSR was below the 85% target value. This was attributed to the agglomeration of "coal-contaminated" pyrite particles, as well as the possible entrainment of liberated pyrite particles during flotation [45].

As part of the investigation of the effectiveness of the selective agglomeration method, the three cleaned coals were analyzed for trace element content. Twelve trace elements and their compounds were identified as potential hazardous air pollutants under Title III of the CAAA. These elements are antimony, arsenic, beryllium, cadmium, chlorine, chromine, cobalt, lead, manganese, mercury, nickel, and selenium [45].

For the Upper Freeport coal, selective agglomeration, combined with conventional cleaning, significantly reduced all elements except beryllium, chlorine, fluorine, and mercury. Arsenic, cobalt, and lead were reduced by more than 75%. The concentration of barium, manganese, nickel, and zinc was more than halved. Similarly, the Pittsburgh No. 8 coal showed a decrease in all trace elements except chlorine and fluorine. Substantial reductions were achieved for arsenic (84%), barium (79%), cobalt (86%), lead (92%), and zinc (81%). For the Illinois No. 5 coal, selective agglomeration reduced all trace elements except cobalt and chlorine. More than 70% of the barium, manganese, and zinc was removed [45].

Economic analysis for the commercial-scale selective agglomeration plant was performed based on the results of the demonstration run. The analysis indicated that pelletization and thermal drying had a significant impact on overall plant costs. As much as 50% of the processing cost was associated with the drying and pelletization of the agglomerates. For the one option producing a dewatered, nonpelletized product having 8% moisture, the cost of $SO_2$ removed decreased to $239/t for the Upper Freeport coal and to $352/t for the Pittsburgh No. 8 coal. For the 25% moisture nonpelletized product option, the cost was reduced even further to $190/t and $284/t for the Upper Freeport and Pittsburgh No. 8 coals, respectively [45].

*Advanced froth flotation.* Advanced froth flotation extends the principles of conventional froth flotation to micrometer-sized particles [45].

Laboratory and bench-scale testing was performed on the three coals using an advanced froth flotation procedure. Here, however, it was found that a pre-cleaning, circuit was necessary for the froth flotation unit. One of the objectives of precleaning is to reject as much mineral matter and pyrite as possible before fine grinding and subsequent advanced processing steps. This is done to mini-mize the quantity of material to be processed in these expensive unit operations [45].

Based on the results obtained for the three coals, it was found that conven-tional precleaning unit operation could be used effectively to reject coarse pyrite from all three coals. The other objective was to produce coarse and intermediate-size coal streams that can be blended with the advanced froth flotation clean coal product to aid in dewatering, transport, and storage [45].

Analysis of the cleaned coals showed that, in all cases, the performance targets of 85% pyritic sulfur rejection, 6% product ash content, and 85% Btu recovery were nearly achieved. Performance was expected to improve in the POC test module due to the better hydrodynamics of the larger scale advanced flotation column. These POC tests for the three coals were currently under way [45].

Economic analysis predictions for the three coals were estimated by ICF-Kaiser for a commercial-scale advanced flotation plant producing clean coal at a rate of 454 t/h based on the bench-scale test results. The commercial plant design did not include a product reconstitution step. The costs of $SO_2$ removed were $241/t, $210/t and $272/t respectively, for the three coals. These costs will be updated when the POC test program results are obtained [45].

*Advanced cycloning.* Processes categorized as advanced cycloning processes involve the separation of two species of different densities using a medium of an intermediate density to the two species. As the process refers to coal cleaning, the heavy mineral matter component of the coal is separated from its lighter organic matter counterpart. Advanced cycloning technology seeks to extend the density-based separating principle to fine and ultrafine coal through the use of alternative media, such as heavy liquids and micronized magnetic, and through the application of centrifugal force to accelerate the separation process.

The dense medium is generally a suspension of fine, high-density particles (magnetite or sand) with the density of the medium varying directly with the solids concentration. The separating performance and downstream recovery of the dense medium suspension require that the difference in size between dense medium particles (magnetite) and coal particles be significant [44, 45].

The Coal Technology Corp. (CTC) in cooperation with ICF-Kaiser, Process Technology Inc. (PTI), and Intermagnetics General Corp. (IGC) was awarded a contract in September 1990 to develop an advanced cycloning process based on the best available medium that could achieve the goals set up by the Acid Rain Control Initiative (ARCI) program [44, 45].

After an extensive study of 50 candidate media for the separation, two were chosen—calcium nitrate and methylene chloride–perchlorethylene solution; these two solutions would be used in bench-scale testing. These two best candidates proved most suitable in a wide range of areas but were specifically well suited in areas of viscosity and nonreactivity. Some of the other criteria used in evaluation of the media were [45]:

- Ability to achieve and adjust specific gravity
- Physical and chemical properties
- Health, safety, and environmental implications
- Cost and availability
- Anticipated separation performance
- Anticipated adverse cleaning plant and boiler impacts
- Anticipated handlability of clean coal and refuse
- Anticipated ease of medium recovery and regeneration

Several pieces of separation equipment were chosen for bench-scale testing and process evaluation. A Krebs cyclone, an Alfa-Laval centrifuge, a Celleco Cleanpac (nested minicyclones), and the Centrifloat (Hydro Processing and Mining, Ltd.) were used in the study. Results of bench-scale testing showed that the organic media using a Krebs 50-mm cyclone and aqueous calcium nitrate using an Alfa-Laval decanter centrifuge were selected based on the process economic evaluation. It was expected that the capital cost associated with centrifuges would be higher than those of cyclones, while the costs for recovery of the medium would be higher for the aqueous media than for the organic media. The proof-of-concept scale testing is still under way [44, 45].

Proof-of-concept and bench-scale testing of selective agglomeration, froth flotation, and advanced cycloning have proved to be capable of significant pyrite rejection (70% to 90%) at high energy (Btu) recovery (80 to 85%). Also, preliminary economic analysis indicated that these performance marks can be achieved at a dollar per ton of $SO_2$ removed that compares favorably with flue gas desulfurization. It is anticipated that advanced coal cleaning could play an important role in bringing coal-fired electric utilities into compliance with acid rain emission requirements while making eastern and midwestern coals more attractive [45].

### 2.2.2 International Programs

**Great Britain.** A major research body that has a well-known reputation for development of clean coal technology is the British Coal Corp. Coal Research Establishment (CRE). They have established themselves as important developers of clean, efficient methods of using coal in the domestic, commercial, industrial, and power generation markets [46].

**Domestic developments.** The CRE, in conjunction with leading boiler manufacturers, has developed a number of boiler designs to handle previously undesired coal feedstocks. Specifically, nonsmokeless, bituminous coals are used in these boilers, where previously only smokeless anthracite coal had been used. The need for the development of a boiler that emitted little or no smoke arose out of the Clean Air Acts of 1956 and 1968, which allowed local authorities to introduce some degree of regulation in high-smoke areas. Since the 1970s, CRE has worked with many boiler designers to develop more efficient boiler systems to reduce smoke emission. Two designs that resulted were the downflow design and the stoker design [46].

The downflow boiler is a 13.2-kW unit designed to draw combustion smoke through a hot firebed and refractory. Additional air is added in this secondary combustion zone, which completes the burning of the smoke from the coal. This principle has been used in open fire, room heater, and boiler forms to make a range of appliances suitable for use in smoke control areas [46].

The second boiler design is a stoker design in which specially sized and prepared coal is mechanically fed to a small combustion zone. Ash and slag material are pushed off the burner to storage or into an automatic removal system. This combination of burner and fuel type provides an ease of coal handling and automation not previously achieved with solid fuel [46].

**Industrial developments.** New industrial developments arose primarily out of the need to reduce $SO_x$ and $NO_x$ emissions to acceptable levels. Also important, however, was the need to develop a more efficient desulfurization design for new power generation facilities. Older facilities, currently being used, employed some FGD method and a low-$NO_x$ burner for emission control. As we have already seen, FGDs, while effective, are associated with high capital and operating costs. In addition, some degree of efficiency loss or derating occurs when retrofitting existing systems with these units [46].

The alternative to the conventional FGD and low-$NO_x$ burner design was an approach based on fluidized bed technology and on the use of gasification. Specifically, the CRE has examined new technologies that utilize combustion in a fluidized bed type application, namely circulating fluidized bed combustion (CFBC), pressurized fluidized bed combustion (PFBC), and integrated gasification combined cycles (IGCC) [46].

**Circulating fluidized bed combustion (CFBC).** Coal is combusted in a fluidized bed mode of operation. Here, entrained solids and offgases are separated in a high-efficiency cyclone, allowing the solids to be returned back to the bed for recombustion. Heat is extracted from the combustor and from a waste heat boiler that cools the combustion gases before final cleanup. Power is generated by superheated steam produced by the boiler, which drives a conventional condensing stream turbine. $NO_x$ emissions are low and $SO_x$ emissions may be further reduced by addition of limestone to the combustor. Approximately 160 units are operating or under construction worldwide, with a maximum unit size of 250 MWe as of February 1993 [46].

**Pressurized fluidized bed combustion (PFBC).** Combustion of coal is performed in a bubbling fluidized bed at 850°C and 12 bar. Air is fed via a compression section of a gas turbine. Off-gases are passed through cyclones and then advanced to the turboexpander inlet. The turbine exit gases are fed to a waste heat boiler, final gas cleanup, and finally the stack. Heat from the waste heat boiler and combustor heat exchangers is used to raise superheated steam for the reheat steam cycle. Limestone injection controls $SO_x$ emissions and $NO_x$ emissions are low [46].

Commercial PFBC systems have been built by ABB Carbon at Stockholm (Sweden), Escatron (Spain), and Tidd (USA). The largest of these is the Stockholm CHP scheme. Two PFBC units, each with a gas turbine and linked by a common steam turbine and to a district heating system, have been installed to provide 135 MW of electricity and 224 MW of heat [46].

**Integrated gasification combined cycles (IGCC).** Coal is first converted to raw fuel gas in a gasification section. Here, coal reacts with oxygen and steam or water. The raw gas then undergoes cleaning to remove any pollutants, hydrogen sulfide, before firing in a gas turbine to generate power. Hot exhaust from the gas turbine is passed through a waste heat boiler to raise all or part of the steam for a conventional condensing steam turbine, which produces additional power [46].

For IGCC systems to achieve efficient performance, a high degree of carbon utilization in the gasifier is essential. This can have implications for reactor size, fuel flexibility, and control. Gasifiers proposed for IGCC have included members of the three generic types—entrained bed (e.g., Shell, Dow, and Texaco), fixed bed (e.g., British Gas/Lurgi), and fluidized bed (e.g., IGT and KRW). The optimum cycle configuration depends on the particular gasification system chosen [46].

The Texaco gasifier has been well proven technically on the 100-MWe IGCC demonstration facility at Cool Water, California, although economic operation was not achieved. Two 100-MWe gas turbines have been operated successfully on gas from Dow's gasification process at Plaquemine, Louisiana. The British Gas/Lurgi gasifier has been demonstrated at up to 550 tons of coal a day and used to power a Rolls Royce SK30 gas turbine, although not in a combined cycle arrangement. An order has been placed for a 250-MWe unit using Shell technology in Buggenum, Holland [46]. More detailed discussion of the IGCC technology can be found in Chapter 5.

**British coal topping cycle.** The use of so-called topping cycles is a means by which high overall efficiency may be obtained without the need for complete gasification. Here, partial coal gasification is performed and the resulting fuel is clean and burned in a gas turbine. The remaining unconverted coal is burned in a combustion system that forms part of the steam cycle. The cycle achieves 90% sulfur removal efficiency with the addition of limestone to the gasifier and combustor. Sulfur recovery may be further improved by using a zinc-ferrite polishing process. $NO_x$ emissions are low. Some advantages of the topping cycle include [46]:

1. Coal with a wider range of properties may be used. Because the cycle achieves partial conversion of coal to fuel gas, a high degree of carbon utilization in the gasifier is not essential as with the IGCC system.
2. Because the cycle achieves 90% sulfur removal with limestone, cooling flue gases before sending to a scrubber is not necessary. As a result, the system can use the high outlet gas temperature and exploit the potential of high inlet temperature gas turbines while minimizing the energy loss incurred in the production of fuel gas.
3. No oxygen production plant is required.

**Germany.** Germany relies heavily on the use of coal, especially lignite, for the generation of electricity. Prior to the unification, East Germany relied heavily on the use of lignite to produce some 90% of the country's 24,000 MW of electricity. West Germany, however, used lignite and other types of coal to produce 40% of its electric capacity, which at the time was 104,000 MW [47].

Since the unification, Germany still relies heavily on lignite and coal; however, stricter regulations apply regarding the emissions of $CO_2$, $SO_x$, and $NO_x$. The government has committed to a reduction in carbon dioxide emissions by 25 to 30% by the year 2005. Electric power plants in Germany were reportedly responsible for 25% of the total $CO_2$ emissions produced. The required reductions will be met in three ways:

1. Construction of new nuclear power plants
2. Operating existing power plants at full capacity
3. Changing the existing fossil fuel power plants from burning high-sulfur to low-sulfur coal

**Emission legislation.** The German government, in an effort to force utilities to operate more efficiently, is considering a proposal for a $CO_2$ emissions tax. The tax could increase the cost of electricity in Germany. Some of the features of the taxation proposal are outlined below [47].

- Under the proposal, an emission source would be fined $(US)6/ton of $CO_2$ emitted. The tax would apply to utilities using coal, oil, and gas-fired power plants.
- A standard net efficiency of 40% would be required for all oil- and coal-fired units and 50% for gas-fired units. The emission tax would decrease for plants operating at higher efficiencies but would increase drastically when a plant's efficiency was lower than standard.
- Coal-fired and heavy oil-fired electric power plants, combined heat and power plants, and district heating plants of 1 MW and above would be subjected to the $CO_2$ emission tax. However, only gas-fired plants of 10 MW or above would be subjected to the tax.

**European Community regulations.** In addition to Germany, the European Community (EC) is expected to propose legislation limiting the maximum emission level of $CO_2$. The proposal is a combined energy-$CO_2$ taxation levy. Under this proposal, 75% of the tax would be levied on the energy used and 25% would be levied on the $CO_2$ emission produced. A cost for this tax has not been calculated, but some say that it could increase the cost of all fuels from $(US)1 to $(US)10/barrel of oil [47].

In the EC, the Large Combustion Plants Directive is responsible for setting limits on sulfur dioxide as well as setting targets for each country. In essence, the directive acts much like the Department of Energy in the United States, although it is not quite clear how membership in the EC is determined. In any case, the directive has set limits on sulfur emissions in the past in an effort to curb pollution from acid rain. In 1988, the 12 EC countries further agreed to a reduction in sulfur dioxide levels of 23% by 1993, 42% by 1998, and 57% by 2003 based on 1980 levels. As well as specifying national tonnage quotas for sulfur dioxide, the directive sets a ceiling of 400 mg/m$^3$ $SO_2$ in flue gas for new plants larger than 500 MW(t), with a sliding scale up to 2000 mg/m$^3$ for smaller plants [48].

The directive sets different targets for each country. Germany, France, Belgium, and the Netherlands, for instance, must make cuts of 40% by 1993 and 70% by 2003, whereas Greece, Ireland, and Portugal are allowed to increase their emissions. The United Kingdom must cut emissions by 20% by 1993, 40% by 1998, and 60% by 2003 [48].

## 2.3 ENVIRONMENTAL ISSUES OF COAL MINING

Coal occurs in stratified deposits interlain with soil and rock in sedimentary basins. Coal mining operations are therefore determined by geologic features or occurrences that took place during peat formation and subsequent coalification [39]. Eight features associated with coal seams are summarized by Speight [39]. They are (1) clastic dikes and clay veins, (2) cleats, (3) concretions, (4) dipping and folded strata, (5) faulting, (6) igneous intrusions and sills, (7) partings or splits, and (8) washout or cutout. This information is crucially important in planning and conducting mining operations [39].

The depth of a deposit beneath the earth's surface widely varies from seam to seam and from location to location. The thickness of the coal strata is also widely varying, from 2 ft to over 100 ft. In parts of the western United States, Wyoming, North Dakota, South Dakota, Montana, and New Mexico, seams 100 ft thick are being surface mined. Coals from these regions generally contain less sulfur than eastern coals and demand for them has been steadily growing.

Even though utilization of coal has been very steady over the past several decades, the mining industries have not gained much deserved attention. Due to the environmental implications and restrictions associated with the product coal and mining profitability, many miners in the eastern United States have had very difficult times. In recent years, its sulfur content has become a critical parameter that determines the market value of a particular coal.

In this section, environmental issues related to coal mining operation are discussed.

### 2.3.1 Abandoned Mining Sites

All over the world, a great number of abandoned mining sites are scattered. The reasons for abandonment are diverse, including exhaustion of recoverable coal, unsafe operation, profitability, high sulfur content, and urban development plans. Many such abandoned mining sites are becoming environmental hazards for various reasons, including groundwater contamination, overburden collapse, huge underground water storage or underground flooding, gob piles, pits contaminated with polycyclic aromatic hydrocarbons (PAHs), and surface water contamination due to rain runoff.

Most mines historically operated their own washing processes. Therefore, the abandoned mining sites are generally rich in mineral matters including trace elements. This mineral matter exists in more leachable forms than those naturally buried in coal deposits.

It should be noted that coal mines are supposed to be closed after operation is either complete or ceased. Proper closing of coal mines is an important issue, because it affects the ecology over a long time. Revegetation of such areas is an equally important subject.

## 2.3.2 Spoil Piles or Gob Piles

Spoil piles are normally rich in pyrites and other mineral matters. Once spoil piles are left discarded, pyrites go through slow air oxidation processes, resulting in ferrous sulfate ($FeSO_4$) and ferric sulfate [$Fe_2(SO_4)_3$]. Both sulfate forms are water leachable. Similar fates are foreseeable from other mineral matter. Spoil piles can catch fire due to the heat generated by chemical reactions occurring inside the pile.

## 2.3.3 Dust

Dust problems are common in both surface mining and underground mining. As discussed earlier, fine coal dusts are pyrophoric and often cause mining fires as well as transportation fires. However, coal dusts can contribute to airborne particulates, especially in the mining areas and nearby towns. Coal fines are evidently different from construction dust or fly ash. Typically, coal fines or coal mining dusts can be characterized distinctly as follows.

1. Coals dusts contain calcite ($CaCO_3$) and dolomite ($CaCO_3.MgCO_3$), whereas fly ash contains more calcium oxide (CaO) and magnesium oxide (MgO). In this regard, coal dusts are similar to construction dusts. Using X-ray powder diffraction (XRD), carbonates and oxides can be readily distinguished.
2. Coal dusts contain more hydrocarbons than fly ash or construction dusts. CHN analysis or elemental analysis can be used to determine the contents of C, H, and N.

Coal dusts in surface mining operations contribute more to air pollution, whereas coal dusts in underground mining affect the safety of mining as well as the miners' health. All three types of surface mining involve dust problems of a similar nature.

## 2.3.4 Water Runoff

Water runoff from coal mining areas can be classified into four types:

1. Surface water runoff. Rainfalls wash off and carry some ingredients of coals and eventually contaminate the water streams.
2. Runoff from underground water storage. This happens in abandoned or poorly sealed coal mines.
3. Waste from blasting streams of water. In a mining method, quite common in Russia and China, low-pressure streams of water are used to wash blasted coal out of the mine. Blasting can be eliminated by the use of high-pressure

streams of water [39]. This method is not used in the United States. It involves a potentially large liability for environmental groundwater contamination.

4. Water from coal washeries. Mined coals are often washed before being shipped to final destinations. Heavy fluid media are used to separate good portions of pyrite and other mineral matter from coal. As a heavy medium, water is most frequently used. Wastewater from this coal washing operation is of environmental concern, requiring a special attention for treatment as well as recycled use.

## 2.3.5 Blasting

Blasting is normally more a safety problem than an environmental problem. However, repeated blasting can affect the mechanical structure of neighboring soil, cause faults, or induce rubblization of neighboring rock structures.

## 2.3.6 Noise

Noise has become a form of pollution in our highly industrialized society, because sound levels have tended to increase with higher population density, more mechanical devices, and amplified sound levels of electronic devices [49]. The average urban noise intensity is now twice what it was in 1955 [49].

Noise from coal mining operation comes from blasting, augering, conveyor belts, trolleys, power shovels, shuttle cars or trains, and motors.

Most coal mines are located in rural areas, where population density is low. If the exposure to excessive noise is infrequent, the effects will be minimal. If a person is exposed day after day, as is an operator in a coal mine, the restoration process stops functioning as well, and hearing loss eventually results [49]. The amount of noise that can be tolerated without risking hearing impairment depends on the frequency band involved. For broadband noise, 120 dB can be tolerated for a minute or two, and the tolerance level ranges down to 85 dB for an 8-hour day.

## REFERENCES

1. Elliot, M. A. *Chemistry of Coal Utilization,* second supplementary volume. New York: Wiley, 1981.
2. Cooper, B. R., and L. Retrakis. *Chemistry and Physics of Coal Utilization.* New York: American Institute of Physics, 1981.
3. Haenel, M. W. Recent progress in coal structure research. *Fuel* 71:1211–1223 (1992).
4. Meyers, R. A. *Coal Structure.* New York: Academic Press, 1982.
5. Gorbaty, M. L., J. W. Larsen, and I. Wender. *Coal Science,* vol. 1. New York: Academic Press, 1982.

6. Speight, J. G. *The Chemistry and Technology of Coal.* New York: Marcel Dekker, 1994.
7. Smith, K. L., L. D. Smoot, T. H. Fletcher, and R. J. Pugmire, *The Structure and Reaction Processes of Coal.* New York: Plenum, 1994.
8. Spiro, C. L., and P. G. Kosky, Space-filling models for coal. 2 . Extension to coals of various ranks. *Fuel* 61:1080–1084, November 1982.
9. Shinn, J. H. From coal to single-stage and two-stage products: A reactive model of coal structure. *Fuel* 63:1187–1196, 1984.
10. Huttinger, K. J., and A. W. Michenfelder, Molecular structure of a brown coal. *Fuel* 66:1987.
11. Wen, C. Y., and E. S. Lee. *Coal Conversion Technology.* Reading, MA: Addison-Wesley, 1979.
12. Schobert, H. H. *Coal, the Energy Source of the Past and Future.* Washington, DC: American Chemical Society, 1987.
13. Baker, T. C. Electric power generation using walloon coals. *Coal Geology,* vol. 1. Surat-Moreton Basin Symposium, Australia, 1979.
14. Klaus, R. and B. Bohm. Overview of best practice. Technological options available for power generation, clean use of coal technologies and meeting environmental goals. Conference Proceeding Clean and Efficient Use of Coal: The New Era for Low-Rank Coal, Budapest, February, 1992.
15. Whitney, R. S. The future for coal in the process industries—the potential and the challenge. 18th Australian Chemical Engineering Conference, New Zealand, August 27–30, 1990.
16. Smith, M. I., A. K. Hjalmarsson, and H. N. Soud. Environmental pollution control for power generation: An overview for low rank coal. Conference Proceedings Clean and Efficient Use of Coal: The New Era for Low-Rank Coal, Budapest, February 1992.
17. Moor, K. R. The world wide strategy in laws and regulations for the control of environmental pollution caused by coal fired power plants. *EKOL Journal* 37(3):63–70, 1992.
18. Hemenway, A., S. Roberts, W. A. Williams, and D. A. Huber. The Clean Air Act Amendments of 1990 and their implications on the clean coal technology program. Proceedings International Conference Coal Slurry Technology, 16th. Washington, DC: Coal Slurry Association, 1991.
19. Awan, A. A., and C. L. Miller. "Clean Coal Technology: An Option to Meet Environmental Requirements, both in near and long-term". Proceedings of the American Power Conference.
20. Gillette, J. L. and C. B. Szpunar. International competitiveness of clean coal technologies. Proceeding the 25th International Society Energy Conversion Engineering Conference, Reno, NV, vol. 4, 1990.
21. Snow, B. L. Reclamation program cuts coal ash disposal costs. *Power Eng.* 97:29–31, 1993.
22. Selle, S. J., G. H. Gronhoud, and E. A. Sondreal. Review of slagging and fouling from low-rank coals: Plant survey, bibliography, and discussion. Final Report to the U.S. Department of Energy, vols. 1 and 2, November 1986.
23. Christensen, J. W. *Energy Resources and Environment.* Dubuque, IA: Kendall/Hunt Publishing Company, 1981.
24. Khemani, L. T., G. A. Momin, P. S. Prakasa Rao, P. D. Safai, G. Singh, and R. K. Kapoor. *Atmos. Environ.* 23(4):757–762, 1989.
25. Pinero, C. E. Can the coal industry survive in a world concerned with global warming?" ASTM Standardization News, February 1992.
26. Smith, M. I., and K. V. Thambimuthu. Greenhouse gas emissions, abatement and control. The role of coal. *Energy and Fuels,* 7(1):7–13, 1993.
27. Springer, K. J. Global what? Control possibilities of $CO_2$ and other greenhouse gases. *J. Eng. Gas Turbines Power* 113: July 1991.
28. Smith, I. *$CO_2$ and climate change.* London: International Energy Agency Coal Research, 1988.
29. McAteer, J. D. Mine safety and health: Are we doing enough? American Mining Congress, Washington, DC 1991.
30. Thomson, L. H. Safety with progress. *Mining Technol.* 70(85):291–294, 1988.
31. OECD. Moving Coal. A study of transport systems by the Coal Industry Advisory Board, Paris, 1985.

32. Meyers, R. A. *Coal Handbook*. New York: Marcel Dekker, 1981.
33. Linday, I. Energy for tomorrow's world. Conference Proceedings, the Clean and Efficient Use of Coal and Lignite: Its Role in Energy, Environment and Life. Hong Kong, November 30–December 3, 1993.
34. Toohey, A. C. The reality of coal and coal technology. Conference Proceedings, The Clean and Efficient Use of Coal and Lignite: Its Role in Energy, Environment and Life. Hong Kong, November 30–December 3, 1993.
35. Williams, J. R, and P. H. Maxwell. The long view: The perspectives of coal uses and suppliers. Conference Proceedings, The Clean and Efficient Use of Coal and Lignite: Its Role in Energy, Environment and Life, Hong Kong, November 30–December 3, 1993.
36. Makansi, J., ed. Clean Air Act Amendments: The engineering response. *Power* June:11–66, 1991.
37. Carter, R. Experts predict clean air compliance strategies, *Coal*, November:45–47, 1991.
38. Kuehn, S. E. Utility plans take shape for Title IV compliance. *Power Eng.*, August:19–26, 1993.
39. Speight, J. G., ed. *Fuel Science and Technology Handbook*, New York: Marcel Dekker, 1990.
40. Bretz, E. A. CAA phase I scrubbers: What utilities chose. *Electrical World* February:23–29, 1994.
41. Kumar, K. S., R. E. Sommerlad, and P. L. Feldman. Know all impacts from switching coals from CAA compliance. *Power* May:31–38, 1991.
42. Rupinskas, R. L., and P. A. Hiller. Considerations for switching to low-sulfur coal. *Power Eng.* November:23–26, 1991.
43. Rupinskas, R. L., and P. A. Hiller. Switching to low-sulfur coal: case histories. *Power Eng.* January:22–26, 1992.
44. Barnett, W. P., and T. J. Feeley III. Status of advanced coal cleaning as a compliance technology. *Mining Eng.* October:1225–1230, 1992.
45. Feeley, T. J. III, and V. L. McLean. Advanced physical coal cleaning as a Clean Air Act compliance technology. *Mining Eng.* October:1253–1257, 1993.
46. Minchener, A. J. Clean coal research. *Mine and Quarry* January/February: 8–10, 1993.
47. Smith, D. J. Germany emphasizes power plant pollution control. *Power Eng.* December: 30–32, 1993.
48. Anon. FGD—the next ten years. *Chem. Eng.* April:26–28, 1991.
49. Kraushaar, J. J., and R. A. Ristinen. *Energy and Problems of a Technical Society*. New York: Wiley, 1988.
50. Torrens, I. M., and J. B. Platt, Electric utility response to the Clean Air Act Amendments. *Power Eng.* 98:43–47, 1994.

## PROBLEMS

1. List the top five coal-producing states in the United States. Was there any change in this list in the past 20 years? What was the reason for such a change?

2. When comparing two coals that have the same total weight percentage of moisture and mineral matter, which coal has the higher heating value, the one with the higher H/C ratio or the one with the lower H/C ratio?

3. Define "synthesis gas."

4. Is this sentence true or false? "Classically, synthesis gas was produced using any of the following methods: electrolysis of water, coke oven gas, off-gas

from catalytic reformer, acetylene process, water gas shift reaction, steam reforming of natural gas, etc."

5. Is the devolatilization of coal faster than the steam gasification reaction at high temperature, say 860°C? From the standpoint of fundamental kinetics, how can you explain this?

6. What kind of pore volume is measured by high-pressure mercury porosimetry?
   (a) Total pore volume
   (b) Superficial pore volume
   (c) Internal pore volume
   (d) Open pore volume
   (e) Absorbed liquid volume

7. What rank of coal is usually the most agglomerating?
   (a) HVC bituminous
   (b) Subbituminous
   (c) Anthracite
   (d) Lignite

8. Suppose a pulverized bituminous coal particle undergoes a pyrolysis reaction. What is the most likely and noticeable change in the particle's structure?
   (a) Swelling
   (b) Cracking
   (c) Shrinking
   (d) Complete deformation

9. Major elements of Montana Rosebud ash are analyzed as shown in Table P-1. Suppose we have residues from complete combustion of Montana Rosebud char. Estimate the density of this residue. Explain all your assumptions.

**Table P-1 Ash analysis of Montana Rosebud char**

| Major elements in ash | Percentage |
|---|---|
| Silica | 43.0 |
| $Al_2O_3$ | 16.6 |
| $Fe_2O_3$ | 8.0 |
| $TiO_2$ | 0.7 |
| CaO | 15.7 |
| MgO | 4.0 |
| $Na_2O$ | 0.7 |
| $K_2O$ | 0.6 |
| Sulfites | 4.8 |

10. Compute the dissociation pressure of calcium carbonate as a function of temperature. Find the temperature at which $K_p = 1$ (atm). Compare your *computed* results with the experimental data (Royster, P. H. *J. Phys. Chem.* 40:435, 1936). Gibbs free energy of formation data can be found in various literature sources including the JANAF table.

$$CaO_3(s) = CaO(s) + CO_2(g)$$

11. Compute the dissociation pressure as a function of temperature. Find the temperature at which $K_p$ equals unity.

$$MgCO_3(s) = MgO(s) + CO_2(g)$$

12. Based on the calculations of Problems 10 and 11, estimate the dissociation pressure of dolomite as a function of temperature. Assume that the dolomite is an approximately 1:1 mixture of calcite and magnecite.

$$MgCO_3 \cdot CaCO_3 = MgO \cdot CaCO_3 + CO_2$$

$$MgO \cdot CaCO_3 = MgO \cdot CaO + CO_2$$

13. Among the trace elements in coal, what species are likely to volatilize during a combustion process?
14. What are the major air pollution problems associated with ex situ coal plants? List the three principal problems.
15. Which one is likely to serve as an active site in coal char?
    (a) Carbon edge
    (b) Dislocated carbon
    (c) Inorganic impurities
    (d) Oxygen and hydrogen functional group
    (e) Carbon in a blind pore
16. What are the trace basic forms of sulfur in coal?
17. Describe schematically the heavy medium separation process of pyritic sulfur from coal. Answer the following questions:
    (a) In terms of pyrite removal, which would be more effective, pulverized coal or lumped coal?
    (b) What ranges of specific gravity of heavy media are considered ideal?
    (c) Is this process equally effective in mineral matter removal?
    (d) Explain the hindered settling theory.
    (e) What are the heavy media typically used for this process?
18. Why is gob pile coal richer in sulfur content? Define gob pile coal.
19. Humic acid is known to be an excellent fertilizer. Humic acid can be obtained by reaction with subbituminous coal. Describe the chemical reaction involved.
20. Coal has good absorbing and adsorbing properties. Alberta Research Council developed a process called the coal agloflotation process that can be used

for cleaning town gas soil by selective process combination between agglomeration and flotation.

(a) Define town gas soil.

(b) Polycyclic aromatic hydrocarbons (PAHs) are known to be polluting contaminants in town gas soils. Discuss the molecular chemical structure, the health effects, and the occurrences in non–coal related areas.

(c) Describe the process fundamentals schematically.

21. Discuss the analysis procedure for organic sulfur content. The ASTM procedure involves the difference between the total sulfur and the inorganic (pyritic and sulfatic) sulfur. Critically, assess the indirect nature of the determination.

22. Discuss the procedure for the determination of the percent oxygen in coal.

23. "America East" coal mine produces coal whose average heating value is 11,200 Btu/lb and sulfur content is 0.7 wt. %. If this coal is burned to generate electricity:

(a) How many pounds of sulfur dioxide (or its equivalent) will be formed per 1,000,000 Btu generated?

(b) Does the $SO_2$ generation level computed in (a) satisfy the level referenced by the Clean Air Act Amendments of 1990? This level is often called the "compliance coal requirement," even though it is somewhat misleading.

24. Which of the following is an unlikely source of oxygen (O) in coal chemical structure?

(a) Phenols

(b) Aliphatic alcohols

(c) Aromatic ketones

(d) Aromatic carboxylics

(e) Quinones

## COAL GASIFICATION: SYNTHESIS OF SYNGAS, FUELS, AND PETROCHEMICALS

### 3.1 INTRODUCTION

Conversion of coal by any of the processes to produce a mixture of combustible gases is termed gasification. In general, gasification of coal entails controlled partial oxidation of the coal to convert it into desired gaseous products. The coal is heated either directly by combustion or indirectly by a heat source. The gasifying media are passed into intimate contact with the heated coal and reacted with carbon or other primary products of decomposition to produce desired gaseous products. Coal gasification is always performed in connection with a downstream process. The primary emphasis may be on electricity generation, on syngas production for pipeline applications, or on synthesis of fuels and chemicals.

Conversion of coal in its solid form to a gaseous fuel is widely practiced today. During earlier years (1920–1940), coal carbonization was being employed to produce manufactured gas in hundreds of plants worldwide. This technology became obsolete in the post–World War II era because of the availability of oil and natural gas in large quantities at attractive prices. But with the advent of the oil embargo and consequent increase in oil prices and shortages of natural gas in the early 1970s, interest in coal gasification was revived. This has resulted in many major coal gasification research, development, and demonstration activities aimed at improving the old technology to make it competitive in modern fuel markets [1].

Coal gasification includes a series of processes that convert coal containing C, H, and O as well as impurities such as S and N into fuel and/or synthesis gas. This conversion is accomplished by introducing a gasification agent (air or oxygen and steam) into a reactor vessel containing coal feedstock under controlled conditions of temperature, pressure, and flow regime (moving, fluidized, or entrained bed). The proportions of the product gases (CO, $CO_2$, $CH_4$, $H_2$, $H_2O$, $N_2$, $H_2S$, etc.) depend on the type of coal and its composition, the gasification agent, and the thermodynamics and chemistry of the gasification reactions as controlled by the operating parameters.

Coal gasification technology can be utilized in four energy systems of potential importance:

• Production of fuel for use in electric power generation units
• Manufacturing substitute or synthetic natural gas (SNG) for use as pipeline gas supplies
• Production of synthesis gas for use as a chemical feedstock
• Generation of fuel gas (low or medium Btu) for industrial purposes

Coal is the largest domestic energy source and a major source of chemicals. Synthesis gas production is the main starting point for production of a variety of chemicals. The success of the Tennessee Eastman Corporation in producing acetic anhydride from coal shows the great potential of coal as a feedstock [2]. The major concern about this technology is related to contaminants in coal. Coals contain an appreciable amount of sulfur, which is of main concern in relation to the downstream processes because many catalysts that might be used in producing chemicals are highly sulfur sensitive. Coals also contain appreciable amounts of alkali metal compounds, which contribute to fouling and corrosion of the reactor vessels in the form of slags. In addition, coals contain a number of trace elements that may also affect downstream processes. If coal gasification technology is used to produce chemicals, the choice of gasification technology is very critical because different processes produce different qualities of synthesis gas (CO + $H_2$).

Synthesis gas (SG) is an important starting material for both fuels and petrochemicals. Syngas can be obtained from various sources including petroleum, natural gas, coal, and biomass. If syngas is used for synthesis of other chemicals, it is typically classified as $H_2$-rich syngas, CO-rich syngas, $CO_2$-rich syngas, and so on. Principal fuels and chemicals directly made from syngas include ammonia, methanol, hydrogen, carbon monoxide, gasoline, diesel fuel, dimethyl ether, methane, isobutane, ethylene, $C_1$–$C_5$ alcohols, ethanol, ethylene glycol, and $C_2$–$C_4$ olefins. Principal fuels and chemicals made via methanol include formaldehyde, acetic acid, gasoline, diesel fuel, methyl formate, methyl acetate, acetaldehyde, acetic anhydride, vinyl acetate, dimethyl ether, ethylene, propylene, *iso*-butylene, ethanol, $C_1$–$C_5$ alcohols, propionic acid, methyl *tert*-butyl ether (MTBE), ethyl *tert*-butyl ether (ETBE), *tert*-amyl methyl ether (TAME), benzene, toluene, xylenes, ethyl acetate, and a methylating agent. The synthesis route via methanol is called indirect synthesis.

## 3.2 HIGH-, MEDIUM-, AND LOW-BTU GASES

Depending on the heating value of the gases produced in the gasification process, there are three types of gas mixtures [3].

1. Low-Btu gas: Consists of a mixture of carbon monoxide, hydrogen, and some other gases with a heating value less than 300 Btu/scf.
2. Medium-Btu gas: Consists of a mixture of methane, carbon monoxide, hydrogen, and various other gases with a heating value in the range of 300–700 Btu/scf.
3. High-Btu gas: Consists predominantly of methane with a heating value of approximately 1000 Btu/scf. It is also referred to as synthetic natural gas.

Coal gasification involves the reaction of coal carbon and other pyrolysis products with oxygen, hydrogen, and water to provide fuel gases.

**Low-Btu gas.** During the production of low-Btu gases, air is used as the combustant. Because oxygen is not separated, the product gas invariably has many undesirable constituents like nitrogen. It results in a low heating value of 150–300 Btu/scf. Sometimes, gasification of coal is carried out in situ where mining by other techniques is not favorable. For such in situ gasification, low-Btu gas is a desired product. Low-Btu gas has five components, with about 50% v/v nitrogen and some quantities of hydrogen and carbon monoxide (combustible), carbon dioxide, and some traces of methane. Such a high content of nitrogen makes the product gas low Btu. The other two noncombustible components ($CO_2$ and $H_2O$) further lower the heating value of the gas. The presence of these components limits the applicability of low-Btu gas to chemical synthesis. The two major combustible components are hydrogen and carbon monoxide, whose ratio varies depending on the gasification conditions employed. One of the most undesirable components is hydrogen sulfide, whose content is proportional to the sulfur content of the feed. It must be removed by washing procedures before product gas can be used for useful purposes.

**Medium-Btu gas.** In the production of medium-Btu gas, pure oxygen rather than air is used as the combustant, which results in an appreciable increase in the heating value by about 300–400 Btu/scf. This gas contains predominantly carbon monoxide and hydrogen with some methane and carbon dioxide. It is used primarily in the synthesis of methanol, higher hydrocarbons via Fischer-Tropsch synthesis, and a variety of chemicals. It can also be used directly as a fuel to raise steam or to drive a gas turbine. The $H_2/CO$ ratio in medium-Btu gas varies from 2:3 (CO rich) to more than 3:1 ($H_2$ rich). The increased heating value is attributed to the higher methane and hydrogen content as well as to lower carbon dioxide contents.

**High-Btu gas.** High-Btu gas consists mainly of pure methane (>95%), and because of this its heating value is around 900–1000 Btu/scf. It is compatible with natural gas and can be used as a substitute for SNG. It is usually produced by catalytic reaction of carbon monoxide and hydrogen.

$$3H_2 + CO \longrightarrow CH_4 + H_2O$$

The large quantities of $H_2O$ produced are removed by condensation and recirculated. The catalyst used for this process is prone to sulfur poisoning, and care must be taken to remove all the hydrogen sulfide before the reaction starts. This results in very pure product gas. The methanation reaction is highly exothermic in nature.

## 3.3 VARIOUS SYNGAS GENERATION PROCESSES

### 3.3.1 Historical Background

It was known as early as the seventeenth century that gas could be produced by heating coal. Around 1750, in England, coal was subjected to pyrolysis to form gases that were used for lighting [4]. With the invention of the Bunsen gas burner (atmospheric pressure), the potential of heating was opened to gas. In 1875 the cyclic carburetted water gas process was developed for gas production. In this process, water gas ($H_2O + CO$) was produced by reacting hot coke with steam. Heat was supplied by introducing air intermittently to burn the portion of coke. The development of coal-to-gas processes was an incentive in Europe during those days because coal was the major fuel available. By the early 1920s, there were at least five Winkler fluid bed processes running, all of which were air blown, yielding producer gas at 10 million scf/hr. Some of them were later converted to use oxygen instead of air to produce nitrogen-free gas.

The Lurgi process was developed to manufacture town gas by complete gasification of brown coal in Germany. In 1936, the first commercial plant based on this process went operational. It produced of town gas from lignite coal at 1 million scf/day. Through 1966, there were at least 10 Lurgi plants in Europe and Asia producing synthesis gas.

In 1942, Heinrich Koppers in Germany developed the Koppers-Totzek suspension gasification process based on the pilot plant work initiated 4 years earlier. The first industrial plant, built in France around 1949, produced synthesis gas at 5.5 million scf/day that was later used to produce ammonia and methanol. By the early 1970s, there were at least 20 K-T plants all over the world. All of them used oxygen as the primary gasification medium.

The Winkler, Lurgi and Koppers-Totzek processes employed steam and oxygen (or air) to carry out gasification. Little development of these processes took place in the United States because of the discovery of natural gas as a fuel, in addition to an abundant supply of oil.

## 3.3.2 General Aspects of Gasification

The rates and degrees of conversion for various gasification reactions are functions of temperature, pressure, gas composition, and the nature of the coal being gasified. The rate of reaction is always higher at higher temperatures, whereas the equilibrium of the reaction may be favored at either higher or lower temperatures. The effect of pressure on the rate depends on the specific reaction. Gasification reactions like the carbon-hydrogen reaction to produce methane are favored at high pressures (70 atm) and relatively low temperatures (760–930°C), whereas low pressures and high temperatures favor the production of carbon monoxide and hydrogen. Most gasification systems can be classified into four categories according to the method of contacting coal and gaseous streams as discussed briefly in an earlier section.

Supply of heat is an essential element in the gasification. Partial oxidation of char with steam and oxygen leads to generation of heat and synthesis gas. Another way to produce a hot gas stream is via the cyclic reduction and oxidation of iron ore. The type of coal being gasified is also important to the operation. Only a suspension-type gasifier can handle any type of coal; if caking coals are to be used in a fixed or fluidized bed, special measures must be taken so that the coal does not agglomerate during gasification. In addition, the chemical composition, volatile content, and moisture content of coal affect its processing during gasification.

## 3.3.3 Classification

There are two ways to classify gasification processes:

1. By the Btu content of the gas that is produced.
2. By the type of reactor vessel and whether or not the system reacts under pressure.

The following processes for conversion to gases are grouped according to the heating value of the gases produced [5].

1. Medium or high Btu with alternative to low Btu
   (a) Lurgi gasifier

   (b) Winkler gasifier
   (c) Synthane gasifier
2. Medium or high Btu only
   (a) Koppers-Totzek process
   (b) Texaco gasification process
   (c) Shell gasification process
   (d) Molten salt process
   (e) $CO_2$ acceptor process
3. Low Btu only
   (a) U-gas process

Based on the reactor configuration and method of contacting gaseous and solid streams, gasification processes can be divided into four categories [3].

1. Fixed or moving bed: In the fixed bed reactor, coal is supported by a grate, the gases (steam, air, or oxygen) pass upward through the supported bed, and the product gases exit from the top of the reactor. Only noncaking coals can be used in fixed bed reactors. In the moving bed reactor, coal and gaseous streams move countercurrently. The temperature at the bottom of the reactor is higher than at the top. Because of the lower temperature of devolatilization, relatively large amounts of liquid hydrocarbons are produced in the gasifier. In the reactor, residence time of the coal is much higher than in a suspension reactor. Ash is removed from the bottom as dry ash or slag. Lurgi and Wellman-Galusha gasifiers are examples of this type of reactor.
2. Fluidized bed: This reactor allows intimate contact between gas and solids and provides longer residence times. It uses finely pulverized coal particles and the gas flows up through the bed. It exhibits liquid-like characteristics. Due to the ascent of particles and separation, a larger coal surface area results, which promotes chemical reaction. Dry ash is removed continuously from the bed or the gasifier is operated at such a high temperature that it can be removed as agglomerates, which results in improved carbon conversion. Such beds also have limited ability to handle caking coals. Winkler and Synthane processes use this type of reactor.
3. Entrained bed: This system uses finely sized coal particles blown into the gas stream before entry into the reactor, with combustion occurring in the coal particles suspended in the gas phase. Due to shorter residence times, very high temperatures are required to obtain good conversion. This is achieved by using excess oxygen. This bed configuration handles both caking and noncaking coals. Examples of commercial gasifiers that use this type of bed are Koppers-Totzek and Texaco.
4. Molten bath reactor: In this reactor, coal is fed along with steam or oxygen in the molten bath of salt or metal operated at 1800–2500°F. Ash and sulfur

are removed as slag. The Kellogg and ATGAS processes are examples of this type of reactor [6].

### 3.3.4 Gasification Processes

**Lurgi gasification.** The Lurgi gasification process is one of the few processes for which the technology has been fully developed [7]. Since its development in Germany before World War II, this process has been used in more than 15 commercial plants throughout the world. This process produces low- to medium-Btu product gases. It is a fixed bed process in which the reactor configuration is similar to a fixed bed reactor. The older version of the Lurgi process is a dry ash gasification, which significantly differs from the more recently developed slagging process.

The dry ash Lurgi gasifier is a pressurized vertical reactor that accepts only crushed noncaking coals [8]. The coal feed is supported at the base of the reactor by a revolving grate, which also introduces a steam and oxygen mixture and removes the ash. This process takes place at around 24 to 31 atm and in the temperature range 620 to 760°C. The residence time in the reactor is about 1 hour. Steam introduced from the bottom of the reactor provides the necessary hydrogen, and the heat is supplied by the combustion of a portion of the product char. The product gas has a relatively high methane content compared with that from a nonpressurized gasifier. If oxygen is used as an injecting medium, the gas coming out has a heating value of approximately 450 Btu/scf. The crude gas leaving the gasifier contains many liquid products, including tar, oil, and phenol, which are separated in a devolatilizer where gas is washed to remove unsaturated hydrocarbons and naphtha. The gas is then subjected to methanation to produce a high-Btu gas (pipeline quality).

A modification of the Lurgi process called the slagging Lurgi gasifier has been developed to process caking coals [3]. The operating temperature of the gasifier is kept higher and the injection ratio of steam is reduced to 1–1.5 mole per mole of oxygen. These two factors cause the ash to melt easily, and the molten ash is removed as a slag. Coal is fed to the gasifier through a lock hopper system and distributor. It is gasified with steam and oxygen injected near the bottom. The upward movement of hot product gases makes the preheating and devolatilization of coal easier. The molten slag formed passes through the slag tap hole. It is quenched with water and removed through a lock hopper. The amount of unreacted steam passing through is minimized in this process. Also, the high operating temperature and fast removal of product gases lead to higher output rates in a slagging Lurgi gasifier than a conventional unit.

Conventional Lurgi gasification is best known for its role as the gasifier technology for South Africa's SASOL complex. A typical product composition for oxygen-blown operation is given in Table 1.

**Table 1 Typical Lurgi gas products**

| | |
|---|---|
| CO | 16.9% |
| $H_2$ | 39.4 |
| $CH_4$ | 9.0 |
| $C_2H_6$ | 0.7 |
| $C_2H_4$ | 0.1 |
| $CO_2$ | 31.5 |
| $H_2S$ + COS | 0.8 |
| $N_2$ + Ar | 1.6 |
| HHV = 285 Btu/scf | |

*Source:* Reference 9.

*Lurgi dry ash gasifier.* This is one of the most important commercial gasifiers and the only nonatmospheric kind in operation today. In this gasifier, coal sized 1.5 to 4 mesh reacts with steam and oxygen in a slowly moving bed. The process is semicontinuous. A schematic of a Lurgi pressure gasifier is shown in Figure 1 [9]. The gasifier is equipped with the following parts [10]:

1. An automated coal lock chamber for feeding coal from a coal bin to the pressurized reactor.
2. A coal distributor through which coal is uniformly distributed into the moving bed.
3. A revolving grate through which the steam and oxygen are introduced and the ash is removed
4. An ash lock chamber for discharging the ash from the pressurized reactor into an ash bin, where the ash is cooled by water quenching
5. A gas scrubber in which the hot gas is quenched and washed before it passes to a waste heat boiler

The gasifier shell is water cooled. Steam is produced from the water jacket. A motor-driven distributor, located at the top of the coal bed, evenly distributes the feed coal coming from the coal lock hopper. The grate at the bottom of the reactor is also driven by a motor to discharge the coal ash into the ash lock hopper. The volume between the inlet and outlet grates has several distinct zones. In the topmost zone the coal is preheated by contact with the hot crude gas leaving the reactor. As the coal is heated devolatilization and gasification proceed at temperatures ranging from 620 to 760°C. Devolatilization of coal is accompanied by gasification of the resulting char. The interaction between devolatilization and gasification is a determining factor in the kinetics of the process.

The bottom of the bed is the combustion zone, where carbon reacts with oxygen to yield mainly carbon dioxide. The exothermic heat generated by this reaction provides the heat for gasification and devolatilization. In other words, there is heat integration within the gasifier. More than 80% of the coal feed is

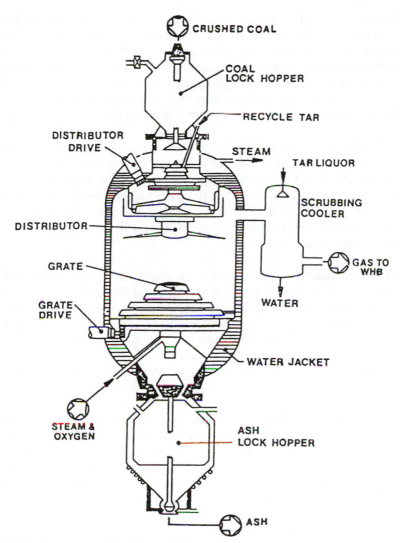

CRUSHED COAL

COAL LOCK HOPPER

RECYCLE TAR

DISTRIBUTOR DRIVE

STEAM

TAR LIQUOR

SCRUBBING COOLER

DISTRIBUTOR

GRATE

GAS TO WHB

GRATE DRIVE

WATER

WATER JACKET

STEAM & OXYGEN

ASH LOCK HOPPER

ASH

**Figure 1** Lurgi pressure gasifier. (Source: Reference 9.)

gasified, the remaining being burned in the combustion zone. The temperature of the combustion zone must be selected in such a way that it is below the ash fusion point but high enough to ensure complete gasification of coal. This temperature is determined by the steam/oxygen ratio.

The material and heat balance of Lurgi gasifier is determined by the following process variables:

1. Pressure, temperature, and steam/oxygen ratio.

2. The nature of coal. The type of coal decides the gasification and devolatilization reaction. Lignite is the most reactive coal, for which reaction proceeds at 650°C. Coke is the least reactive, for which the minimum temperature required for chemical reaction is around 840°C. Therefore, more coal is gasified per unit of oxygen for lignite compared with other types of coal.
3. The ash fusion point, which decides the steam/oxygen ratio.
4. The volatile matter of the coal, which influences the quality of tar and oils.

The Lurgi gasifier has relatively high thermal efficiency because of its medium pressure operation and the countercurrent solid-gas flow. At the same time, it consumes a great deal of steam, and the concentration of hydrogen and carbon dioxide in the product gas is low. Also, the crude gas leaving the gasifier contains carbonization products such as tar, oil, naphtha, and ammonia. This is passed through a scrubber, where it is washed by circulating gas liquor and cooled down by a waste heat boiler.

***Slagging Lurgi gasifier.*** This gasifier is an improved version of the Lurgi dry ash gasifier. A schematic [9] of the slagging Lurgi gasifier is shown in Figure 2. The temperature of the combustion zone is kept higher than the ash fusion point. This is achieved by using a smaller amount of steam. The ash is removed from the bottom as slag. The main advantage of this gasifier over the conventional dry ash one is that the yield of carbon monoxide and hydrogen is high and the coal throughput increases many times. Also, steam consumption is minimized [11].

**Koppers-Totzek gasification.** This gasification process uses the entrained flow technology whereby finely pulverized coal is fed into the reactor with steam and oxygen [12, 13]. The process operates at atmospheric pressure. The gasifier itself is a cylindrical, refractory-lined coal burner with at least two burner heads through which coal, oxygen, and steam are charged. The burner heads are spaced either 180° or 90° apart (representing a two- or four-headed arrangement) and are designed so that steam covers the flame and prevents the reactor refractory walls from getting excessively hot. The reactor typically operates at a temperature of about 1500°F and atmospheric pressure. About 90% of carbon is gasified in a single pass, depending on the type of coal. Lignite is the most reactive coal, with a reactivity approaching 100% [3].

In contrast to moving bed and fluidized bed reactors, this gasifier has few limitations on the nature of feed coal in terms of caking behavior and mineral matter properties. Because of the very high operating temperatures, the ash agglomerates and drops out of the combustion zone as molten slag and is subsequently removed from the bottom of the reactor. The hot effluent gases are quenched and cleaned. This gas product contains no tars, ammonia, or condensable hydrocarbons and is predominantly synthesis gas. It has a heating value

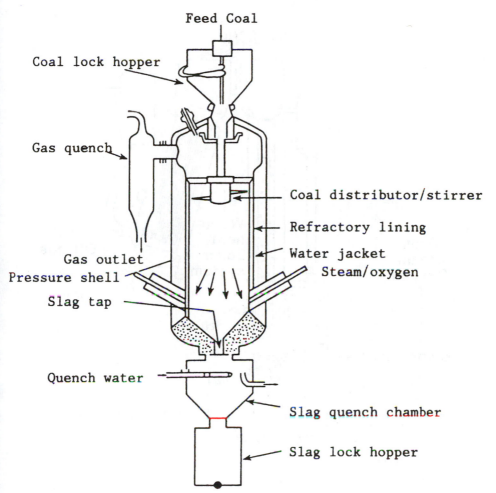

Feed Coal

Coal lock hopper

Gas quench

Coal distributor/stirrer

Refractory lining

Water jacket
Steam/oxygen

Gas outlet
Pressure shell

Slag tap

Quench water

Slag quench chamber

Slag lock hopper

**Figure 2** Schematic representation of a slagging Lurgi gasifier. (Source: Reference 3.)

of about 280 Btu/scf and can be further upgraded by reacting with steam to form additional hydrogen and carbon dioxide.

*Koppers-Totzek gasifier.* This gasifier is one of the most important entrained bed gasifiers in commercial operation today. It accepts almost any type of coal, including caking coal, without any major restrictions. It has the highest operating temperature (around 1400–1500°C) of all the gasifiers. There are two versions of it in terms of design: a two-headed and a four-headed gasifier. A schematic of the Koppers-Totzek two-headed gasifier[103] is shown in Figure 3. The original version, designed in 1948 in Germany, is two headed, with the heads mounted

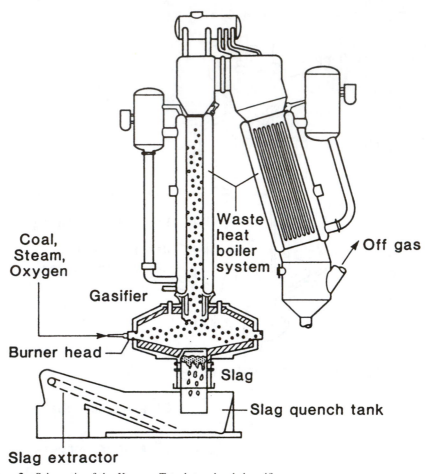

**Figure 3** Schematic of the Koppers-Totzek two-headed gasifier.

at the ends. The gasifier as such is ellipsoidal and horizontal. Each head contains two burners. The shell of the gasifier is water jacketed and has an inner refractory lining. Design of the four-headed gasifier began in India around 1970. In this case, burner heads are spaced at 90° instead of 180° as in the two-headed design. All the burner heads are installed horizontally. The capacity of a four-headed burner gasifier is larger than that of its two-headed counterpart [14].

*Features of the Koppers-Totzek process*

1. High capacity: These units are designed for coal feed rates up to 800 tons per day or about 42 million scf/day of 300-Btu gas.

2. Versatility: The process is capable of handling a variety of feedstocks, including all ranks of solid fuels and liquid hydrocarbons, pumpable slurries containing carbonaceous materials. Even feedstocks with high sulfur and ash contents can be readily used in this process.
3. Flexibility: The changeover from solid to liquid fuels involves only a change in the burner heads. Multiple feed burners permit wide variations in turn-down ratio. This process is capable of instantaneous shutdown with full production resumed in 30 minutes.
4. Simplicity of construction: No complicated mechanical equipment or pressure scaling device is required. The only moving parts in the gasifiers are the moving screw feeders for solids or pumps for liquid feedstocks.
5. Ease of operation: Control of the gasifiers is achieved primarily by maintaining carbon dioxide concentration in the clean gas at a reasonably constant value. Slag fluidity may be visually monitored. Gasifiers display a good dynamic response.
6. Low maintenance: Simplicity of design and a minimum number of moving parts result in little maintenance between the scheduled annual events.
7. Safety and efficiency: The process has over 20 years of safe operation. The gasifier's overall thermal efficiency is 85 to 90%. The on-stream time or availability is better than 95%.

*Process description.* The Koppers-Totzek gasification process, whose schematic is shown in Figure 4, employs partial oxidation of pulverized coal in suspension with oxygen and steam. The gasifier is a refractory-lined steel shell equipped with a steam jacket for producing low-pressure process steam. A two-headed gasifier is capable of handling 400 tons of coal per day. Coal, oxygen, and steam are brought together in opposing gasifier burner heads spaced 180° apart. In the case of four-headed gasifiers, these burners are 90° apart. They can handle up to 850 tons of coal per day. Exothermic reactions produce a flame temperature of approximately 1930°C. Gasification of coal is almost complete and instantaneous. The carbon conversion depends on the reactivity of the coal, approaching 100% for lignites.

Gaseous and vapor hydrocarbons evolving from coal at a moderate temperature are passed through a zone of very high temperature in which they decompose so readily that there is no coagulation of coal particles during the plastic stage. Thus, any coal can be gasified regardless of the caking property, ash content, or ash fusion temperature. As a result of the endothermic reactions occurring in the gasifier between carbon and steam and radiation to the refractory walls, the flame temperature decreases from 1930 to 1500°C. Under these conditions, only gaseous products are produced, with no tars, condensable hydrocarbons, or phenols. The process therefore is essentially pollution free. Typical compositions of Koppers-Totzek gas products are shown in Table 2.

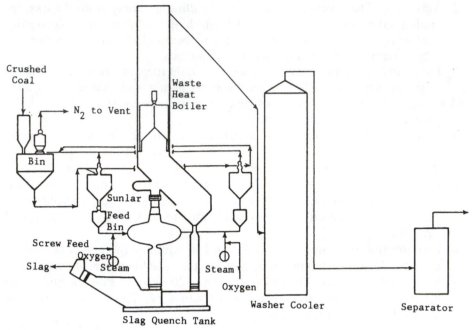

**Figure 4**   Schematic representation of the Koppers-Totzek process.

Ash in the coal feed is liquefied in the high-temperature zone. Approximately 50% of the coal ash drops out as slag into a slag quench tank below the gasifier. The remaining ash is carried out of the gasifier as fine fly ash. The gasifier outlet is equipped with water sprays to lower the gas temperature below the ash fusion temperature. This prevents slag particles from adhering to the tubes of the waste heat boiler, which is mounted above the gasifier. Ash fusion characteristics can be adjusted by addition of the flux to the coal feed.

The raw gas from the gasifier passes through the waste heat boiler where high-pressure steam, up to 100 atm, is produced. After leaving the waste heat

**Table 2  Typical Koppers-Totzek raw gas products (oxygen-blown)**

| | |
|---|---|
| CO | 52.5 |
| $H_2$ | 36.0 |
| $CO_2$ | 10.0 |
| $H_2S + COS$ | 0.4 |
| $N_2 + Ar$ | 1.1 |
| HHV = 286 Btu/scf | |

*Source:* Reference 9.

boiler; the gas at 175–180°C is cleaned and cooled in a high-energy scrubbing system, which reduces the entrained solids to 0.002–0.005 grains/scf and lowers the temperature from 175°C to 35°C. If the gas coming out of the Koppers-Totzek process is to be compressed to high pressures for chemical synthesis, electrostatic precipitators are used for further cleaning. Several gasifiers can share common cleaning and cooling equipment.

The cool, cleaned gas leaving the gas cleaning system contains sulfur compounds, which must be removed to meet gas specifications. The type of system chosen depends on the end uses and the pressure of the product gas. For low-pressures and low-Btu gas applications, there are chemical reaction processes such as amine and carbonate processes. At higher pressures, physical absorption processes, such as Rectisol are used. The choice of the process is also dependent on the desired purity of the product gas and its selectivity with respect to the concentration of carbon dioxide and sulfides.

**Shell gasification.** The Shell coal gasification process is a relatively new process developed by Royal Dutch/Shell group in the early 1970s. It uses a pressurized, slagging entrained flow reactor for gasifying dry pulverized coal [15]. It has the potential to gasify a wide range of coals including low-rank lignites with high moisture content. Unlike other gasifying systems, it uses pure oxygen as the gasifying medium. A schematic of Shell gasification process is given in Figure 5. The process has the following advantages [16]:

1. Almost 100% conversion of a wide variety of coals, including high-sulfur coals, lignites, and coal fines
2. High thermal efficiency in the range of 75 to 80%
3. Efficient heat recovery through production of high-pressure superheated steam
4. Production of clean gas without any significant amount of by-products
5. High throughput
6. Environmental compatibility

Coal, before feeding to the gasifier vessel, is crushed and ground to a size less than 90 μm. This pulverized dried coal is fed through diametrically opposite diffuser guns into the reaction chamber [17]. The coal is then reacted with the pure oxygen and steam where the flame temperature reaches as high as 1800–2000°C. A typical operating pressure is about 30 atm. Raw gas typically consists mainly of carbon monoxide (62–63%) and hydrogen (28%) with some quantities of carbon dioxide and nitrogen. A water-filled bottom compartment is provided where molten ash is collected. Some amount of ash is entrained with the synthesis gas and is then recycled along with the unconverted carbon. A quench section is provided at the reactor outlet. Removal of particulate matter

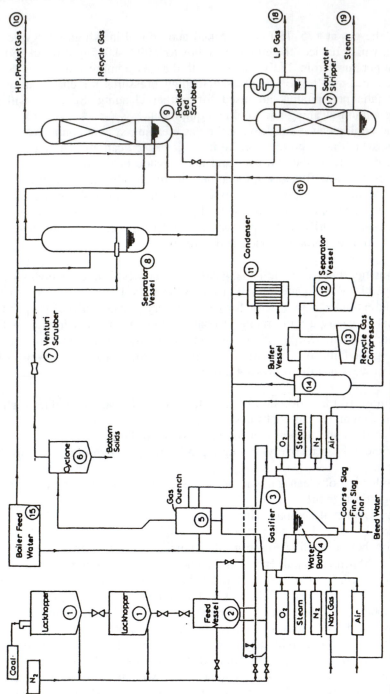

**Figure 5** Schematic of the Shell Gasification Process (6 tpd PDU). (Source: Reference 8.)

from the raw gas is integrated with the overall process. This removal system consists of cyclones and scrubbers. Its main advantage is elimination of solids containing wastewater, thus omitting the need for filtration.

**Texaco gasification.** The Texaco process also uses the entrained bed technology for gasification of coal. It gasifies coal under relatively high pressure by injection of oxygen (or air) and steam with concurrent gas-solid flow. Fluidized coal is mixed with either oil or water to make it into a slurry. This slurry is pumped under pressure into a vertical gasifier, which is basically a pressure vessel lined inside by refractory walls. The slurry reacts with either air or oxygen at high temperature. The product gas contains primarily carbon dioxide and hydrogen with some quantity of methane. Because of the high temperature, oil and tar are not produced. This process is basically used to manufacture synthesis gas [3]. A schematic of the Texaco gasification process is shown in Figure 6.

This gasifier evolved from the commercially proven Texaco partial oxidation process [8] used to gasify crude oil and hydrocarbons. Its main feature is the use of a coal slurry feed that simplifies the coal feeding system and operability

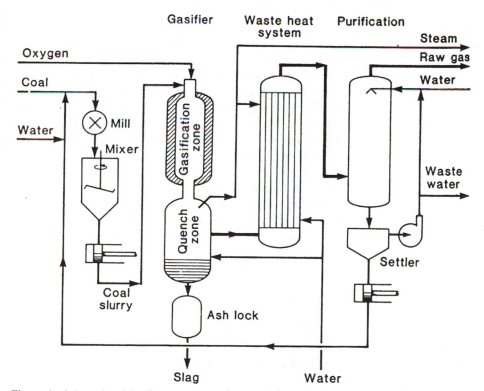

**Figure 6** Schematic of the Texaco process. (Source: Reference 18).

of the gasifier. The gasifier is a simple, vertical, cylindrical pressure vessel with refractory linings in the upper partial oxidation chamber. It is also provided with a slag quench zone at the bottom where the resulting gases and molten slag are cooled down. In the later operation, large amounts of high-pressure steam can be obtained, which result in a large increase in thermal efficiency. Another important factor that affects the gasifier thermal efficiency is the water content of the coal slurry. This water content should be minimized because a large amount of oxygen must be used to supply the heat required to vaporize the water. This gasifier favors high-energy dense coals, so the water-to-energy ratio in the feed is small. Eastern bituminous coals are preferable to lignites for this gasifier. The gasifier operates at about 1100–1370°C and a pressure of 20–85 atm.

The product gases and molten slag produced in the reaction zone pass downward through a water spray chamber and a slag quench bath where the cooled gas and slag are then removed for further treatment. The gas after being separated from slag and cooled is treated to remove carbon fines and ash. These are then recycled to the slurry preparation system. The gas is then treated for acid gas removal and elemental sulfur is recovered from the hydrogen sulfide ($H_2S$)–rich stream.

**In situ gasification.** In situ gasification or underground gasification is a technology for recovering the energy contents of coal deposits that it is not feasible to recover economically or technically by conventional mining technologies. Coal reserves that are suitable for in situ gasifications are the ones with low heating value, thin seam thickness, great depth, high ash or excessive moisture content, large seam dip angle, or undesirable overburden properties. A considerable amount of investigative work has been performed on underground coal gasification (UCG) in the former USSR, but it is only in recent years that the concept has been revived in Europe and North America as a means of fuel gas production. In addition to its potential for recovering deep, low-rank coal reserves, the UCG process offers many advantages with respect to its resource recovery, minimal environmental impact, safety, process efficiency, and economic potential. The aim of in situ gasification of coal is to convert coal into combustible gases by combustion of coal seam in the presence of air, oxygen, or steam.

The basic concepts of underground coal gasification may be illustrated by Figure 7 [18]. The principles of in situ gasification are very similar to those involved in the aboveground (ex situ) gasification of coal. The combustion process itself could be forward or reverse. In forward combustion the movement of the combustion front and injected air is in the same direction, whereas in reverse combustion the front moves in the opposite direction to the injected air. The process involves drilling and subsequent linking of the two boreholes so that the gas passes between the two. Combustion is initiated at the bottom of one borehole, called the injection well, and is maintained by continuous injection of

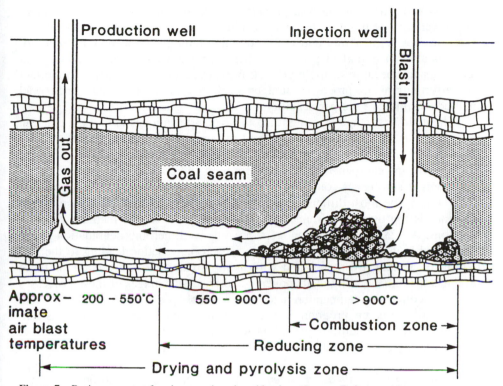

**Figure 7** Basic concepts of underground coal gasification. (Source: Reference 18.)

air. In the initial reaction zone, carbon dioxide is generated by reaction of oxygen (air) with the coal and further reacts with the coal to produce carbon monoxide in the reduction zone. In addition, at such a high temperature, moisture present in the seam may react with carbon to form carbon monoxide and hydrogen. In addition to all these basic gasification reactions, coal decomposes in the pyrolysis zone due to high temperatures to produce hydrocarbons and tars, which contribute to the product gas mix. The heating value from the air-blown in situ gasifier is roughly 100 Btu/scf. The low heat content of the gas makes it necessary to use the product gas on site. A detailed discussion on in situ gasification can be found in in-depth reviews by Thompson [19] and Gregg and Edgar [20].

Juntgen [21] has explored the possibilities of using microbiological techniques for in situ conversion of coal into methane. Microorganisms have been found that develop on coal as a sole carbon source. Both forms of sulfur, organic and inorganic (pyritic and sulfatic), can be removed by biochemical techniques and microorganisms are able to grow, in principle, in narrow pore systems of solids. Conversion of aromatics is also potentially possible. An important precondition to developing such new techniques for in situ coal conversion in deep seams is knowledge of the coal properties, both physical and chemical, under the prevailing conditions. The two most important coal properties for in situ processes are permeability and rank. For microbiological conversion, microporosity is also an important parameter. The permeability of coal seams at great depth is small because of the high rock pressure, and accessibility is important for performing in situ processes. Juntgen's review article mentioned some model methods to improve the permeability.

The main advantage of using microbiological techniques is that the reaction takes place at ambient temperatures, and the progress in developing these processes is remarkable. A remarkable effect of such reactions in coal is that the microorganisms can penetrate the fine pores of the coal matrix and can create new pores if substances contained in the coal matrix are converted into soluble compounds. The most difficult and complex problem for microorganism-based reactions is the transition from solely oxidative processes to methane-forming reactions. At least three reaction steps are involved: aerobic degradation to biomass and higher molecular products; an anaerobic reaction leading to the formation of acetate, hydrogen, and carbon monoxide; and conversion of these products to methane using methanogenic bacteria. Advantages of these processes are lower conversion temperature and more valuable products. At the same time, an intensive research program is necessary to adapt reaction conditions and product yields to conditions prevailing in coal seams at depth, where transport processes play a significant role in the overall reaction.

*The underground gasification system.* The underground gasification system involves three distinct sets of operations: pregasification, gasification, and utilization. Pregasification operations provide access to the coal deposit and prepare it for gasification. Connection between the inlet and outlet through the coal seam is achieved via shafts and boreholes. Linking can be achieved by several means, such as pneumatic linking, hydraulic or electrolinking, and explosives. Sometimes part of linking may also be due to natural permeability. Among all the linking methods, only directionally drilled boreholes provide positive connections between inlet and outlet sections; all other methods permit a certain degree of uncertainty to creep into the system. A plan view of a linked-vertical-well underground gasification plant operated near Moscow [18] is shown in Figure 8.

The gasification operations are those that result in reliable production of low-Btu gas. They consist of input of gasification agents like air or oxygen-steam or alternating air-steam, followed by ignition. Ignition is achieved by electrical means or by burning solid fuels. Ignition results in contact between gasification agents and coal seam at the flame front. The flame front may advance in the direction of gas flow (forward burning) or in the direction opposite to the gas flow (backward burning). During these operations, the major problems are in the area of process control.

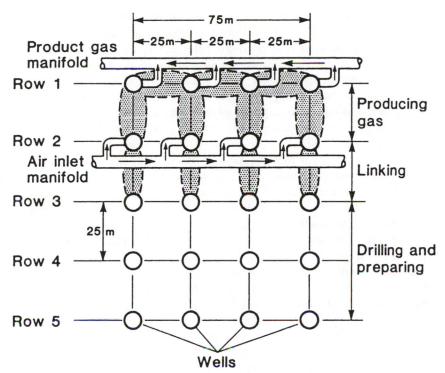

**Figure 8** Plan view of linked-vertical-well underground gasification plant operated near Moscow. (Source: Reference 18.)

The next and most important operation is the utilization of the product gas and requires a coupling between the gas source and the energy demand. The product gas can be used as an energy source to produce electricity on site or can be upgraded to a high-Btu pipeline quality gas for transmission. In some other applications, it could be utilized near the deposit as a hydrogen source, as a reducing agent, or as a basic raw material for chemicals manufacture. No major technical problems are involved in the utilization area.

***Methods for underground gasification.*** Two main types of methods have been tried successfully: shaft methods and shaftless methods (and combinations of the two) [22–24]. Selection of the method to be used depends on such parameters as the permeability of the coal seam; the geology of the coal deposit; the seam thickness, depth, and inclination; and the amount of mining desired. Shaft methods involve driving of shafts and other large-diameter openings, which requires underground labor; shaftless methods use boreholes for gaining access to the coal seam and do not require underground labor.

**Shaft methods** The *chamber or warehouse method* requires the preparation of underground galleries and the isolation of coal panels with brickwood. The blast of air for gasification is applied from the gallery at the previously ignited face of one side of the panel, and the gas produced is removed through the gallery at the opposite side of the panel. This method relies on the natural porosity of coal for flow through the system. Gasification and combustion rates are usually low, and the product gas sometimes has a variable composition. Coal seams are charged with dynamite to crush them in advance of the reaction zone by a series of explosions.

The borehole producer method requires the development of parallel underground galleries located about 500 ft apart within the coal bed. From these galleries, 4-in.-diameter boreholes are drilled about 15 ft apart from one gallery to the opposite. Electric ignition of the coal in each borehole is achieved by remote control. This method was designed to gasify substantially flat-lying seams. Variations on this technique utilized hydraulic and electric linking as alternatives to the use of boreholes.

The stream method can be applied to steeply pitched coal beds. Inclined galleries following the dip of the coal seam are constructed parallel to each other and are connected at the bottom by a horizontal gallery or "fire-drift." A fire in the horizontal gallery starts the gasification, which proceeds upward with air coming down one inclined gallery and gas departing through the other. One obvious advantage of the stream method is that ash and roof material drop down, tend to fill void space, and do not tend to choke off the combustion zone at the burning coal face. This method is less successful for horizontal coal seams because of roof collapse problems.

**Shaftless methods** In shaftless methods, all development and gasification are carried out through a borehole or a series of boreholes drilled from the surface into the coal seam. A general approach has been to make the coal bed more permeable between the inlet and outlet boreholes by some linking method, ignite the coal seam, and then gasify by passing air or other gasifying agents from the inlet borehole to the outlet one.

Percolation or filtration methods are the most direct approach to shaftless gasification of a coal seam using multiple boreholes. The distance between boreholes depends on the seam permeability. Most lower rank coals such as lignites have considerable natural permeability and can be gasified without overt linking. Higher rank coals are less permeable, in which case it becomes necessary to connect boreholes by some linking technique that will increase the permeability and/or fracture the coal seam so that an increased rate of gas flow can be obtained. Air or air and steam are blown through one hole, and gas is removed from the second one. Forward or reverse combustion is permitted in this method. As the burning progresses, the permeability of the seam increases and compressed air blown through the seam also enlarges cracks in the seam. When the combustion of a zone nears completion, the process is transferred to another pair of boreholes. In this operation, coal ash and residue are strong enough to prevent roof collapse.

***Potential problem areas in in situ gasification.*** There are several reasons why the foregoing methods cannot produce a high quality and constant quantity of gases, recover a high percentage of coal in the ground, and control groundwater contamination. The technical problem areas are discussed in this section.

1. *Combustion control.* Combustion control is essential to the control of gas quality and the level of coal recovery. The contact between the coal and the reacting gas should be such that the coal is completely gasified in situ, all free oxygen in the inlet gas is consumed, and the production of fully burned carbon dioxide and water is minimized. In this method, as the time goes by, the heating value of the product gas decreases. This is due to increasingly poor contact of gas with the coal face resulting from large void volumes and from roof collapse. The problem of efficient contacting has not been solved satisfactorily in these methods.
2. *Roof control.* After the coal is burned away, a substantial roof area is left unsupported. Uncontrolled roof collapse causes problems in the combustion control and hinders successful operation of the gasification process. It also results in the leakage of reactant gases, the seepage of groundwater into the coal seam, the loss of product gas, and surface subsidence above the coal deposit.
3. *Permeability, Linking, Fracturing.* A coal bed usually does not have sufficiently high permeability to permit the passage of oxidizing gases through it without a high pressure drop. Also, intentional methods like pneumatic, hydraulic, electrolinking, and fracturing with explosives do not result in a uniform increase in permeability. They also tend to disrupt the surrounding strata and worsen the leakage problems. The use of boreholes provides a positive predictable method of linking and is a preferred technique.
4. *Leakage Control.* This is one of the most important problems, because the loss of a substantial amount of product gas can adversely affect the quality and the extraction costs. Influx of water can also affect the control of the process. Leakage varies from site to site and is dependent on geological conditions, depth of coal seam, permeability of the coal bed, and so forth. It is imperative that in situ gasification never be attempted in a severely fractured area, in shallow seams, and in coal seams adjoining porous sedimentary layers, and it is essential to prevent roof collapse and to properly seal inlet and exit boreholes.

***Monitoring of underground processes.*** Proper monitoring and measurement of the underground processes are necessary components of an underground gasification system but are not sufficient for successful operation. Detailed and thorough knowledge of all the parameters affecting the gasification is required so that adequate process control philosophy can be evolved for controlling the operation. These factors are the location and shape of the fire front; the temperature distribution along the fire front; the extent and nature of collapsed roof debris; the permeability of coal seam and debris; the leakage of reactants, products, and groundwater; and the composition of the product gases.

***Criteria for an ideal underground gasification system.*** The following are the criteria for an ideal underground coal gasification system for successful operation:

- It must lend itself to operation on a large scale.
- The processing features must be such that no big deposits of coal are left ungasified or partially gasified.
- The processing features must be controllable so that desired levels of product gases are consistently obtained.
- The mechanical features must to some extent be able to control the undesirable factors such as groundwater inflows and leakage of reactants and products.
- The process should not require any underground labor, either during operation or preferably even during the installation of the facilities.

## Other commercial gasifiers

***Winkler process.*** The Winkler process is the oldest commercial process employing the fluidized bed technology [25]. The process was developed in Europe in the 1920s. More than 15 plants are in operation today all over the world, with the largest having an output of 1.1 million scf/day. In this process, crushed coal is dried and fed into a fluidized bed reactor by means of a variable-speed screw feeder. The gasifier operates at atmospheric pressure and a temperature of 815–1000°C. Coal reacts with oxygen and steam to produce off-gas rich in carbon monoxide and hydrogen. The high operating temperature leaves very little tar and liquid hydrocarbons in the product gas stream. The gas stream, which may carry up to 70% of the ash, is cleaned by water scrubbers, cyclones, and electrostatic precipitators. Unreacted carbon carried over by the gas is converted by secondary steam and oxygen in the space above the fluidized bed. As a result, a maximum temperature occurs above the fluidized bed. To prevent ash particles from melting and forming deposits in the exit duct, gas is cooled by a radiant boiler section before it leaves the gasifier. Raw gas leaving the gasifier is passed through a further waste heat recovery section. Gas is then compressed and shifted. It has a heating value of about 275 Btu/scf. The thermal efficiency of the process runs about 75%.

**Process description.** In early 1920s, Winkler, working for Davy-Power Gas Inc., conceived the idea of using a fluidized bed for gasifying the coal. The first commercial process was built in 1926 at Leuna. Since then, more than 30 producers and 15 installations have put into operation this process for coal gasification.

In the earlier facilities, dryers were used prior to the entry of coal in the gas generator to drive off the moisture to less than 8%. But later, with experience, it was realized that as long as the feed coal could be sized, stored, and transported without plugging, the dryers could be omitted. Without dryers, moisture in the coal is vaporized in the generator with the heat provided by using additional oxygen. Also drying the coal in the generator offers one advantage, elimination of an effluent stream, the dryer stack, which requires further particulate and sulfur removal treatment.

**Gasifier (generator).** A schematic of a Winkler fluidized bed gasifier [18] is shown in Figure 9. Coal is fed to the gasifier through variable-speed screws. These screws not only control the coal feed rate but also serve to seal the gasifier, preventing steam from wetting the coal and blocking the pathway. The high-

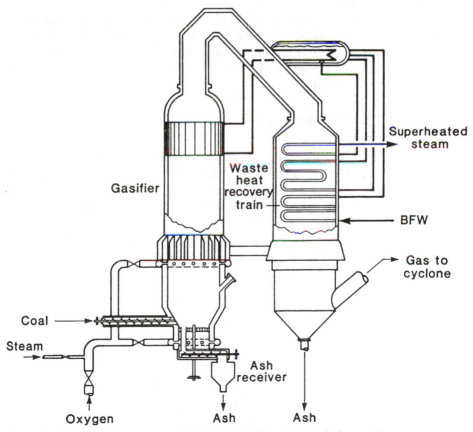

**Figure 9**   Schematic of the Winkler fluidized bed gasifier. (Source: Reference 18.)

velocity gas stream flows up from the bottom of the gasifier. This gas stream fluidizes the bed of coal and intimately mixes the reactants. This helps in attaining uniform temperatures between the solid and the gas stream, which further facilitates the reactions approaching equilibrium in the shortest possible time. Gasification in the Winkler gasifier is a combination of combustion and water gas shift reaction and due to relatively high temperatures, all of the tars and heavy hydrocarbons are reacted [26].

As a result of the fluidization, the ash particles are segregated according to size and specific gravity. About 30% of the ash leaves the bottom, while 70% is carried overhead. The lighter particles carried up along with the producer gas are further gasified in the space above the bed. The quantity of gasifying medium injected into this bed must be proportional to the amount of unreacted carbon being carried over. If it is too small, ungasified carbon is carried out of the generator, resulting in lower thermal efficiency, and if it is too large, product gas is unnecessarily consumed by combustion. The maximum temperature in the bed generator occurs in the space above the fluidized bed because of this secondary gasification.

Just above the bed, a radiant boiler is installed that cools the gas 150–205°C before it leaves the generator. This helps prevent sintering of the fly ash on the refractory walls of the exit duct. The sensible heat recovered generates superheated steam and preheats the boiler feed water (BFW). Typical Winkler gas products are shown in Table 3.

### Features of the Winkler process

1. It can gasify a variety of coal feeds ranging from lignite to coke. Lignite, being younger in geological age, is more reactive than bituminous and anthracite. With more reactive coal, the gasification temperature decreases and the overall gasification efficiency increases. For less reactive coals, the heat losses in unburned solids increase.
2. It can gasify coal with a high ash content. Although a high ash content leads to increased residues and uncombusted materials, usually coals with high

**Table 3 Typical Winkler gas products**

|  | $O_2$-blown | Air-blown |
|---|---|---|
| CO | 48.2 | 22.0 |
| $H_2$ | 35.3 | 14.0 |
| $CH_4$ | 1.8 | 1.0 |
| $CO_2$ | 13.8 | 7.0 |
| $N_2$ + Ar | 0.9 | 56.0 |
| HHV | 288 | 126 |

*Source:* Reference 9.

ash contents are cheaper and sources of feed coal are greatly expanded. The Winkler gasifier is not sensitive to variations in the ash content during operation.

3. The Winkler gasifier can also gasify liquid fuels in conjunction with coal gasification. Addition of supplementary liquid feeds results in an increase in production and heating value of the product gas.
4. Winkler generation is very flexible in terms of the capacity and turn-down ratio. It is limited at the lower end by the minimum flow required for fluidization and at the upper end by the minimum residence time for complete combustion of residues.
5. Start-up and shutdown are very easy just by stopping the flows of oxygen, coal, and steam. This can be achieved within minutes. Even for hard coals that are difficult to ignite, the heat loss during shutdown may be reduced by brief injection of air into the fuel bed.
6. Maintenance of the generator is easy, since it consists only of a brick-lined reactor with removable injection nozzle for the gasification medium.

**Wellman-Galusha process.** This process has been in commercial use for more than 40 years. It can produce low-Btu gas using air for fuel gas or synthesis gas using oxygen. There are two types of gasifiers, the standard type without agitator and the agitated type. The rated capacity of the agitated type is about 25% more than that of a standard gasifier of the same size. The agitated type can handle volatile caking bituminous coals [3]. A schematic of a Wellman-Galusha agitated gasifier [9] is shown in Figure 10.

This gasifier falls under the fixed or moving bed type of the reactor. The gasifier shell is water jacketed and hence the inner wall of the reactor vessel does not require a refractory lining. The gasifier operates at around 540–650°C and at atmospheric pressure. Crushed coal is fed to the gasifier through a lock hopper and vertical feed pipes, and steam and oxygen are injected at the bottom of the bed through tuyeres. The fuel valves are operated to maintain constant flow of coal to the gasifier, which also helps in stabilizing the bed and therefore the quality of the product gas. The injected air or oxygen passes over the water jacket and generates the steam required for the process. A rotating grate is located at the bottom of the gasifier to remove ash from the bed uniformly. An air-steam mix is introduced underneath the grate and is distributed through the grate into the bed. This gasifying medium passes through the ash, combustion, and gasifying zone in that order. The product gas contains carbon dioxide, hydrogen, carbon monoxide, and nitrogen (if air is used as injecting medium) and, being hot, dries and preheats the incoming coal before leaving the gasifier. A typical product composition from a Wellman-Galusha gasifier is shown in Table 4.

The product gas is passed through a cyclone separator, where char particles and fine ash are removed. It is then cooled and scrubbed in a direct contact countercurrent water cooler and treated for sulfur removal. If air is used as

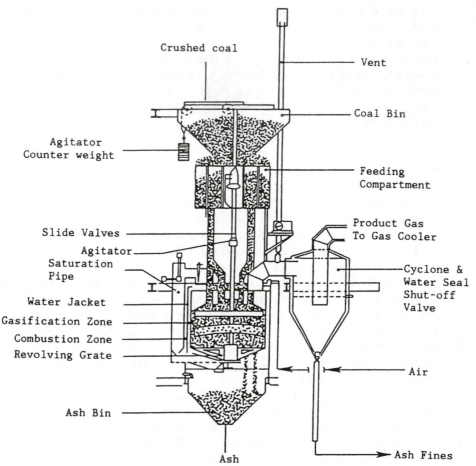

**Figure 10** Schematic representation of an agitated Wellman-Galusha gasifier. (Source: References 3 and 9.)

### Table 4 Typical Wellman-Galusha products (air-blown)

| | |
|---|---|
| CO | 28.6% |
| $H_2$ | 15.0 |
| $CH_4$ | 2.7 |
| $N_2$ | 50.3 |
| $CO_2$ | 3.4 |
| HHV (dry) = 168 Btu/scf | |

*Source:* Reference 9.

combustant, low-Btu gas is obtained, whereas if oxygen is used then medium-Btu gas is produced.

Unlike the standard gasifier, the agitated version is equipped with a slowly revolving horizontal arm that spirals vertically below the surface of the coal bed to minimize channeling. This arm also helps in providing a uniform bed for gasification.

**Catalytic gasification.** In recent years the study of catalytic gasification has received a great deal of attention because it needs less heat and yields high carbon conversion. Studies of the catalysis of coal gasification have two purposes: (1) to understand the kinetics of coal gasification that involves active mineral matter and (2) to design possible processes using these catalysts. The use of catalysts lowers the gasification temperature, which favors product composition under equilibrium conditions as well as high thermal efficiency. However, under normal conditions a catalytic process cannot compete with a noncatalytic process unless the catalyst is very inexpensive and highly active at low temperatures. Research on catalysis covers mainly three subjects: basic chemistry, application problems, and engineering. An extensive review article has been published by Juntgen [27]. Nishiyama [28] has published a review article that features some possibilities in well-defined catalytic research. The article presents the following remarks:

1. Salts of alkali and alkaline earth metals as well as transition metals are active catalysts for gasification.
2. The activity of a particular catalyst depends on the gasifying agent as well as the gasifying conditions.
3. The main mechanism of catalysts using alkali and alkaline earth metal salts in steam and carbon dioxide gasification involves the transfer of oxygen from the catalyst to carbon through the formation and decomposition of a C—O complex.

The mechanism of hydrogasification catalyzed by iron or nickel is still not very clear. But a possible explanation is that active catalyst seems to be in the metallic state and there are two main steps in the mechanisms: hydrogen dissociation and carbon activation [29–33]. For the latter case, carbon dissolution into and diffusion through catalyst particles seems logical. Gasification proceeds in two stages, which have different temperature ranges and thermal behaviors, so that a single mechanism cannot explain the whole reaction. Thus the catalyst is still assumed to activate the hydrogen.

Calcium as a catalyst has also been studied by several researchers [34–42]. This catalyst has very high activity in the initial period when it is well dispersed in the other promoter catalyst, but with increasing conversion, or so-called burn-off, the catalytic activity drops. The chemical state and dispersion are studied by chemisorption of carbon dioxide, X-ray diffraction, and some other techniques. They confirmed the existence of two or more states of a calcium compound as well as the formation of a surface oxygen complex.

Compared to other heterogeneous catalytic systems, the catalysis in gasification is complex: the catalyst is very short lived and effective only while in contact with the substrate, which itself changes. Hence the definition of activity for such systems is very equivocal. For an alkali metal catalyst, the rate increases due to the change in catalyst dispersion and the increase in the ratio of catalyst to carbon in the later stage of gasification. Other possible explanations for the rate increase could be the change in surface area by pore opening and the change in chemical state of the catalyst. At the same time, there are some changes that deactivate the catalyst—for example, agglomeration of catalyst particles, coking, and reaction with sulfur or other trace elements.

The activity of the catalyst depends on the nature of the substrate and gasifying conditions. The main properties of the substrate related to the activity are (1) reactivity of the carbonaceous constituents, (2) catalytic effect of minerals, and (3) effect of minerals on the activity of added catalyst. The following general trends have been observed in reference to the factors affecting the activity of the catalysts:

1. Nickel catalysts are more effective toward lower rank coals because they are more dispersed on them. The efficacy of potassium is independent of the rank. In any case, the coal rank as given by the carbon content is not an appropriate parameter with which to predict the catalyst activity.
2. The surface area of coal char is related to the activity of the catalyst. It can be related to the number of active sites in cases in which the amount of catalyst is large enough to cover the available surface area. For an immobile catalyst, the conversion is almost proportional to the initial surface area.
3. Sometimes, pretreatment of coal before the catalytic action helps in achieving higher rates. Although the pretreatment cannot be directly applied to coal as a practical process, a suitable selection of coal species or processing could enhance the activity of catalysts.
4. The effect of mineral matter on the catalysis is twofold. Some minerals like alkali and alkaline earth metals catalyze the reaction, whereas some like silica and alumina interact with catalyst and deactivate it. In general, demineralization results in enhancement of the activity for potassium and only slightly for calcium and nickel.

The method of catalyst loading is also important for the activity. It should be loaded in such a way that definite contact with both solid and gaseous reactants is maintained. It was observed that when the catalyst was loaded from an aqueous solution, a hydrophobic carbon surface resulted in finer dispersion of the catalyst compared with a hydrophillic surface.

The most common and effective catalysts for steam gasification are the salts, namely oxides and chlorides, of alkali and alkaline earth metals, separately or in combination [43]. Xiang et al. studied the catalytic effects of the Na-Ca composite on the reaction rate, methane conversion, steam decomposition, and product gas composition at reaction temperatures of 700–900°C and pressures from 0.1 to 5.1 MPa. A kinetic expression was derived and the reaction rate constants and the activation energy at elevated pressures were determined. Alkali metal chlorides like NaCl and KCl are very inexpensive and are attractive as raw materials of catalytic gasification. However, their activities are quite low compared with the corresponding carbonates because of the strong affinity between alkali metal ion and chloride ion. Takarada et al. [44] have attempted to make Cl-free catalyst from NaCl and KCl by an ion exchange technique. The authors ion-exchanged alkali metals to brown coal from an aqueous solution of alkali chloride using ammonia as a pH-adjusting agent. Chloride ions from alkali chloride were completely removed by water washing. This Cl-free catalyst promoted the steam gasification of brown coal markedly. It was found to be catalytically as active as alkali carbonate in the steam gasification. During gasification, the chemical form of active species was found to be in the carbonate form and was easily recovered. Sometimes an effective way to prepare the catalyst is physical mixing of K-exchanged coal with the higher rank coals [45]. This direct contact between K-exchanged coal and higher rank coal resulted in rate enhancement in gasification. Potassium was found to be a highly suitable catalyst for catalytic gasification by the physical mixing method. Weeda et al. [46] studied the high-temperature gasification of coal under product-inhibited conditions using potassium carbonate as a catalyst to enhance the reactivity. They performed temperature-programmed experiments to characterize the gasification behavior of different samples relative to each other. Some researchers [47] have recovered the catalysts used in the form of a fertilizer of economic significance. They used a combined catalyst consisting of potassium carbonate and magnesium nitrate in the steam gasification of brown coal. The catalysts along with coal ash were recovered as potassium silicate complex fertilizer.

In addition to the commonly used catalysts like alkali and alkaline earth metals for catalytic gasification, some less known compounds made out of rare earth metals, as well as molybdenum oxide ($MoO_2$), have been successfully tried for steam and carbon dioxide gasification of coal [48–50]. Some of the rare earth compounds used were $La(NO_3)_3$, $Ce(NO_3)_3$, and $Sm(NO_3)_3$. The catalytic activity of these compounds decreased with increasing burn-off of the coal. To alleviate this problem, coloading with a small amount of Na or Ca was attempted and the loading of rare earth complexes was done by the ion exchange method.

It is a well-known fact that the coal gasification technology could benefit by the development of suitable catalysts that will help catalyze steam decomposition and carbon-steam reaction. Batelle [51] has developed a process whereby calcium in the form of oxide was used to catalyze the hydrogasification reaction. It was shown that a reasonably good correlation exists between the calcium content and the reactivity of coal chars with carbon dioxide. Other alkali metal compounds, notably chlorides and carbonates of sodium and potassium, can also raise the gasification rate by as much as 35–60%. Apart from oxides of calcium, iron, magnesium, and zinc oxides are also fairly potent, accelerating gasification by 20–30%.

Some speculative mechanisms have been proposed by Murlidhara and Seras [51] concerning the role of calcium oxide in enhancing the reaction rate. For instance, hydrogen may be a donor and then may be abstracted by calcium oxide by the given mechanism.

*Scheme 1*:

$$\text{Organic} \longrightarrow \text{organic*} + H_2$$

$$\text{CaO} + 2H_2 \longrightarrow \text{CaH}_2 + H_2O$$

$$\text{Organic*} + CO_2 \longrightarrow 2CO$$

$$CO_2 + \text{CaH}_2 \longrightarrow \text{CaO} + CO + H_2$$

*Scheme 2*:

$$\text{CaO} + CO_2 \longrightarrow \text{CaO(O)} + CO$$

$$\text{CaO(O)} \longrightarrow \text{CaO} + (O)$$

$$(O) + C \longrightarrow CO$$

*Scheme 3*:

$$\text{CaO} + 2C \longrightarrow \text{CaC}_x + CO$$

$$\text{CaC} + \text{organic (oxygen)} \longrightarrow \text{CaO} + \text{organic*}$$

Exxon [52] has reported that impregnation of 10–20% potassium carbonate lowers the optimum temperature and pressure for steam gasification of bituminous coals from 1800 to 1400°F and from 1000 to 500 psi. In their commercial-scale plant design, the preferred form of makeup catalyst was identified as potassium hydroxide. This catalyst adds several advantages to the overall process. It increases the rate of gasification, it prevents swelling and agglomeration when handling caking coals, and, most importantly, it promotes the methanation equilibrium. Therefore, the production of methane is favored in comparison to synthesis gas. A catalyst recovery unit is provided after the gasification stage to recover the used catalyst.

**Molten medium gasification.** When salts of alkali metals and iron are used as a medium to carry out coal gasification, it is referred to as molten medium gasification. The molten medium not only catalyzes the gasification reaction but also supplies the necessary heat and serves as a heat exchanger medium [3, 53].

Several main commercial processes have been developed over the years: the Kellogg-Pullman salt process, the Atgas molten iron gasification process, Rockwell molten salt gasification, and Rummel-Otto molten salt gasification. Schematics of the Rockwell molten salt gasifier and Rummel-Otto single shaft gasifier are shown in Figures 11 [9] and 12 [18].

*Kellogg molten salt process.* In this process, gasification of coal is carried out in a bath of molten sodium carbonate through which steam is passed [54]. The molten salt offers the following advantages:

1. The steam-coal reaction, being basic in nature, is strongly catalyzed by sodium carbonate, permitting complete gasification at a relatively low temperature.
2. Molten salt disperses coal and steam throughout the reactor, thereby permitting direct gasification of caking coals without carbonization.
3. A salt bath can be used to supply heat to the coal undergoing gasification.
4. Due to the uniform temperature throughout the medium, the product gas obtained is free of tars and tar acids.

Crushed coal, which is stored in lock hoppers, is picked up by a stream of preheated oxygen and steam into the gasifier. In addition, sodium carbonate recycle from the ash rejection system is metered into the transport gas stream and the combined coal, salt, and carrier are admitted to the gasifier. The main portion of the preheated oxygen and steam is admitted into the bottom of the reactor for passage through the salt bath to support the gasification reactions. Along with the usual gasification reactions, sulfur entering with the coal accumulates as sodium sulfide to an equilibrium level. At this level, it leaves the reactor according to the following reaction:

$$Na_2S + CO_2 + H_2O \longrightarrow Na_2CO_3 + H_2S$$

Ash accumulates in the melt and leaves along with the bleed stream of salt, where it is rejected and sodium carbonate recycled. The bleed stream of salt is quenched in water to dissolve sodium carbonate and permit rejection of coal ash by filtration. The dilute solution of sodium carbonate is further carbonated for precipitation and recovery of sodium bicarbonate. The filtrate is recycled to

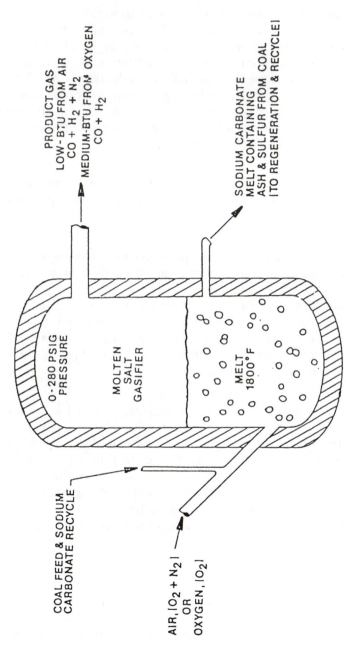

**Figure 11** Rockwell molten salt gasifier. (Source: Reference 9.)

PRODUCT GAS
LOW-BTU FROM AIR
$CO + H_2 + N_2$
MEDIUM-BTU FROM OXYGEN
$CO + H_2$

SODIUM CARBONATE
MELT CONTAINING
ASH & SULFUR FROM COAL
(TO REGENERATION & RECYCLE)

0 - 280 PSIG
PRESSURE

MOLTEN
SALT
GASIFIER

MELT
1800°F

COAL FEED & SODIUM
CARBONATE RECYCLE

AIR,$(O_2 + N_2)$
OR
OXYGEN,$(O_2)$

128

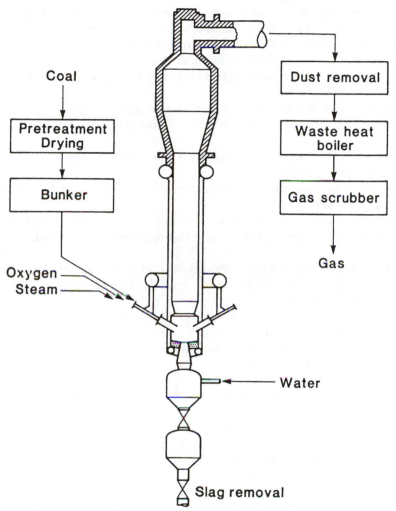

**Figure 12** Rummel-Otto single-shaft gasifier. (Source: Reference 18.)

quench the molten salt stream leaving the reactor. The sodium bicarbonate filtrate cake is dried and heated to regenerate sodium carbonate for recycle to the sodium carbonate from the gasifier. The gas stream leaving the gasifier is processed to recover the entrained salt, and the heat is then further processed for conversion to the desired product gas such as syngas, pipeline gas, or substitute natural gas.

*Atgas molten iron coal gasification.* This process is based on the molten iron gasification concept, in which coal is injected with steam or air into a molten iron bath. Steam dissociation and thermal cracking of coal volatile matter generate hydrogen and carbon monoxide. The coal sulfur is captured by the iron and transferred to a lime slag from which elemental sulfur can be recovered as a by-product. The coal dissolved in the iron is removed by oxidation to carbon monoxide with oxygen or air injected near the molten iron surface. The Atgas process uses coal, steam, and oxygen to yield product gases with heating values around 900 Btu/scf.

The Atgas molten iron process has several inherent advantages over gas-solid contact gasification in either fixed or fluidized bed reactors [55].

1. Gasification is carried out at low pressures, hence the mechanical problem of coal feeding is eliminated.
2. Coking properties, ash fusion temperatures, and fines generation of coal are not problematic.
3. The sulfur content of coal does not create any environmental problem because sulfur is retained in the system and recovered as elemental sulfur from the slag.
4. The system is very flexible with regard to the physical and chemical properties of the coal. Relatively coarse particles can be handled without any special pretreatment.
5. Formation of tar is eliminated because of high-temperature operation.
6. The product gas is essentially free of sulfur compounds.
7. Shutdown and start-up procedures are greatly simplified compared with fixed or fluidized bed reactors.

Coal and limestone are injected into the molten iron through tubes using steam as the carrier. The coal devolatilizes with some thermal cracking of the volatile constituents, leaving the fixed carbon and sulfur to dissolve in iron, whereupon carbon is oxidized to carbon monoxide. The sulfur in both organic and pyritic forms migrates from the molten iron to the slag layer, where it reacts with lime to produce calcium sulfide.

The product gas, which leaves the gasifier at approximately 1425°C, is cooled, compressed, and fed to a shift converter, where a portion of carbon monoxide is reacted with steam to attain a $CO/H_2$ ratio of 1:3. The carbon dioxide produced is removed, and gas is cooled again and enters a methanator, where carbon monoxide and hydrogen react to form methane. Excess water is removed from the methane-rich product. The final gas product has a heating value around 900 Btu/scf.

### 3.3.5 Modeling of Different Coal Gasifiers

As research and development continue on new coal gasification concepts, mathematical models are needed to gain insight into their operation and commercial potential. The influence of design variables and processing conditions on the gasifier performance must be determined. Such models are then used as tools for design modifications, scaling, and optimization.

Coal gasification is performed in different types of reactors in which, depending on the type of gas-solid contact, the bed can be moving, fluidized, entrained, or molten salts. Of these, the moving bed is the most widely used because of its high coal conversion rates and thermal efficiency. Different approaches have been used to model this and other types of reactors. There are mainly two kinds of models. Thermodynamic or equilibrium models are relatively easy to formulate but generate only certain restrictive information, like off-gas compositions. The other type of model is the kinetic model, which predicts the behavior inside the reactor. This helps in both design and simulation. Adanez and Labiano [56] have developed a mathematical model of an atmospheric moving bed countercurrent coal gasifier and studied the effect of operating conditions on the gas yield and composition, the process efficiency, and the longitudinal temperature profiles. The model has been developed for adiabatic reactors. It assumes the gasifier consists of four zones with different physical and chemical processes. These are the coal preheating and drying zone, the pyrolysis zone, the gasification zone, and the combustion zone followed by the ash layer, which acts as a preheater of the reacting gases. In reality, there is no distinction between them and the reactions occurring in each zone vary considerably. The model uses an unreacted shrinking core model to define the reaction rate of the coal particles. The most critical parameter in the operation of these moving bed gasifiers with dry ash extraction is the longitudinal temperature profile, because the temperature inside the reactor must not exceed the ash softening point at any time. The model also takes into account the effect of coal reactivity, particle size, and steam/oxygen ratio. As a partial check of the validity of the model, data predicted on the basis of the model were compared with real data on the product gas composition for various coals, and good agreement was found. The authors have concluded that the reactivity of the coals and the emissivity of the ash layer must be known accurately, as they have a strong influence on the temperature profiles and the maximum temperature in the reactor and on its capacity for processing coal.

Lim et al. [57] have constructed a mathematical model of a spouted bed gasifier based on simplified first-order reaction kinetics for the gasification reactions. The spouted bed gasifier has been under development in Canada and Japan over the last 12 years [58, 59]. This model treats only the motion of gases and not the motion of solids. The spout is treated as a plug flow reactor of a fixed diameter with cross-flow into the annulus. The annulus is treated as a series of steam tubes, each being a plug flow reactor with no dispersion. The model calculates the composition profile of various product gases in the spout as a function of the height, radial composition profiles, and average composition in the annulus at different heights; the average compositions exiting the spout and annulus; and the flow rates and velocities in the spout and annulus.

Johnson et al. [60] have developed a similar model for simulating the performance of a cross-flow coal gasifier. Gasification in a cross-flow gasifier is analogous to the batch gasification in a combustion pot. Therefore the model equations for kinetics and mass and energy balances formulated were based on a batch process. In the cross-flow coal gasifier concept, operating temperatures are much higher than 1000°C and the diffusion through the gas film and ash layer is a critical factor. The model assumes a shrinking unreacted core model for kinetic formulations. Simulation results of the model were compared with the experimental data obtained in batch and countercurrent gasifications and the agreement was found to be quite good. It was also concluded that the performance of the gasifier depends on the gas-solid heat transfer coefficient and the particle size and the bed voidage affected the time for complete gasification.

Watkinson et al. [61] have formulated a model to predict the gas composition and the yield from coal gasifiers. Gas composition depends on the contacting pattern of blast and fuel, the temperature and pressure of the operation, the composition of the blast, and the form of fuel feeding. The authors presented a calculation method and compared the predicted data with the data from nine different types of commercial and pilot-scale gasifiers including Texaco, Koppers-Totzek, and Shell gasifiers, which are entrained bed types; the Winkler fluidized bed gasifier; and the Lurgi dry ash as well as slagging Lurgi moving bed gasifier. The model consists of elemental mass balances for C, H, O, N, and S; equilibria for four key reactions; and an optional energy balance. Predictions were best for entrained bed systems, slightly poor for fluidized bed gasifiers, and more uncertain for moving bed gasifers. This is due to the lower temperatures and uncertain volatile yields in the latter ones. This resulted in deviation between the calculated and reported values.

Lee et al. [62] developed a single-particle model to interpret kinetic data on coal char gasification with $H_2$, $CO_2$, and $H_2O$. Their model yields an analytical solution taking into account all the major physical factors that affect the gasification rate. Some of the factors taken into account were changing magnitudes of surface area, porosity, activation energy, and effective diffusivity. Their model closely describes the characteristic shape of the conversion versus time curves as determined by $CO_2$ gasification studies. The shape under certain restrictions leads to a "universal" curve of conversion versus an appropriate dimensionless time. The model developed is mathematically very simple, and all the parameters have physical significance. The number of adjustable parameters in this model is only two.

In their review, Agarwal et al. [63] have critically examined all the mathematical models for fluidized bed reactors. The review is concerned primarily with the modeling of bubbling fluidized bed coal gasifiers. It starts with a discussion of rate processes occurring in a fluidized bed reactor. They have also reviewed some of the reported models in the literature with the presentation of their own.

When a coal particle is fed into a gasifier, it undergoes several physico-chemical transformations: drying, devolatilization, and gasification of the residual char in different gas atmospheres. These heterogeneous phenomena are accompanied by a number of supplementary reactions that are homogeneous in nature. Detailed kinetic studies are an important prerequisite for the development of the mathematical model. Mathematical models for a bubbling fluidized bed coal gasifier can be broadly classified as thermodynamic and rate models. Thermodynamic models predict the equilibrium compositions and temperature of the product gas based on a given set of steam/oxygen feed ratios, the operating pressure, and the desired carbon conversion. These models are independent of the type of gasifier and based on the assumption of complete oxygen consumption. They cannot be used to investigate the influence of operating variables on the gasifier performance. The kinetics or the mathematical model, on the other hand, predicts the composition and the temperature profiles inside the gasifier for a given set of operating conditions and reactor configurations and hence can be used to evaluate the performance of a gasifier. They are developed by combining a suitable hydrodynamic model for the fluidized bed with appropriate kinetic schemes for the processes occurring inside the gasifier. Agarwal et al. [63] have classified various rate models into four groups on the basis of hydrodynamic model used:

1. Simplified flow models
2. Davidson-Harrison models
3. Kunii-Levenspiel models
4. Kato-Wen models

The review [63] critically examines and compares the types of models. Although many investigators have compared their model predictions with experimental data, a detailed evaluation of the influence of model assumptions on its predictions has not been reported. Efforts have been made to compare the predictions of different models, but an attempt to evaluate the model with experimental data from different sources has not been made.

The authors of the review article [63] have developed a model of their own for a bottom-feeding bubbling fluidized bed coal gasifier based on the following assumptions:

1. The bubble phase is in plug flow without any particles; the emulsion phase is completely mixed and contains the particles in fluidized conditions.
2. Excess gas generated in the emulsion phase passes into the bubble phase. The rate of this excess per unit bed volume is constant.
3. The coal particles in the feed are spherical, homogeneous, and uniform in size.
4. Only the water-gas shift reaction occurs in the homogeneous gas phase.

5. External diffusion and intraparticle diffusion are assumed to be negligible in the char gasification reactions.
6. Entrainment, abrasion, agglomeration, or fragmentation of the bed particles is assumed negligible.
7. The gasifier is at a steady state and is isothermal.

All the model equations are derived on the basis of these assumptions. The model predictions were compared with the experimental data from three pilot-scale gasifiers reported in the literature [63]. It was concluded that the predictions were more sensitive to the assumptions regarding the combustion or decomposition of the volatiles and the products of char combustion than to the rate of char gasification. Hence, in pilot-scale gasifiers, due to the short residence time of coal particles, the carbon conversion and the product gas yields are determined mainly by the fast coal devolatilization, volatiles combustion-decomposition, and char combustion rather than by the slow char gasification reactions. This explains why models based on finite rate char gasification reactions are able to fit the same pilot-scale gasification data.

A better understanding of coal devolatilization, decomposition of the volatiles, and char combustion under conditions prevailing in a fluidized bed coal gasifier is important for the development of a model with good predictive capability. There is a strong need to investigate the kinetics of gasification of coal char in synthesis gas atmospheres and to obtain experimental data for the same coal char in a pilot-scale plant.

It is well known that there are many physical changes occurring when the coal char particles are gasified. There have been many attempts to unify these dynamic changes through various normalizing parameters such as half-life, reactivity, or surface area. According to some studies [64], the experimental char conversion versus time data from different experiments can be unified into a single curve where time is considered to be normalized time, $t/t_{1/2}$, $t_{1/2}$ being the half-life of the char gas reaction. This unification curve with only one parameter is fitted into the rate models commonly used, such as the grain model and the random pore model. With the aid of reported correlations for unification curves, a master curve is derived to approximate the conversion-time data for most of the gasification systems. Also, because the half-life is simply related to the average reactivity, it can be generally used as a reactivity index for characterizing various char-gas reactions and conversions up to 70% can be predicted with reasonable accuracy over a wide range of temperatures.

## 3.4 VARIOUS CHEMICAL REACTIONS IN GASIFICATION

### 3.4.1 Steam Gasification

Coal conversion reactions are classified into two types: devolatilization or primary gasification and gasification of char or secondary gasification. Both kinds of reactions occur in different parts of the gasifier.

Primary gasification involves thermal decomposition of the raw coal and gives char and volatiles.

$$Coal \longrightarrow char + volatiles$$

Volatiles include tars, phenols, methane, oils, naphtha, $H_2S$, and some CO and $H_2$. This process itself requires heat other than that required to raise the coal to devolatilizaiton temperatures. In general, devolatilization and pyrolysis reactions are endothermic in nature. Steam gasification involves the reaction of char with water vapor (steam) to form gaseous products, mainly carbon monoxide, carbon dioxide, and hydrogen. There are many ways to write the reactions that occur in steam gasification, but char for all practical purposes is written as pure carbon.

$$C + H_2O = CO + H_2 \qquad \Delta H^{\circ}_{298} = 131.3 \text{ kJ/mol}$$

$$C + 2H_2O = CO_2 + 2H_2 \qquad \Delta H^{\circ}_{298} = 90.1 \text{ kJ/mole}$$

Both steam gasification reactions are endothermic in nature. Both reactions have thermodynamic constraints that influence the thermal efficiency of the gasifier. Usually, a substantial amount of steam in excess of the stoichiometric requirement is needed. A high temperature is essential for promotion of the gasification reactions and also minimizes methane formation. High-temperature, low-pressure operation produces mainly syngas, whereas low-temperature, high-pressure operation increases methane formation. The energy associated with use of excess steam usually cannot be recovered as useful energy. This thermal efficiency loss increases substantially with an increase of the steam/oxygen ratio in the feed gas, which corresponds to an increase in the $H_2/CO$ ratio in the synthesis gas. This leads to the conclusion that synthesis with a low $H_2/CO$ ratio is thermally most efficient under the thermodynamic equilibrium constraint.

Alkali metal salts are among the oldest additives known to increase the rate of steam gasification of carbonaceous materials. The reactivities of coal increase with increased catalyst dispersion, higher alkali/metal ratio, and order of catalytic activity of alkali metals (Cs > Rb > K > Na > Li). Li and Heiningen [65] have studied the kinetics of black liquor char (BLC) by steam gasification. Because the chemical compositions of BLC and alkali metal carbonate–impregnated chars are similar, extensive studies of the latter serve as a useful reference. The kinetic studies were carried out in a thermogravimetric analysis reactor, where a Langmuir-Hinshelwood type of kinetic expression was developed. A gasification mechanism was also proposed whereby $CO_2$ was established as one of the primary gas products.

## 3.4.2 Boudouard Reaction

The Boudouard reaction is a gasification reaction in which the carbon dioxide formed reacts with carbon to form carbon monoxide.

$$C + CO_2 = 2CO \qquad \Delta H^\circ_{298} = 172.5 \text{ kJ/mol}$$

This is also an endothermic reaction favored at high temperatures ($T > 680°C$). Unlike the gasification reactions mentioned so far, combustion reactions are highly exothermic, so the heat generated in combustion is usually balanced by the steam gasification reaction to attain heat integration. An idealized char gasifier with all inlet and outlet streams is used at the same temperature (i.e., 370°C). The energy constraint can be represented by $\Sigma X_i H_i = 0$, where $X_i$ is the fractional amount of carbon converted to products by the $i$th reaction and $\Delta H_i$ is the heat of reaction for the same reaction.

Because the above reaction, also known as the $CO_2$ gasification reaction, is one of the primary reactions taking place in gasification processes, there have been many studies of the kinetics of the reaction [66–68]. Most of the studies have been performed in a thermogravimetric apparatus in a fixed or fluidized bed with laminar flow. In a thermogravimetric analyzer (TGA), the use of small particles does not present any difficulty, but the velocity of gas is restricted to laminar region as high velocities may lead to a discrepancy in weight measurements. The mechanism of this reaction is not generally agreed upon, but it is proposed that $CO_2$ adsorption on the surface of char (oxygen transfer) and then CO desorption (carbon removal) are the main steps determining the apparent rate.

Several different reaction mechanisms have been suggested for carbon–carbon dioxide reaction. Even though the coal or char reaction with carbon dioxide cannot be automatically assumed to be identical to the carbon reaction, their analogous interpretations may be quite useful.

**Walker's interpretation.**

$$C_f + CO_2(g) = C(CO_2)$$

$$C_f + C(CO_2) = C(O) + C(CO)_A$$

$$C_f + C(O) = C(CO)_B$$

$$C(CO)_A = CO(g) + C_f$$

$$C(CO)_B = CO(g) + C_f$$

$$CO(g) + C_f = C(CO)_c$$

$C_f$ is an unoccupied (free) carbon site, and $C(CO)_A$, $C(CO)_B$, $C(CO)_C$ are adsorbed species. Experimentally it has been observed that carbon monoxide is an intermediate product of the chemisorption of $CO_2$ on C and that the adsorption of $CO_2$ is not reversible to give immediate desorption of $CO_2$. Consequently,

$$C_f + CO_2(g) = C(O) + CO(g)$$

$$C_f + C(O) = C(CO)_B$$

$$C(CO)_B = CO(g) + C_f$$

$$CO(g) + C_f = C(CO)_C$$

There is a further possibility that the transition of $C_f + C(O) \rightarrow C(CO)_B$ is either slow (case 1) or very fast (case 2) in comparison with $C(CO)_B \rightarrow CO(g) + C_f$. It is conceivable that each case will be operative but in a different temperature range. Assuming that case 1 holds, the general expressions given above can be simplified to:

$$C_f + CO_2(g) = C(O) + CO(g) \tag{A}$$

$$C(O) = CO(g) + C_f \tag{B}$$

$$CO(g) + C_f = C(CO) \tag{C}$$

At least two mechanisms can be shown to be consistent with experimental data.

## Mechanism A

$$C_f + CO_2(g) \xrightarrow{i_1} C(O) + CO(g)$$

$$C(O) \xrightarrow{j_3} CO(g) + C_f$$

$$CO(g) + C_f \underset{j_2}{\overset{i_2}{\rightleftharpoons}} C(CO)$$

In the equations, $i_1$, $i_2$, $j_2$, and $j_3$ are the rate constants. This mechanism is called the CO adsorption mechanism or inhibition by CO adsorption.

**Mechanism B.** This mechanism is based on the assumptions that the rate of the backward reaction of (B) is negligible and reaction (C) is not important.

$$C_f + CO_2(g) \underset{j_1}{\overset{i_1}{\rightleftharpoons}} C(O) + CO(g)$$

$$C(O) \xrightarrow{j_3} CO(g) + C_f$$

This mechanism is called Ergun's mechanism or the oxygen-exchange mechanism.

It should also be recognized that the chemisorption reaction may be, in reality, a dual-site adsorption process composed of fundamental steps, for example,

$$CO_2 + 2C_f = C'(CO) + C(O)$$

$$C'(CO) = CO + C_f$$

If one assumes $C'(CO)$ is a fleeting intermediate whose life is very short, then the pseudo-steady-state approximation can be applied and the following reaction equation can result:

$$C_f + CO_2 = C(O) + CO$$

As such, the mechanism reduces to a single-site chemisorption case, under approximate assumptions.

Both mechanisms A and B state that CO retards the gasification of C by $CO_2$ by decreasing the fraction of the surface that is covered by oxygen atoms under steady-state conditions.

It was observed that the $CO_2$ gasification rate of char is approximately first order with respect to partial pressure of $CO_2$ at low pressures and then becomes almost independent above 10 atm. The only mechanism investigation at higher pressures (4–36 atm) is that of Blackwood [69]. They find higher reactivity at 790–870°C than can be attributed to the Ergun mechanism. The volume reaction model and the grain model were examined to interpret the experimentally obtained conversion-time curves. The experimental results were well expressed by the volume reaction model. The Langmuir-Hinshelwood type of rate equation was derived from the slope of the conversion-time curve.

$$\text{Rate} = \frac{k_1 P_{CO_2}}{1 + k_2 P_{CO} + k_3 P_{CO_2}}$$

where $P_{CO}$ and $P_{CO_2}$ are the partial pressures of CO and $CO_2$, respectively, and $k_1$, $k_2$, and $k_3$ are the rate constants. The rate constants of this equation were obtained from the various levels of the carbon conversion. Activation energies obtained from the calculation were compared with values published in the literature and found to be in good agreement. They were identical for both the volume and grain reaction models. It was also observed that the entropy of disintegration of the carbon-ash matrix in char was highly negative, indicating that carbon and ash in char formed a highly interactive phase.

### 3.4.3 Hydrogasification

This reaction involves direct addition of hydrogen to coal under pressure to form methane.

$$\text{Coal} + H_2 \longrightarrow CH_4 + C$$

$$C + 2H_2 = CH_4 \qquad \Delta H^\circ_{298} = -74.8 \text{ kJ/mol}$$

The hydrogen necessary for this reaction is manufactured from steam by using the char that leaves the hydrogasifier [3]. The product gas consists mainly of methane, unreacted hydrogen, and some amount of liquid by-products. The main advantage of direct hydrogasification is that it maximizes the formation of methane in the hydrogasification unit, thereby minimizing the amount of methane that must be formed by the methanation reaction,

$$CO + 3H_2 = CH_4 + H_2O(g) \qquad \Delta H^\circ_{298} = -206.1 \text{ kJ/mol}$$

On an overall basis, direct hydrogasification minimizes coal utilization per unit of methane produced. The final product has a methane concentration in excess of 80% and heating value over 900 Btu/scf without hydrogen separation. The most critical parameters in the hydrogasification of coal are the temperature and hydrogen partial pressure [70]. A higher temperature at a high partial pressure of hydrogen results in a higher carbon conversion rate and high methane concentration in the raw product gas.

Hydrogasification of coal and carbon has not been studied as extensively as steam or $CO_2$ gasification. This is because the reaction occurs much more slowly than in any oxidizing environment. It also requires higher pressures to overcome thermodynamic limitations and is further limited as the temperature is increased. The rate of hydrogasification also decreases with conversion of coal char. But since it is a direct source to methane where a significant partial pressure of hydrogen exists, it is one of the primary coal conversion processes. The reaction of carbon with pure hydrogen offers a unique opportunity to study the role of oxygen in gasification. Toomajian et al. [71] have studied the effect of oxygen and other treatments on the hydrogasification rate of coal char. They showed that addition of 0.1% oxygen to the hydrogen reactant gas increases the rate of methane formation by an order of magnitude. The decrease in hydrogasification rate with conversion was overcome by studying the effects of oxidation (partial burn-off in air at 375°C), demineralization, and heat treatment at 1000°C followed by $K_2CO_3$ addition on hydrogasification. They observed that oxidation of partially hydrogasified and heat-treated chars resulted in a threefold increase in subsequent hydrogasification rate. Addition of $K_2CO_3$ catalyst eliminates the decay in rate for mineral-free chars. The catalyst also removes strongly chemisorbed hydrogen, resulting in an active surface structure.

Among the many catalysts used for hydrogasification, nickel is more suitable for producing a high concentration of methane in one pass using pressurized steam. However, nickel is very expensive. Iron is promising as an alternative to nickel [72] and iron chlorides and sulfates are easily available as acid waste from the steel industry. A large increase in hydrogasification rate was observed when Cl-free iron catalyst was used at comparatively low temperatures under pressure.

It is known that iron group metals catalyze coal gasifications in hydrogen. This hydrogasification of carbons catalyzed by Fe group metals is greatly promoted by calcium, as noted by Haga and Nishayama [73]. Calcium as such is not a catalyst for hydrogasification. They studied the promotion of calcium salts between 1 and 30 atm and between 750 and 900°C. It was observed that the promotion increased with an increase in the amount of added calcium, temperature, and pressure. The most important factor dominating the activities of nickel and ferrous-based catalyst is the increased catalyst dispersion due to the addition of calcium. It also increases the carbon-catalyst interaction. But it is still unclear whether this interaction results in catalyst activation or deactivation. The effectiveness of Ca in promoting catalysis also depends on the kind of char.

### 3.4.4. Water Gas Shift Reaction

The gaseous products from a gasifier usually contain a large amount of carbon monoxide and hydrogen plus some quantities of other gases. If the objective of the gasification reactions is to obtain a high yield of methane, then carbon monoxide and hydrogen must be in a mole ratio of 1:3. Usually some adjustment to this ratio is required, and to attain this carbon monoxide formed is reacted with steam to convert it to carbon dioxide and hydrogen according to the water gas shift reaction:

$$CO + H_2O(g) = CO_2 + H_2 \qquad \Delta H^\circ_{298} = -41.2 \text{ kJ/mol}$$

Unlike most gasification reactions, the water gas shift reaction is slightly exothermic in nature. With this reaction, a desired 1:3 mole ratio of carbon monoxide and hydrogen may be obtained. If the water gas shift reaction is carried out with oxygen and steam rather than air and steam, the product gas obtained is essentially nitrogen-free raw synthesis gas, which consists almost entirely of carbon dioxide, carbon monoxide, and hydrogen and, after removal of carbon dioxide, offers an important feedstock for a wide variety of chemical syntheses. The equilibrium constant of the water gas shift reaction varies relatively little over a wide range of temperatures [74]. However, at higher temperatures, carbon dioxide formation is somewhat lower because the equilibrium moves to the left.

The water gas shift reaction is one of the principal reactions taking place during steam gasification, the practical importance of which lies in the production of hydrogen or in tuning of the CO/H$_2$ ratio for hydrocarbon synthesis on an industrial scale. Therefore many investigators have studied the kinetics of the water gas shift reaction over iron, copper, nickel, chromium, and molybdenum catalysts [75]. But few have investigated the role of this reaction in the alkali metal–catalyzed reaction. Meijer et al. [76] have investigated the kinetics of the water gas shift reaction and proposed a model based on a series of elementary processes pertaining to the overall reaction. They observed that the water gas shift oxygen exchange is catalyzed by the alkali metal cluster, which is anchored to the carbon surface. This oxygen exchange is dependent on partial pressures of the reducing and oxidizing agents. The oxygen exchange rate is low if carbon dioxide is present in the gas mixture. On the basis of measurements of the forward and backward reaction rates as a function of experimental conditions, a three-step kinetic model was developed for the water gas shift reaction in which $CO_2$, CO, and oxygen exchange proceeds through a $CO_2^*$ intermediate. The estimated parameters of this model obeyed all the thermodynamic constraints, providing additional support for the model selected.

### 3.4.5 Partial Combustion

Combustion of coal is its reaction with air or oxygen to form carbon monoxide and eventually carbon dioxide with release of a substantial amount of heat:

$$C + O_2 \longrightarrow CO_2 \qquad \Delta H^\circ_{298} = -393.5 \text{ kJ/mol}$$

Given sufficient air or $O_2$, combustion proceeds sequentially through vapor-phase oxidation and ignition of volatile matter to eventual ignition of the residual char. Partial combustion involves a complex reaction mechanism that depends on how fast and efficiently combustion progresses.

Detailed theoretical and experimental studies [24] have shown that combustion of carbon involves at least four carbon-oxygen interactions:

$$C + \tfrac{1}{2}O_2 \longrightarrow CO$$

$$CO + \tfrac{1}{2}O_2 \longrightarrow CO_2$$

$$C + CO_2 \longrightarrow 2CO$$

$$C + O_2 \longrightarrow CO_2$$

The stoichiometric reactions are quite simple, but because of the heterogeneous nature of the reactions (both solid and gas phases involved) two main hypotheses are envisaged. According to one hypothesis, the combustion process begins with chemisorption of oxygen at "active" sites on char surfaces to form surface oxides. This decomposition generates mainly carbon dioxide, which then reacts with more oxygen to form carbon dioxide in a gaseous boundary zone around the char particle. The other hypothesis advocates the formation of carbon monoxide only.

Thus, what appears to be a simple stoichiometric equation between a carbonaceous material and oxygen can be explained adequately only through heterogeneous effects involving heat and mass transfer reaction at a surface.

Little effort has been made to understand the combustion reactions chemically, in spite of their importance from the gasification point of view. Most of the reactions are limited by gas diffusion, and therefore complete conversion is quite difficult to achieve even at very high temperatures. Several studies have tried to analyze the factors influencing the reaction rate at high temperatures, but most of the studies have been done in a narrow temperature range for a limited number of coals. A recent study in Japan [77] determined reaction rates of combustion for five coals in a very wide temperature range between 500 and 1500°C to examine the effects of coal rank and catalysis by mineral matter. Then the rates were correlated with various char characteristics. It was found that the catalytic effect of mineral matter determined the reactivity of coal in the region where chemical reaction controlled the rate. For the high-temperature region, where external mass diffusion was the controlling phenomenon, the reactivity decreased with increasing coal rank.

### 3.4.6 Kinetics of Gasification

Fundamental understanding of chemical and physical processes is extremely important before a rational design of a coal gasifier can be made. Two main processes occur: (1) transition of coal to coal char and (2) subsequent gasification of coal char itself.

During the initial stages of gasification, coal pyrolysis takes place. The overall process is divided into three stages. First, the coal undergoes a sort of depolymerization reaction that leads to formation of a metastable intermediate product. Second, the product then undergoes cracking and recondensation, giving out primary gases, oils, and semichar. Finally, semichar is converted to char with evolution of hydrogen. At elevated temperatures, this initial gasification stage is completed in seconds. This is why earlier kinetic investigations of coal gasification revealed fairly high (up to 40%) conversions in the initial moments. Some authors referred to these as initial conversions. The subsequent gasification of coal chars is much slower, and it takes a longer time to obtain significant conversions under practical conditions. Because the reactor design criteria and volume requirements for commercial gasifiers are largely dependent on coal char gasification, the kinetics of coal char gasification is more important and is discussed in detail here.

An advantage of studying coal char gasification kinetics is that it is comparatively easy to conduct experiments with different gas conversions, in which the composition of gas in contact with coal char is known and remains constant. For experimental purposes, fixed bed flowing gas systems have been most frequently used to obtain detailed kinetic information. On the other hand, temperature control is a greater problem in fixed bed operation due to potentially large radial and axial temperature gradients as well as to lower heat transfer rates between gas and solids.

There are some other factors to be considered in interpreting data on coal char gasification. The results obtained differ not only with the type of coal used but also with the procedures employed in preparing the coal char from coals.

Coal char formed depends on the pretreatment temperature, treatment gas environment, and residence time.

Studies of char gasification have found that measured reaction rates change during the conversion and depend not only on the reaction-conditions and chemical composition of the char but also on the physical characteristics of the char. Bhatia and Perlmutter [78, 79] derived a random pore model that interprets conversion data as a combination of chemical rates and pore size distributions, in terms of three dimensionless parameters that characterize pore structure, particle size, and kinetics. Su and Parlmutter [80] applied this model to the char-air reaction system and found good agreement between two independent evaluations of the structure parameters. They concluded that rate variations with conversion are controlled by pore structural changes, even though overall rates also include contributions of intrinsic reactivity. Chars with different pore structures can be generated from the same coals by varying pyrolysis conditions: slow heating and a higher pyrolysis temperature generate char with a more compact structure. The difference in pore structure of the chars affects their subsequent oxidation rates. The pore structure parameters obtained by fitting the kinetic data were compared with the values determined by direct measurements of the initial pore structures of the chars. The agreement among these independent evaluations demonstrated the degree of applicability of the random pore model.

For the model evaluations, Wei and Perlmutter obtained surface areas and pore size distributions for various partially reacted char samples. Surface areas followed the reaction rates as a function of conversion, supporting the interpretation that reaction of porous char with steam is governed by the structural changes that occur as the conversion changes. A pyrolytic change of char mass resulted in a significant increase in surface area and decrease in pore structure parameter. It was shown that this pore structure parameter was very sensitive to the changes in conversion for low conversion and large parameter values.

Identification of the appropriate total surface area for carbon gasification is critical for calculation of intrinsic reactivities and finding the factors that determine carbon gasification reactivities. Hurt et al. [81] studied the gasification in oxygen and carbon dioxide of two high-surface area synthetic carbons to understand the role of microporous surface area in uncatalyzed carbon gasification. For these preselected carbons, they measured low-temperature gasification reactivities, vapor adsorption isotherms, and adsorption equilibrium times at various stages of conversion. They also examined surface features of individual char particles as a function of conversion. These results were then interpreted and correlated with the actual coal char reactivities.

Fott and Straka [82] studied the effect of pore diffusion on the reactivity of lump coke. They observed that because of pore diffusion, the reactivity decreases considerably compared with that of fine-grained coke under conditions that are otherwise the same. They applied a simplified mathematical model to express the gasification of lump coke considering the effect of pore diffusion. Kinetic parameters were determined from experiments with fine-grained coke, and effective diffusivity was evaluated from experiments with a single coke particle. They also found a correlation between effective diffusivity and the fraction of largest pore volume.

Goyal et al. [83] have developed a correlation for determining the rate of gasification of coal char produced from bituminous coal. The correlation was developed on the basis of anticipated data to be used in commercial fluidized bed reactor. They used different gaseous media including steam, hydrogen, a mixture of the two, and a synthesis gas mixture. The rate of gasification was found to be the highest with steam. The lower gasification rate with steam-hydrogen and synthesis gas mixtures can be explained by the retardation of steam-carbon reaction by hydrogen and by both hydrogen and carbon monoxide in the gasification with the synthesis gas mixture. The overall rate constant was correlated with the process variables.

The following coal char gasification reactions are the most important ones in developing commercial processes to convert coal to high- or low-Btu gases:

$$C + 2H_2 \rightleftharpoons CH_4 \tag{1}$$

$$C + CO_2 \rightleftharpoons 2CO \tag{2}$$

$$C + H_2O \rightleftharpoons CO + H_2 \tag{3}$$

Reaction (1) has a very strong favorable effect on the pressure and is usually studied at elevated pressures. This reaction is important to the production of high-energy gas. Reaction (1) is exothermic, whereas reactions (2) and (3) are endothermic. Reactions (2) and (3) are important to the production of both high- and low-energy gases. Whereas reactions (1) and (2) each occur alone in the system according to the stoichiometry indicated, reaction (3) usually occurs with the water gas shift reaction:

$$CO + H_2O \rightleftharpoons CO_2 + H_2 \tag{4}$$

Carbon dioxide formed in the shift reaction reacts competitively with the steam-carbon reaction according to reaction (2) at high or low pressures. Because hydrogen is a product of the steam-carbon reaction as well as the shift reaction, methane can also be formed in such systems. A variety of other reactions can also be postulated that are stoichiometric combinations of reactions (1), (2), and (3). In complex gasification systems containing steam, hydrogen, and carbon dioxide, most experiments determine only the overall stoichiometry.

Gasification reactions have usually been studied under isothermal conditions, often with differential gas conversion. Sometimes, to study the temperature effect in the same experiment, a constant heat-up rate is applied. The carbon gasification rate, $G$, can be defined by the expression

$$G = \frac{dx/dt}{1 - x}$$

where $x$ is the initial carbon conversion fraction (mol/mol initial carbon) and $t$ is time. This expression implies that if the absolute carbon gasification rates are proportional to the amount of carbon present at any time, $G$ remains constant with increasing $x$. This is usually true until the carbon conversion reaches 80%. After this stage, specific reaction rates decrease to some extent. If any catalyst is used during the process, the approximation of relatively constant specific gasification rates cannot be valid. The activity of catalyst significantly varies the rates even under constant environmental conditions. If the coal used has agglomerating tendencies during the initial devolatilization process, it can lead to inhibition of gaseous diffusion during the subsequent gasification. When this occurs, very low initial specific reaction rates are obtained. But they increase with increasing conversion as restrictions are removed and internal char surface becomes more available to the reacting atmosphere.

### Kinetics of various reactions occurring in coal char gasification

*Coal char–$H_2$–$CH_4$ system.*

$$C + 2H_2 \rightleftharpoons CH_4$$

For the gasification of coal char and other carbons in pure hydrogen, specific carbon gasification rates have been correlated with two types of equations:

$$G = \frac{k_H P_{H_2}^2}{1 + K_H P_{H_2}} \tag{1}$$

or
(2)

$$G = k_H^* P_{H_2}$$

where $k_H$, $K_H$, and $K_H^*$ are kinetic parameters that can be expressed by an Arrhenius form of expression. These parameters depend on the temperature and the nature of char. $P_{H_2}$ is the partial pressure of hydrogen. At very high hydrogen partial pressures, equation (1) reduces to (2) with $k_H^* = k_H/K_H$. It can be concluded that with either of the two equations the gasification rates increase significantly with increasing hydrogen partial pressure.

There are two main causes of variations in the reactivity of coal chars during gasification with hydrogen. One is related to the original coal from which chars are derived and the other to the conditions employed, especially the temperature used in preparing them. It has been found that char reactivity decreases with increasing preparation temperature relative to the temperature of subsequent gasification in hydrogen.

*Coal char–CO–CO₂ system.* The reaction of coal char with $CO_2$ to form CO (Boudouard reaction) is studied from both gasification and combustion points of view. There are inherent differences between gasification of carbon with carbon dioxide and gasification with hydrogen, discussed earlier. At low pressures, near atmospheric conditions, gasification rates in carbon dioxide are higher by about two to three orders of magnitude than the rates in hydrogen. But at higher pressures, over a wide range, the rates in hydrogen increase in direct proportion to the pressure whereas the rates in carbon dioxide reach a plateau at approximately 10 atm. Such an effect is reflected in the following expression, which describes the gasification kinetics of carbon in carbon dioxide and carbon monoxide mixtures at conditions away from equilibrium:

$$G = \frac{dx/dt}{1 - x} = \frac{k_C P_{CO_2}}{1 + k_{C_1} P_{CO} + k_{C_2} P_{CO_2}}$$

where $k_C$, $k_{C_1}$, and $k_{C_2}$ are kinetic parameters that depend on the temperature and the nature of char and $P_{CO}$, and $P_{CO_2}$ are the partial pressure of CO and $CO_2$, respectively. This equation shows that gasification rates decrease with increase in carbon monoxide partial pressure because of an inhibiting effect, not related to thermodynamic stability. At very high pressures, the preceding equation simplifies to

$$G = \frac{k_C^*}{1 + K_C^*(P_{CO}/P_{CO_2})}$$

where $k_C^* = k_C/k_{C_2}$ and $k_C^* = k_{C_1}/k_{C_2}$.

*Coal char–synthesis gas system.* The reaction kinetics for the gasification of carbonaceous solids in systems containing carbon, hydrogen, and oxygen is very complex. This is because the reaction stoichiometry in such systems is a function of reaction conditions, which is not the case in systems containing only carbon and oxygen or carbon and hydrogen. But under some limiting conditions, the reaction stoichiometry in such carbon-hydrogen-oxygen systems is simplified because of the nature of the relative kinetics of the gasification reaction that occurs. At lower pressures, methane formation is inhibited, and since steam and hydrogen predominate in the system, the only reaction that occurs is

$$C + H_2O \rightleftharpoons CO + H_2$$

At higher pressures, methane becomes an important product and the overall reaction can be written as

$$2C + H_2 + H_2O \rightleftharpoons CO + CH_4$$

Under such conditions, where both carbon monoxide and methane are primary products of gasification, there are greater variations in kinetic parameter evaluations than for simpler systems. Hence the kinetic correlations developed to describe the gasification at high pressures are largely empirical in nature with little theoretical basis.

A wide variety of rate expressions have been used to correlate the gasification rates of carbon to form carbon oxides in steam-hydrogen mixtures. Some of these are

$$G = k_{OH}P_{H_2O} \tag{1}$$

$$G = \frac{k_{OH}P_{H_2O}}{1 + k_{OH_1}P_{H_2} + k_{OH_2}P_{H_2O}} \tag{2}$$

$$G = k_{OH}\left(\frac{P_{H_2O}}{P_{H_2}}\right)^{1/2} \tag{3}$$

$$G = \frac{k_{OH}P_{H_2O}^2}{1 + k_{OH.1}P_{H_2} + k_{OH.2}P_{H_2O}} \tag{4}$$

$$G = \frac{k_{OH}P_{H_2O}}{(1 + k_{OH.1}P_{H_2} + k_{OH.2}P_{H_2O})^2} \tag{5}$$

where $k_{OH}$, $k_{OH.1}$, and $k_{OH.2}$ are kinetic rate constants and $P_{H_2}$ and $P_{H_2O}$ are the hydrogen and steam partial pressures. Equation (1) can be applied only over a very narrow range of temperature, pressure, and gas composition. All the other rate expressions predict that increasing the hydrogen partial pressure lowers the gasification rates, which is consistent with mechanistic observations.

**Particle size effects on coal char gasification.** When the coal particles undergo devolatilization before undergoing secondary gasification, the surface structure becomes porous. In such cases, transport of reactant and product gases between the external bulk gas phase and the internal particle surface becomes a critical factor. Under these conditions, these exists a concentration gradient between the inner particles and the outer surface. Hence, mass transport and diffusion processes affect the overall gasification rate. The rate generally decreases with increasing particle size. The gaseous concentration gradient exists between either (1) the bulk gas phase and the external particle surface due to film diffusion resistance or (2) external and internal particle surfaces due to diffusion limitations within the particle pore structures. In either case, these effects must be considered in interpreting results and applying kinetic models to anticipate behavior in large-scale systems.

A great deal of research has been done on the kinetics of gasification by steam. Alkali metal salts are among the oldest known additives that markedly increase the rate of steam gasification of carbonaceous materials. The effect of steam and hydrogen concentration on the gasification rate can be described by

Langmuir-Hinshelwood kinetics [84]. The authors studied black liquor char kinetics in a thermogravimetric analysis reactor. They also proposed a gasification mechanism whereby $CO_2$ is one of the primary gasification products. A reaction scheme was formulated in which an alkali metal oxide acts as an oxygen transferring medium. The reducing agents in this redox mechanism are either C (producing CO) or C(O) (producing $CO_2$). The suggested mechanisms can also be related to the Langmuir-Hinshelwood type of kinetic expression used to describe the influence of gas composition.

Gasification rates of various coal chars differ significantly. Several methods have been developed to interpret and measure this difference in rate including parameters like the total surface area, the active surface area, and the amount of metals in the ash. The most popular method is the transient kinetic method [85–88]. In this method, a char sample is gasified in a differential fixed bed flow reactor to a specified conversion level. The flow of the gaseous reactant ($CO_2$) is then interrupted and switched to that of an inert gas at the reaction temperature. The steady-state and transient responses of CO concentration versus time are monitored by an analyzer. The integrated area under the transient portion of the curves is used to determine the reactive surface area of the char at that conversion level. In this study, the main focus is the distinction between stable carbon-oxygen complexes (C—O) and reactive surface intermediate [C(O)]. The transient kinetics technique allows this distinction to be made. It has made possible the measurement of the conversion dependence of the reactive surface area of the carbon and chars. It is also helpful in elucidating some of the mechanistic aspects of catalyzed carbon gasification. The concept of reactive surface area (RSA) has been also applied to several catalyzed gasification processes and a good correlation between the reactivity and the RSA was observed for K-catalyzed reactions. It was concluded that desorption of the reactive C(O) intermediate was the rate-determining step in the overall mechanism of both catalyzed and uncatalyzed reaction. For the Ca-catalyzed reaction, the quality of the correlation depended on the catalyst dispersion, suggesting that in addition to the decomposition of the reactive intermediate, another process contributed to the transient evolution of C(O). No good correlation was found for the Ni-catalyzed reaction. Mixed catalyst systems were also studied. Nozaki, Radovic, Furusawa, and Kyotani et al. [85–88] remarked after the studies that the transient kinetics technique can be useful for determining the reactive surface areas of catalyzed chars. It could also provide useful information not only on the dispersion of the catalyst but also on the mechanism of the catalysis of carbon gasification.

### 3.4.7 Thermodynamics of Coal Gasification

Thermodynamic considerations are important in evaluating the limiting performance characteristics of individual processes and practical design of experimental or commercial reactors in coal gasification systems. But for this purpose, the thermodynamic properties of all the systems must be known in order to define the heat effects and equilibrium behavior. Because coal is complex and hetero-geneous, measurement of such properties is difficult.

**Equilibrium effects in various gasification reactions.** First, to study the equi-librium effects, it becomes necessary to define the standard free energies of formation of reactants and products. Because of the lack of accurate experimen-tal information for coal and coal chars, relatively crude techniques of interpo-lation or extrapolation from the properties of pure solid aromatic compounds have been used. But it is difficult to evaluate the uncertainties that result from this technique even after taking into account the heterogeneous and amorphous nature of coal. Equilibrium constants for various gas-carbon and associated re-actions are summarized in Table 5.

*Hydrogasification reaction*

$$C + 2H_2 \rightleftharpoons CH_4$$

For this reaction, experimental yields obtained during the reaction have been interpreted in terms of pseudoequilibrium constants by assigning activity coef-ficients to the reacting carbon [89]. The equilibrium constant for the preceding reaction is generally expressed as

$$K_{CH_4} = \frac{P_{CH_4}}{a_c P_{H_2}^2}$$

where $P$ is the partial pressure of the respective species and $a$ is the activity of the particular carbon referred to graphite (for graphite $a_c = 1$).

Several investigators [90, 91] have evaluated experimental values of $P_{CH_4}/P_{H_2}^2$ obtained at pseudoequilibrium conditions during the gasifications. Based on their experimental results, some conclusions can be drawn as follows:

1. Values of $a_c$ as high as 20 have been obtained for data obtained at a total pressure of 1 atm. Such high values have been attributed to the suggestions that amorphous carbons have excessive free energies compared with graphites.
2. Some investigators [92] have argued that for gasifications of carbons in gases containing both steam and hydrogen, it is not valid to interpret carbon activities only on the basis of the preceding reaction (hydrogasification) because methane can be formed by other reactions involving steam and hydrogen. They showed that for high-pressure gasification with hydrogen the ratio $P_{CH_4}/P_{H_2}^2$ in product gases never exceeded the equilibrium value where $a_c = 1$ and concluded that the coal-char gasification systems are

**Table 5 Equilibrium constants**

| Temperature (K) | $\log_{10}K_p$ | | | | | |
|---|---|---|---|---|---|---|
| | $C + \tfrac{1}{2}O_2 = CO$ | $C + O_2 = CO_2$ | $C + H_2O = CO + H_2$ | $C + CO_2 = 2\,CO$ | $CO + H_2O = CO_2 + H_2$ | $C + 2\,H_2 = CH_4$ |
| 300 | +23.93 | +68.67 | −15.86 | −20.81 | +4.95 | +8.82 |
| 400 | 19.13 | 51.54 | −10.11 | −13.28 | 3.17 | 5.49 |
| 500 | 16.26 | 41.26 | −6.63 | −8.74 | 2.11 | 3.43 |
| 600 | 14.34 | 34.40 | −4.29 | −5.72 | 1.43 | 2.00 |
| 700 | 12.96 | 29.50 | −2.62 | −3.58 | 0.96 | 0.95 |
| 800 | 11.93 | 25.83 | −1.36 | −1.97 | 0.61 | 0.15 |
| 900 | 11.13 | 22.97 | −0.37 | −0.71 | 0.34 | −0.49 |
| 1000 | 10.48 | 20.68 | +0.42 | +0.28 | 0.14 | −1.01 |
| 1100 | 9.94 | 18.80 | +1.06 | +1.08 | −0.02 | −1.43 |
| 1200 | 9.50 | 17.24 | +1.60 | +1.76 | −0.16 | −1.79 |
| 1300 | 9.12 | 15.92 | +2.06 | +2.32 | −0.26 | −2.10 |
| 1400 | 8.79 | 14.78 | +2.44 | +2.80 | −0.36 | −2.36 |
| 4000 | 5.84 | 5.14 | — | — | — | — |

equilibrium limited systems in which carbon activities are approximately the same as for graphite.

For some experimental systems where coal devolatilization reactions are occurring in addition to the subsequent gasification of coal char, apparent high values of $a_c$ can result. This occurs because coal decomposition reactions are generally irreversible and the methane yields obtained are a result of relative kinetic factors rather than equilibrium or pseudoequilibrium characteristics in the presence of hydrogen at elevated pressures. The gas composition obtained under equilibrium conditions for coal char gasification can be qualitatively estimated by assuming that solid carbon has the same free energy of formation as graphite. For the hydrogasification reaction given above, the equilibrium gas composition can be expressed only as a function of the temperature and the total pressure. Gas-phase species are assumed to be ideal. The equilibrium methane mole fraction decreases with decreasing pressure and increasing temperature.

***Boudouard reaction.*** For the carbon–carbon dioxide–carbon monoxide system,

$$C + CO_2 \rightleftharpoons 2CO$$

the equilibrium gas compositions are a function only of the temperature and the pressure. The equilibrium carbon monoxide mole fractions for this reaction increase with increasing temperature and deceasing pressure, which is opposite to the behavior exhibited by the carbon-hydrogen-methane system.

***Steam gasification reactions.*** For the carbon-hydrogen-oxygen gasification system, where all the components are present, the following species are assumed to exist in the gas phase: $CO$, $CO_2$, $H_2$, $H_2O$, and $CH_4$. For such a system, the equilibrium composition can be expressed as a function of the temperature, pressure, and H/O ratio in the gas phase. The selection of reactions for the equilibrium calculations is arbitrary because a variety of equivalent combinations of reactions are possible.

$$CO + H_2O \rightleftharpoons CO_2 + H_2$$

$$C + 2H_2 \rightleftharpoons CH_4$$

$$C + CO_2 \rightleftharpoons 2CO$$

The equilibrium trends are as shown in Table 6 [86]. These characteristics are constructed on the assumption that solid carbon phase has the same thermodynamic properties as graphite.

**Heat effects.** The effects of heat on the coal gasification system can be expressed in terms of the sensible heats of the reacting substances and the heats of reactions. For calculating sensible heats of individual components, their spe-

**Table 6  Effects on equilibrium mole fractions**

| Mole fraction, $X_i$ | Temp. ↑ | Pressure ↑ | $H_2/O_2$ ratio ↑ |
|---|---|---|---|
| $X_{CO}$ | ↑ | ↓ | ↓ |
| $X_{CO2}$ | ⌢ | ↑ | ↓ |
| $X_{H2}$ | ↑ | ↓ | ↑ |
| $X_{CH4}$ | ⌢ | ↑ | ↑ |
| $X_{H2O}$ | ↓ | ↑ | ⌢ |

cific heats must be evaluated. For coal and coal char, a variety of experimental techniques have been used below 400°C to avoid discrepancy due to heats of coal decomposition. A useful correlation has been proposed for estimating the actual specific heats of coal. According to this correlation, the specific heat of coal char is assumed to be a weighted summation of the specific heats of major coal components:

$$C_p = X_v C_v + X_a C_a + X_w C_w + X_c C_c$$

where $X_v$, $X_a$, $X_w$, and $X_c$ are the mass fractions of potential volatile matter, ash, water, and fixed carbon, respectively, with $C_v$, $C_a$, $C_w$, and $C_c$ their specific heats and $C_p$ is the average specific heat of coal or coal char. The specific heats are expressed as functions of temperature.

**Pressure effects.**  A recently developed coal gasification technology features the use of pressurized gasification in which operating pressures as high as 6 MPa have been applied for effective operation. For such high-pressure operations, correlation of the operating pressure with coal pyrolsis, coal caking, and coal char reactivity is important. A study in China [93] has shown all the required parameters correlated with respect to the pressure. The studies led to the following conclusions:

1. Gas product yields obtained as a result of pyrolsis increase with increasing pressure regardless of whether the atmosphere is $H_2$ or $N_2$. The yields are much higher in the $H_2$ atmosphere.
2. The reactivities of coal char obtained under pressure decrease with increasing pressure.
3. The steam gasification and carbon dioxide gasification rates increase with increasing partial pressure of the respective gasifying agent up to 1–1.5 MPa and then tend to have zero order with respect to the reacting gases.
4. Coal hydrogasification is very sensitive to the partial pressure of hydrogen, but first-order behavior is obtained for a wide range of operating pressures.
5. Caking properties of caking coals increase throughout the pressure range examined, but for noncaking coals, they remain constant above a certain characteristic pressure.

Design and operation of coal gasification and in general coal conversion processes require appropriate thermodynamic and thermophysical property data. The most important data required are the VLE data. Ramanujan et al. [94] have developed a computational technique to predict vapor-liquid equilibrium properties for coal gasification systems. The technique developed is applicable at elevated pressures and temperatures for gas as well as liquid mixtures. Gas-phase behavior was described by the Redlich-Kwong equation of state with modification by Prausnitz-Chueh, and the liquid-phase behavior was described by the regular solution theory of Hildebrand and Scatchard. The technique involved the introduction of a solubility parameter and was applied successfully to several binary and some ternary and quaternary systems. The principal limitation of the described methodology was the use of a hypothetical liquid phase for a supercritical component, because it did not have any theoretical basis. The technique is recommended for the computation of phase behavior of complex hydrocarbon mixtures related to the coal conversion processes provided pure component information is available.

## 3.4.8 Chemical Mechanisms

It is generally recognized that coal gasification proceeds through the surface oxide complex as an intermediate species [95]. By evaluating the amount of surface oxide complex and its rate of formation and desorption, the gasification of coal by steam and carbon dioxide can be analyzed. It was observed that the desorption rate of surface oxide complex formed through steam gasification is almost the same as that of the complex formed during carbon dioxide gasification. A linear relationship was obtained between the gasification rate and the amount of surface oxide complex formed.

General controlling mechanisms in the gasification of pulverized coal were studied by Smoot and Brown [96]. They investigated the mechanisms with oxygen at atmospheric pressure through analysis of experimental data and by comparison with predictions of a comprehensive model. Data were obtained for various oxygen-steam-coal ratios, and effects of coal feed rate, particle size, and flame type were determined. Their results showed that initial devolatization occurs rapidly at high temperatures and about 70% of coal is consumed. This is followed by reaction of coal volatiles and oxygen to produce high concentrations of carbon dioxide. The addition of steam played a very insignificant role in the coal reaction process. The surface reactions are controlled in high-temperature regions by diffusion to the char surface. But as the temperature decreases as a result of the endothermic reactions, heterogeneous reaction near the external char particle surface with the oxidizer takes the center stage, which depends on the coal type.

Some studies have also been undertaken to understand the mechanism of catalytic coal char gasification by alkali metal and alkaline earth metal catalysts

[97]. They have described an oxygen transfer mechanism in which reaction takes place in three steps: (1) adsorption on metal or carbon active sites, (2) activation of gas, and (3) diffusion of the activated atom to the reaction site. The reactivity of the char depends on both the catalyst and the carbon active sites. The authors did a series of microcalorimetric studies to obtain information regarding the catalyst dispersion and the identity of the active phase. The interaction between oxygen and coal chars promoted with either sodium or magnesium carbonate was studied. Their results suggested that the active phase of sodium is a highly dispersed metallic phase, in contrast to earlier studies in which it was shown that potassium was in a partially oxidized state. This led to the conclusion that not all alkali metals promote coal char gasification through the same mechanism. It was also found that magnesium did not follow the same mechanism for catalytic action as it failed to adsorb a significant quantity of oxygen.

For the three main reactions discussed in the earlier section on kinetics, several workers [98–101] have proposed mechanisms to rationalize the kinetic correlation forms.

**Coal char–hydrogen–methane system:**

$$C + 2H_2 \rightleftharpoons CH_4$$

For this reaction, the following fundamental reaction steps are assumed

Step 1 $$C_f + H_2 \underset{k_1'}{\overset{k_1}{\rightleftharpoons}} C(H_2)$$

Step 2 $$C(H_2) + H_2 \underset{k_2'}{\overset{k_2}{\rightleftharpoons}} CH_4 + C_f$$

where $C_f$ is the active center on the carbon surface and $C(H_2)$ is the surface complex formed by the chemisorption of hydrogen at the active center.

According to this mechanism, gaseous hydrogen is chemisorbed by any active center on the carbon surface. This chemisorbed active surface then further reacts with hydrogen to form methane and regeneration of the active center. Both reaction steps are reversible. Different reactivities of coal chars are attributed to variations in the total number of active centers. This implies that relative behavior of different coal chars should be similar. An alternative set of mechanistic steps have been proposed to describe the hydrogen chemisorption process.

Step 1(a) $$C_f + H_2 \underset{k_3'}{\overset{k_3}{\rightleftharpoons}} (CH)(H)$$

Step 1(b) $$CH(H) \underset{k_4'}{\overset{k_4}{\rightleftharpoons}} C(H_2)$$

According to this mechanistic step, the active center on the carbon surface splits the hydrogen molecule, and one atom is attached to the active center while the other atom reacts with either the original active site or any other active site to form a $-CH_2$ group. This requirement for the formation of surface hydrogen atoms as a precursor to the formation of methane is supported by alternative experiments in which various carbons are reacted directly with hydrogen atoms produced in microwave or radio frequency fields. The results obtained in terms of the gasification rates proved similar to those obtained in molecular hydrogen under actual experimental conditions.

**Boudouard reaction.** This reaction is important in both gasification and combustion processes. The reverse reaction is a carbon deposition reaction that often causes catalyst deactivation by fouling in syngas conversion processes.

$$C + CO_2 \rightleftharpoons 2CO$$

For this reaction the following mechanism steps have been postulated [102]:

Step 1 $$C_f + CO_2 \underset{k_1'}{\overset{k_1}{\rightleftharpoons}} C(O) + CO$$

Step 2 $$C(O) \underset{k_2'}{\overset{k_2}{\rightleftharpoons}} CO + C_f$$

In step 1, the carbon dioxide molecule dissociates at an active site, $C_f$, on the carbon surface, releasing a molecule of carbon monoxide and a complex containing carbon and oxygen, C(O). This reaction results in exchange of oxygen with the solid but the actual gasification occurs in step 2, where the carbon-oxygen surface complex dissociates leading to the formation of carbon monoxide and generation of a new active center on the solid carbon surface. The process then continues. In the mechanism proposed above, it is assumed that the total number of active sites plus the carbon-oxygen complex remains constant. Also, the net rate of formation of these complexes is zero under the steady-state assumption. The mechanistic studies of this particular reaction are done using isotopic gases and/or solid carbon. Isotopes have been a particularly useful tool for studying the mechanism of the carbon–carbon dioxide reaction. Oxygen exchange is an essential step in the above postulation and is proved using isotopic techniques.

Different investigators have postulated other reaction steps different from that described earlier to account for results obtained on gasification of a coconut char in carbon monoxide–carbon dioxide mixtures at elevated pressures (up to 40 atm). This is one of the few mechanisms reported for high-pressure gasification. According to this mechanism, the following reaction steps take place.

Step 1. (forward) $$C_f + CO_2 \xrightarrow{k_1} CO + C(O)$$

Step 2. (forward) $$C(O) \xrightarrow{k_2} CO + C_f$$

Step 3 $$CO + C_f \underset{k_3'}{\overset{k_3}{\rightleftharpoons}} C(CO)$$

Step 4. (forward) $$CO_2 + C(CO) \xrightarrow{k_4} 2CO + C(O)$$

Step 5. (forward) $$CO + C(CO) \xrightarrow{k_5} CO_2 + 2C_f$$

Sams and Shadman [103] have studied the mechanism of the Boudouard reaction, which was potassium catalyzed by a temperature- and concentration-programmed technique. The knowledge of the reaction mechanism is of primary significance in understanding how alkali metal catalysis works and how it can be utilized for process applications. The experimental observations were explained by a redox type of mechanism containing three surface complexes: $-CO_2K$, $-COK$, and $-CK$. The following sequence was suggested for the gasification process after the initial rapid transient stage is over:

$$(-CO_2K) + C = (-COK) + CO$$

$$(-COK) + CO_2 = (-CO_2K) + CO$$

The rate of oxidation is fast enough that the rate-limiting step in the overall process is the reduction of the oxidized surface group, $-CO_2K$, via the first reaction stated above. The presence of CO reduces the observed rate of gasification. The completely reduced group CK was readily decomposed to free potassium metal, which was easily vaporized at gasification temperatures. This explains the mechanism of catalyst loss and its significance under reducing conditions. The proposed mechanism was supported by a complete oxygen balance.

### 3.4.9 Design Criteria

Even though gasification processes of all types are in operation in industries worldwide, the research and development work now in progress to produce advanced environmentally acceptable gasification systems is considerable. There is a wide range of coal gasification research projects, ranging from bench scale to demonstration projects in progress. Even though the stabilization of oil prices in the 1980s contributed to reassessment for priorities for commercialization of processes, numerous demonstration studies have been completed and commercial plants have been installed for production of substitute natural gas. Successful translation of demonstration and pilot-scale plant to commercial operation requires careful consideration of the matter of design of gasifiers. Such translation for a complex technology such as coal conversion is a broad and intricate subject.

**Commercial plant objectives.** Objectives should be designed early in the design program to provide guidance for scale-up criteria and design. Examples of factors to be considered in setting objectives are as shown [104].

Examples of factors to be included in definition of commercial plant objectives:

1. Capacity
2. Reliability
3. Economic goal
4. Product: (a) specification; (b) product slate flexibility
5. Location factors
6. Raw material: (a) availability; (b) characteristics; (c) logistics
7. Product storage capacity
8. Plant logistics: (a) raw material supply; (b) product distribution

**Basic data requirements.** The main consideration here is that pilot operation should provide a detailed basic understanding of the chemical, mechanical, and control engineering factors required for design and operation of the coal conversion facility. The factors to be considered here are:

1. Accuracy and precision of data.
2. Analysis and control accuracy.
3. Material and stream properties.
4. Basic chemical engineering. These include
   Reaction kinetics
   Thermodynamics
   Fluid regimes
   Mass transfer
   Physical equilibrium data
   Catalyst performance

The availability of accurate chemical engineering data provides a sound foundation for the application of translation to scale-up techniques.

5. Process performance.
6. Equipment performance.
7. Materials of construction.
8. Safety and hygiene.
9. Environmental factors.

**Design or scale-up criteria.** Defining the criteria to be used for scale-up is the first step in scaling up a process. These design criteria are usually either performance based or economics based. Sometimes a trade-off between two same types of criteria, such as conversion and ultimate yield, may be involved. Design criteria should always be specific.

Some examples of a number of equipment categories pertinent to scale-up of coal conversion processes are:

1. High-capacity and high-pressure coal slurry pumps
2. Coal slurry preheat furnaces
3. Coal slurry pressure letdown valves
4. Pressure recovery turbines
5. Solids drying equipment
6. Large-capacity gasifiers
7. Solid feed devices for high-pressure gasifiers
8. Gas-solid separation equipment suitable for high pressure and temperature
9. Large-scale synthesis reactors
10. Compressors suitable for gas-solid mixtures and high pressures.

One factor that is a prerequisite for scale-up criteria is the time frame for design, construction, and operation of the facility. For example, in the case of a gasifier, shop fabrication and testing of the gasifiers and other vessels are prudent before a purchase. The design, fabrication, erection, testing, reliability, and economic factors must all be carefully evaluated and sufficient lead time allowed to develop and implement a program to achieve the goals. Included in this evaluation should be the installation, operating, and maintenance costs of ancillary equipment items such as the multiple pumps, heat exchangers, and knockout drums.

## REFERENCES

1. Howard-Smith, I., and G. Werner. Coal Conversion Technology. *Chem. Technology Rev.* 66: 1–133 (1976). Park Ridge, NJ: Noyes Data Corp.
2. Wilson, J., J. Halow, and M. R. Ghate. *Chemtech* February: 123–128, 1988.
3. Speight, J. G. *The Chemistry and Technology of Coal*, pp. 461–516. New York: Marcel Dekker, 1983.
4. Vyas, K. C., and W. W. Bodle. Clean Fuels from Coal—Technical Historical Background & Principles of Modern Technology, pp. 53–84. Symposium II papers, IIT Research Institute, 1975.
5. Bodle, W. W., and K. C. Vyas. *Clean Fuels from Coal*, Symposium Papers, pp. 49–91, Chicago: Institute of Gas Technology, September 10–14, 1973.
6. Braunstein, H. M., and H. A. Pfuderer. Environmental Health and Control Aspects of Coal Conversion: An Information Overview. Report ORNL-EIS-94, Oak Ridge National Laboratory, Oak Ridge, TN, 1977.
7. Rudolph, P. E. H. The Lurgi process—the route to SNG from coal. Presented at the Fourth Pipeline Gas Symposium, Chicago, October 1972.

8. Cooper, B. R., and W. A. Ellingson. *The Science and Technology of Coal and Coal Utilization*, pp. 163–230. New York: Plenum, 1984.

9. Llyod, W. G. Synfuels technology update, In *The Engineeting Synthetic Fueld Industry*, ed. A. Thumann, pp. 19–58. Atlanta: Fairmont Press, 1981.

10. Moe, J. M. SNG from coal via the Lurgi gasification process. In *Clean Fuels from Coal, Symposium Papers*, pp. 91–110. Institute of Gas Technology, Chicago, Illinois, September 10–14, 1973.

11. Johnson, B. C., et al. The Grand Forks slagging gasifier. In *Coal Processing Technology*, vol. IV, a CEP technical manual, pp. 94–98. New York: AIChE, 1978.

12. Michels, H. J., and H. F. Leonard. *Chem. Eng. Prog.* 74(8):85, 1978.

13. Van der Bergt, M. J. *Hydrocarbon Processs* 58(1):161, 1979.

14. Farnsworth, J., H. F. Leonard, and M. Mitsak. Production of gas from coal by the Koppers-Totzek process. In *Clean Fuels from Coal*, Symposium Papers, pp. 143–163. Institute of Gas Technology, Chicago, September 10–14, 1973.

15. *EPRI J.*, December:41–44, 1983.

16. *Oil and gas J.* April 29:51, 1985.

17. *Hydrocarbon Process.* April:96, 1984.

18. Probstein, R. F., and R. E. Hicks. *Synthetic Fuels.* New York: McGraw-Hill, 1982.

19. Thompson, P. N. *Endeavor* (new series) 2:93, 1978.

20. Gregg, D. W., and T. F. Edgar. *AIChE J.*, 24:753, 1978.

21. Juntgen, H. *Fuel* 66:443–453, April 1987.

22. Nadkarni, R. M. Underground gasification of coal. In *Clean Fuels from Coal, Symposium Papers*, pp. 611–638. Institute of Gas Technology, Chicago, September 10–14, 1973.

23. Krantz, W. B., and R. B. Gunn, Underground gasification: The state of the art. *AIChE Symp. Ser. No. 226*, 79:1–185 (1983).

24. Berkowitz, N. *An Introduction to Coal Technology*, pp. 251–298. New York: Academic Press, 1979.

25. Odell, W. U.S. Bureau Mines Information, 7415, 1974, I.N. Banchik, Symp. Coal Gasif. & Liquif., Pittsburgh, 1974; Davy Powergas Inc. Publ., 1974.

26. Banchik, I. N. The Winkler process for the production of low Btu gas from coal. In *Clean Fuels From Coal*, Symposium II Papers, pp. 359–374. IIT Research Institute, 1975.

27. Juntgen, H. Application of catalysts to coal gasification processes. Incentives and perspectives. *Fuel* 62:234–238, (1983).

28. Nishiyama, Y. *Fuel Proc. Tech.* 29:31–42, 1991.

29. Asami, K., and Y. Ohtuska. *Ind. Eng. Chem. Res.* 32:1631–1636, 1993.

30. Yamashita, H., S. Yoshida, and A. Tomita. *Energy Fuels* 5:52–57, 1991.

31. Matsumoto, S. *Energy & Fuels* 5:60–63, 1991.

32. Srivastava, R. C., S. K. Srivastava, and S. K. Rao. *Fuel* 67:1205–1207, 1988.

33. Haga, T., and Y. Nishiyama. *Fuel* 67:748–752, 1988.

34. Oshtuka, Y., and K. Asami. Steam gasification of high sulfur coals with calcium hydroxide. *Proc. 1989 Int. Conf. Coal Sci.* 1:353–356, 1989.

35. Salinas Martinez, C., and A. Lineras-Solano. *Fuel* 69:21–27, 1990.

36. Joly, J. P., A. Martinez-Alonso, and N. R. Marcilio. *Fuel* 69:878–894, 1990.

37. Muhlen, H. J. *Fuel Proc. Tech.* 24:291–297, 1990.

38. Levendis, Y. A., S. W. Nam, and G. R. Gravalas. *Energy Fuels* 3:28–37, 1989.

39. Zheng, Z. G., T. Kyotani, and A. Tomita. *Energy Fuels* 3:566–571, 1989.

40. Muhlen, H. J. *Fuel Proc. Tech.* 24:291–297.

41. Haga, T., M. Sato, Y. Nishiyama. *Energy Fuels* 5:317–322, 1991.

42. Pareira, P., G. A. Somorajai, and H. Heinemann. *Fuels* 6:407–410, 1992.

43. Xiang, R., W. You, and L. Shu-fen. *Fuel* 66:568–571, 1987.

44. Takarada, T., T. Nabatame, and Y. Ohtuska. *Ind. Eng. Chem. Res.* 28:505–510, 1989.

45. Takarada, T., M. Ogirawa, and K. Kato. *J. Chem. Eng. Japan* 25:44–48, 1992.
46. Weeda, M., J. J. Tromp, and J. A. Moulijn. *Fuel* 69:846–850, 1990.
47. Chin, Ge., G. Liu, and Q. Dong. *Fuel* 66:859–863, 1987.
48. Carrasco-Marin, F., J. Rivera, and C. Moreno. *Fuel* 70:13–16, 1991.
49. Lopez, A., F. Carrasco, and C. Moreno. *Fuel* 71:105–108, 1992.
50. Toshimitsu, S., S. Nakajima, and Y. Watanabe. *Energy Fuels* 2:848–853, 1988.
51. Murlidhara, H. S., and J. T. Seras. Effect of calcium and gasification. In *Coal Processing Technology*, vol. IV, pp. 22–25. CEP, AIChE, 1978.
52. Gallagher, J. E., and H. A. Marshall. SNG from coal by catalytic gasification. In *Coal Processing Technology*, vol. V, pp. 199–204. CEP, AIChE, 1979.
53. Cover, A. E., and W. C. Schreiner. In *Clean Fuels from Coal*, Symposium Papers, pp. 273–279. Chicago: Institute of Gas Technology, September 10–14, 1973.
54. Cover, A. E., W. C. Schreiner, and G. T. Skaperdas. Kellogg's coal gasification process. *Chem. Eng. Prog.* 69(3):31–36, 1973.
55. LaRosa, P., R. J. McGarvey. In *Clean Fuels from Coal*, Symposium Papers, pp. 285–300. Chicago: Institute of Gas Technology, September 10–14, 1973.
56. Adanez, J., and F. G. Labiano. *Ind. Eng. Chem. Res.* 29:2079–2088, 1990.
57. Lim, C. J., J. P., Lucas, and A. P. Watkinson. *Can. J. Chem. Eng.* 69:596–606, 1991.
58. Foong, S. K., G. Cheng, and A. P. Watkinson. *Can. J. Chem. Eng.* 59:625–630, 1981.
59. Watkinson, A. P., C. J. Lim, and G. Cheng, *Can. J. Chem. Eng.* 65:791–798, 1987.
60. Monazam, E., E. Johnson, and J. Zondlo. *Fuel Sci. Tech. Int.* 10(1):51–73, 1992.
61. Watkinson, A. P., J. P. Lucas, and C. J. Lim. *Fuels* 70:519–527, April 1991.
62. Lee, S., J. Angus, N. C. Gardner, and R. V. Edwards. *AIChE J.* 30(4):583–593, 1984.
63. Gururajan, V. S., P. K. Agarwal, and J. B. Agnew. *Trans IChemE.* 70 (Part A): 211–238, 1992.
64. Raghunathan, K., and Y. K. Yang. *Ind. Eng. Chem. Res.* 28:518–523, 1989.
65. Li, J., and A. R. Heiningen. *Ind. Eng. Chem. Res.* 30:1594–1601, 1991.
66. Matsui, I., and T. Furusawa. *Ind. Eng. Chem. Res.* 26:91–100, 1987.
67. Prasad, K. H. V., and Y. V. Subbarao. 69:1517–1521, 1990.
68. Kovacik, G., A. Chambers, and B. Ozum. *Can. J. Chem. Eng.* 69:811–815, 1991.
69. Blackwood, J. D. *Aust. J. Chem.* 13:194, 1960.
70. Feldmann, H. F., and S. P. Chauhan. *Coal Processing Technology*, vol. V, pp. 205–206. Chemical Engineering Progress, AIChE, 1979.
71. Toomajian, M., M. G. Lussier, and D. Miller. *Fuel* 71:1055–1061, 1992.
72. Ohtuska, Y., K. Asami, and T. Yamada. *Energy Fuels* 6:678–679, 1992.
73. Haga, T., and Y. Nishayama. *Ind. Eng. Chem. Res.* 28:724–728, 1989.
74. Lee, S. *Methanol Synthesis Technology.* Boca Raton, FL: CRC Press, 1990.
75. Newsome, D. S. The water gas shift reaction. *Catal. Rev. Sci. Eng.* 21:275, 1980.
76. Meijer, R., M. Sibeijn, and J. Moulijn. *Ind. Eng. Chem. Res.* 30:1760–1770, 1990.
77. Kyotani, T., K. Kubota, and A. Tomita. *Fuel Proc. Tech.* 36:209–217, 1993.
78. A random pore model for fluid-solid reactions. *AIChE J.* 26(3):379, 1980.
79. *AIChE, J.,* 27(2):247, 1981.
80. Su, J. L., and D. D. Perlmutter. *AIChE J.* 31(6):973–981, 1985.
81. Hurt, R. H., A. F. Sarofim, J. P. Longwell. *Energy Fuels* 5:290–299, 1991.
82. Fott, P., and P. Straka. *Fuel* 66:1281–1288, 1987.
83. Goyal, A., R. Zabransky, and A. Rehmat. *Ind. Eng. Chem. Res.* 28:1767–1778, 1989.
84. Li, Jian, and A. R. Heiningen. *Ind. Eng. Chem. Res.* 30:1594–1601, 1991.
85. Nozaki, T., T., Adschiri, and T. Furusawa. *Fuel Proc. Tech.* 24:277–283, 1990.
86. Lizzio, A., and L. R. Radovic. *Ind. Eng. Chem. Res.* 30:1735–1744, 1991.
87. Adschiri, T., T. Nozaki, and T. Furusawa. *AIChE J.* 37(6):897–903, 1991.
88. Kyotani, T., H. Yamada, H. Yamashita, and R. Tomita. *Energy Fuels* 6:865–867, 1992.
89. Johnson, J. L., *Kinetics of Coal Gasification*, pp. 2–20. New York: Wiley, 1979.

90. Wen, C. Y., O. C. Abraham, and A. T. Talwalkar, *Am. Chem. Soc. Div. Fuel. Chem. Prepr.* 10(4):168–185, 1966.
91. Squires, A. M. *Trans. Inst. Chem. Eng. (Lond.)* 39:3–9, 1961.
92. Blackwood, J. D., and D. J. McCarthy. *Aust. J. Chem.* 19:797–812, 1966.
93. Sha, Xing, Yi-Gong Chen, J. Cao, and De-Qing Ren. *Fuel* 69:656–659, 1990.
94. Ramanujan, S., S. Leipzlger, and S. Well. *Ind. Eng. Chem. Process. Des. Dev.* 24:658–665, 1985.
95. Green, T., J. Ball, and K. Conkright. *Energy Fuels* 5:610–611, 1991.
96. Smoot, D., and B. Brown. *Fuel* 66:1249–1256, 1987.
97. Gowe, A., and J. Phillips. *Energy Fuels* 6(4):526–532, 1992.
98. Blackwood, J. D., and D. J. McCarthy. *Aust. J. Chem.* 19:797–813, 1966.
99. Blackwood, J. D., and D. J. McCarthy. *Aust. J. Chem.* 20:2525–2528, 1967.
100. Blackwood, J. D., B. D. Cullis, and D. J. McCarthy. *Aust. J. Chem.* 20:1561–1570, 1967.
101. Blackwood, J. D. *Aust. J. Chem.* 12:14–28, 1959.
102. Menster, M., and S. Ergun. *US. Bur. Mines, Bull, 644,* 1973.
103. Sams. D. A., and F. Shadman. *AIChE J.* 32(7):1132–1137, 1986.
104. O'Hara, J. B., N. E. Jentz, and G. H. Harvey. Commercial coal conversion plant design. In *Clean Fuels from Coal,* Symposium II Papers, pp. 818–837. IIT Research Institute, 1975.

## PROBLEMS

1. When we gasify coal, we normally observe the first-stage pyrolysis followed by the slow gasification reaction. In this case, which step is rate controlling?
2. What are the morphological differences between coal and char, if both are derived from the same seam of coal?
3. What is the inhibitor in the steam gasification reaction of coal?
4. How many of the following reactions are stoichiometrically independent?

   $2C + O_2 = 2CO$

   $C + O_2 = CO_2$

   $C + CO_2 = 2CO$

   $CO + H_2O = CO_2 + H_2$

   $C + 2H_2 = CH_4$

   $CH_4 + H_2O = CO + 3H_2$

   $C + H_2O = CO + H_2$

   $2CO + O_2 = 2CO_2$

   $CH_4 + 2O_2 = CO_2 + 2H_2O$

   $2 H_2 + O_2 = 2H_2O$

5. Gauss elimination technique can be used to determine the rank of a matrix.
   (a) Convert the stoichiometric system presented in Problem 4 into a matrix equation of linear systems.
   (b) Apply the Gauss elimination method to determine the number of stoichiometrically independent reactions.
6. Consider the steam gasification of coal. At sufficiently high steam decomposition, is it possible to exceed water gas shift equilibrium? In other words, is it possible that the reaction tends to operate in the reverse direction?

7. Determine the temperature at which $K_p$ equals unity, for the following reactions:
   (a) $C + CO_2 = 2CO$
   (b) $C + H_2O = CO + H_2O$
   (c) $C + 2H_2 = CH_4$

8. At atmospheric pressure, is there any temperature range for which all the above reactions are thermodynamically favored? If so, determine the temperature range.

9. Are the following statements true or false?
   (a) In chemisorption, the surface atoms must have free valence electrons in order to form strong chemical bonds with gas molecules or atoms.
   (b) In carbon-oxygen reactions, the primary product is carbon monoxide, not carbon dioxide. Carbon dioxide is formed solely via oxidation of carbon monoxide.
   (c) There exist unpaired electrons in various types of carbons and this can be confirmed using electron paramagnetic resonance technique.

10. If coal char and synthetic graphite are gasified in a $CO_2$ environment at the same nominal operating conditions, which will have a faster reaction rate? What is the reasoning behind your answer?

11. Many researchers have found that apparent activation energies of heterogeneous char reactions are functions of char conversions when Arrhenius-type rate concepts are applied. Do you expect the activation energy of the char-$CO_2$ reaction to increase with char conversion? Why or why not?

12. The Langmuir-Hinshelwood type of kinetics has been frequently applied to coal gasification reactions. The following three major assumptions must be added to the usual ones (select the correct term):
   (a) The surface is (homogeneous, heterogeneous); that is, (uniform, non-uniform) average activity can be defined for the entire surface.
   (b) Interactions (do, do not) occur among adsorbed species.
   (c) Surface migration is either nonexistent or so rapid that only adsorption or desorption can be rate controlling.

13. According to Walker's interpretation, the C-$CO_2$ reaction involves the following mechanistic steps:
   $$C_f + CO_2(g) = C(CO_2)$$
   $$C_f + C(CO_2) = C(O) + C(CO)_A$$
   $$C_f + C(O) = C(CO)_B$$
   $$C(CO)_A = CO(g) + C_f$$
   $$C(CO)_B = CO(g) + C_f$$
   $$CO(g) + C_f = C(CO)_C$$
   How many of these mechanistic reactions are stoichiometrically independent?

14. Derive a generalized rate expression for the following mechanism:

$$C_f + CO_2(g) \xrightarrow{i_1} C(O) + CO(g)$$

$$C(O) \xrightarrow{j_2} CO(g) + C_f$$

$$CO(g) + C_f \underset{j_2}{\overset{i_2}{\rightleftharpoons}} C(CO)$$

In this case, the number of free sites may be assumed to remain constant with burnoff. If you label oxygen, say $^{17}O$, what kind of conclusion can you draw from the experiment?

15. Derive a generalized rate expression for the following mechanism:

$$C_f + CO_2(g) \underset{j_1}{\overset{i_1}{\rightleftharpoons}} C(O) + CO(g)$$

$$C(O) \xrightarrow{j_3} CO(g) + C_f$$

16. Is the following mechanism a dual-site mechanism or a single-site mechanism?

$$H_2O + 2C_f = C(H) + C(OH)$$
$$C(OH) + C_f = C(O) + C(H)$$
$$C(H) + C(H) = H_2 + 2C_f$$
$$C(O) \longrightarrow CO + C_f$$

17. Which of the following coal reactions at 900°C has the highest possibility for a diffusion rate-limiting case. A diffusion rate-limiting case means that the diffusional rate is the slowest step of all transport-reaction steps.
    (a) Combustion
    (b) $CO_2$ gasification
    (c) $H_2O$ gasification
    (d) Hydrogasification

18. At what temperature is the steam gasification reaction of coal likely to be controlled by the diffusion rate?
    (a) 680°C
    (b) 800°C
    (c) 950°C
    (d) 1200°C

19. The liquid-phase methanol process (LPMeOH) was developed based on the use of CO-rich syngas. What gasifiers produce CO-rich syngas? (Note: For more information regarding the liquid phase methanol process, refer to Lee, S., *Methanol Synthesis Technology*, CRC Press, 1990.)

20. Discuss the distinguishing characteristics of the following reactor types:
    (a) Fixed bed
    (b) Packed bed
    (c) Fluidized bed
    (d) Entrained flow
    (e) Moving bed
21. Consider a laboratory-scale fixed bed gasifier of coal. The sample bed dimensions are 2 in. inner diameter $\times$ 5 in. height. This reactor has an external heating device that keeps the reactor inside surface at a constant temperature of 900°C. Assume a coal sample of $-35 + 60$ mesh has been instantaneously injected into the bed and the bed is packed with this coal sample. Estimate the time required for a coal particle situated at the center point of this bed to "feel" the high-temperature environment, that is, the high temperature of the reactor internal wall. In other words, how long will it take for a center particle to reach the sothermal condition?

    Assume the coal sample has:

    $\varepsilon$(porosity) = 0.3
    $k_e$(effective thermal conductivity) = $(1 - \varepsilon)^2 k$
    $k$(thermal conductivity) = $4 \times 10^{-3}$ cal/cm s°C
    $\rho$(density) = 0.9 g/cm³
    $C_p$(heat capacity)= 0.5 cal/g°C

    Show all your calculations and assumptions.

22. An equilibrium mixture of $N_2$ and $O_2$ is kept at 3300 K and 1 atm. Compute the equilibrium mixture compositions of N, O, $N_2$, $O_2$, $N_2O$, NO, $N_2O_3$, $N_2O_4$, $NO_2$, and $N_2O_5$. Data for Gibbs free energies of formation for each species are available from various thermodynamic data packages including JANAF tables.
    (a) Discuss the method of minimization of total Gibbs free energy of the system.
    (b) Compute the equilibrium mixture composition at this temperature.
    (c) Discuss the numerical method to be used.
    (d) Comment on the formation of $NO_x$ caused by lightning.
    (e) Comment on the efficiency of a "low $NO_x$ burner" in controlling $NO_x$.

# FOUR

# COAL LIQUEFACTION FOR ALTERNATIVE LIQUID FUELS

## 4.1 LIQUEFACTION OF COAL: BACKGROUND

There are three principal routes by which liquid fuels can be produced from coal: coal pyrolysis, direct liquefaction, and indirect liquefaction.

As repeatedly mentioned throughout this book, coal is a very useful energy source and is abundantly available in the world. Even though coal has a reasonably high heating value of approximately 8000–14000 Btu/lb, its solid state has been a major reason for inconvenience as a fuel. In order to make this solid fuel more user-friendly, constant efforts have been made to convert the solid fuel into pipeline quality gaseous fuels or clean liquid fuels. During World War II in the early to middle 1940s, the production of liquid fuels at approximately 100,000 bbl/day was recorded for the German war effort. Their process used a high-temperature, high-pressure technology, and the product liquid fuels were of environmentally poor quality.

The current objectives of coal liquefaction are focused more on easing operating process conditions, minimizing the hydrogen requirement, and making the product liquid more environmentally acceptable. With the recent trends of stable petroleum prices in the world market, the process economics of coal liquefaction has been seriously looked at.

## 4.2 COAL PYROLYSIS FOR LIQUID FUEL

Pyrolysis of coal yields, through destructive distillation, condensable tar, oil and water vapor, and noncondensable gases. The solid residue of pyrolysis is char. The condensed pyrolysis product must be further hydrogenated to remove sulfur and nitrogen species and to improve the liquid fuel quality. Nitrogen and sulfur species not only generate environmentally hazardous gases ($NO_x$ and $SO_x$) when combusted but also poison the upgrading catalyst.

Coal pyrolysis processes are well developed and commercially available. Table 1 presents various coal pyrolysis processes and their operating conditions and yields [1].

A great number of factors affect the process efficiency and product yield. These factors include the coal rank, processing temperature, residence time, reactor pressure, coal particle size, and process mechanics.

A quick glance at Table 1 indicates that higher liquid yields are obtained with shorter residence times and also that a hydrogen atmosphere helps the liquid product yield [2]. A shorter residence time does not allow sufficient reaction time for thermal cracking of the liquid hydrocarbons and leaves more liquid hydrocarbons in the product stream. Adding hydrogen to coal hydrocarbons increases the H/C ratio enough to produce a liquid fuel. A high tar content on supercritical extraction may be attributed to its excellent low temperature solubility that extracts and dissolves the tar.

### 4.2.1 COED Process

The COED (char-oil-energy-development) process was developed by the FMC Corporation. The process has been improved and has become the COED/COGAS process. It is based on a fluidized bed process that is carried out in four successive fluidized bed pyrolysis stages at progressively higher temperatures. The temperatures of the stages are selected to be just below the maximum temperature to which the particular feed coal can be heated without agglomerating and plugging the fluid bed [1]. Figure 1 shows a schematic of the COED/COGAS process [1]. The optimal temperatures for four stages vary depending on the properties of the feed coal. Typical operating temperatures are 315–345°C in the first stage, 425–455°C in the second stage, 540°C in the third stage, and 870°C in the fourth stage [2]. Heat for the process is provided by burning a portion of the product char with a steam-oxygen mixture in the fourth stage. Hot gases flow countercurrent with respect to the char and provide the hot fluidizing medium for pyrolysis stages. The gases leaving both the first and second stages are passed to cyclones, which remove the fines, but the vapors leaving the cyclones are quenched in a venturi scrubber to condense the oil, and the gases and oil are separated in a decanter. The gas is desulfurized and then steam reformed to produce hydrogen and fuel gas. The oil from the decanter is

**Table 1 Summary of several pyrolysis and hydropyrolysis processes**

| Process | Developer | Reactor type | Reaction temperature | | Reaction pressure (psi) | Coal residence time | Yield (wt. %) | | |
|---|---|---|---|---|---|---|---|---|---|
| | | | °C | °F | | | Char | Oil | Gas |
| Lurgi-Ruhrgas | Lurgi-Ruhrgas | Mechanical mixer | 450–600 | 840–1110 | 15 | 20 sec | 55–45 | 15–25 | 30 |
| COED | FMC Corp. | Multiple fluidized bed | 290–815 | 550–1500 | 20–25 | 1–4 hr | 60.7 | 20.1 | 15.1 |
| Occidental coal pyrolysis | Occidental | Entrained flow | 580 | 1075 | 15 | 2 sec | 56.7 | 35.0 | 6.6 |
| Toscoal | Tosco | Kiln-type retort vessel | 425–540 | 795–1005 | 15 | 5 min | 80–90 | 5–10 | 5–10 |
| Clean coke | U.S. Steel Corp. | Fluidized bed | 650–750 | 1200–1380 | 100–150 | 50 min | 66.4 | 13.9 | 14.6 |
| Union Carbide Corp. | Union Carbide Corp. | Fluidized bed | 565 | 1050 | 1000 | 5–11 min | 38.4 | 29.0 | 16.2 |

*Source:* Reference 1.

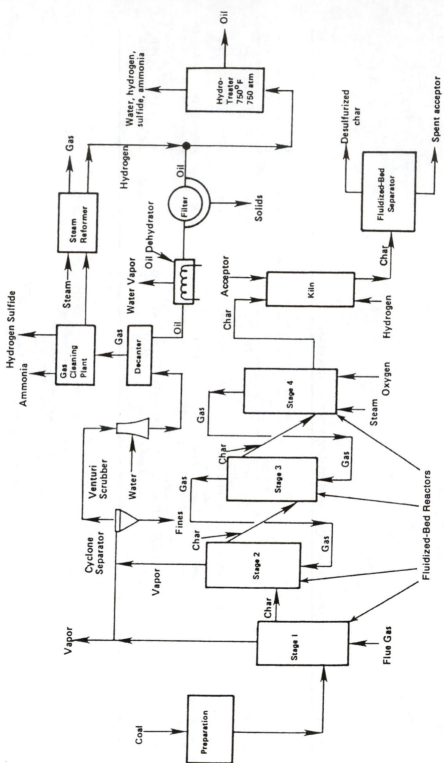

**Figure 1** Schematic representation of the COED/COGAS process. (Source: Reference 1.)

168

dehydrated, filtered, and hydrotreated to remove nitrogen, sulfur, and oxygen to form a heavy synthetic crude oil of approximately 25° API. The properties of synthetic crude oils made from coal by the COED process are shown in Table 2 [1].

The char is desulfurized in a shift kiln, where it is treated with hydrogen to produce hydrogen sulfide, which is subsequently absorbed by an acceptor such as dolomite or limestone [1]. The COGAS process involves the gasification of the COED char to produce a synthetic gas ($CO + H_2$). This COED/COGAS process is significant from both liquefaction and gasification standpoints.

## 4.2.2 TOSCOAL Process

A schematic of the TOSCOAL process is shown in Figure 2 [1]. In the process, crushed coal is fed to a rotating drum that contains preheated ceramic balls at temperatures between 425 and 540°C. The hydrocarbons, water vapor, and gases are drawn off, and the residual char is separated from the ceramic balls in a revolving drum with holes in it. The ceramic balls are reheated in a separate furnace by burning some of the product gas [2]. The TOSCOAL process is analogous to the TOSCO process for producing overhead oil from oil shale. In this analogy, the char replaces the spent shale [1].

Table 3 shows the properties of liquids from Wyodak coal as mined by the TOSCOAL process. As can be seen, the recovery is nearly 100%.

**Table 2 Product properties of synthetic crude oils from coal by the COED process**

| | Coal | |
| Properties | Illinois No. 6 | Utah King |
| --- | --- | --- |
| Hydrocarbon-type analysis (vol. %) | | |
|     Paraffins | 10.4 | 23.7 |
|     Olefins | | |
|     Naphthenes | 41.4 | 42.2 |
|     Aromatics | 48.2 | 34.1 |
| API gravity | 28.6 | 28.5 |
| ASTM distillation (°F) | | |
|     Initial boiling point (IBP) | 108 | 260 |
|     50% distilled | 465 | 562 |
|     End point | 746 | 868 |
| Fractionation yields (wt. %) | | |
|     IBP–180°F | 2.5 | |
|     180–390°F | 30.2 | 5.0 |
|     390–525°F | 26.7 | 35.0 |
|     390-650°F | 51.0 | 65.0 |
|     650–EP | 16.3 | 30.0 |
|     390–EP | 67.3 | 95.0 |

*Source:* Reference 1.

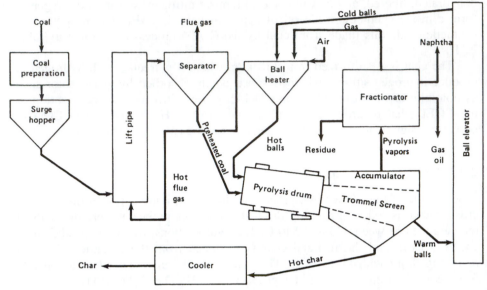

**Figure 2**  Schematic of the TOSCOAL process. (Source: References 1 and 46.)

### 4.2.3 Lurgi-Ruhrgas Process

The Lurgi-Ruhrgas (L-R) process was developed as a low-pressure process for liquid production from lower rank coals. The process, developed in Europe, is currently in commercial use. A schematic of this process is given in Figure 3 [1].

In the L-R process, crushed coal is fed into a mixer and heated rapidly to 450–600°C by direct contact with hot recirculating char particles that were previously heated in a partial oxidation process in an entrained flow reactor [1].

**Table 3  Liquids from Wyodak coal as-mined by the TOSCOAL process**

| Temperature | | | |
|---|---|---|---|
| °C | 425 | 480 | 520 |
| °F | 800 | 900 | 970 |
| Yield (wt. %) | | | |
| Gas ($\leq C_3$) | 6.0 | 7.8 | 6.3 |
| Oil ($\geq C_4$) | 5.7 | 7.2 | 9.3 |
| Char | 52.5 | 50.6 | 48.4 |
| Water | 35.1 | 35.1 | 35.1 |
| Recovery (wt. %) | 99.3 | 100.7 | 99.1 |

*Source:* Carlson, F. B., L. H. Yardumian, and M. T. Atwood. Reprints, Clean Fuels from Coal II Symposium, Institute of Gas Technology, Chicago, 1975, p. 504.

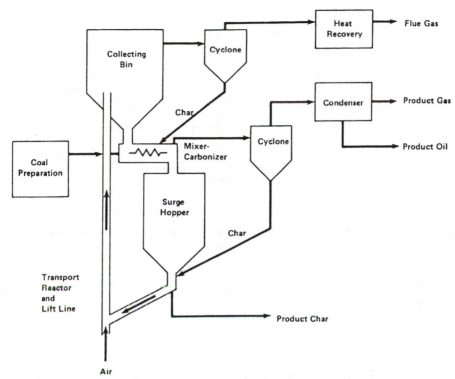

**Figure 3** Schematic of the Lurgi-Ruhrgas process. (Source: References 1 and 46.)

Cyclone removes the fines from the product gases and the liquid products are collected by a series of condensers. The liquid products are hydrotreated to yield stable products. The high gas yield is due to the relatively long residence time, and the gaseous products include both primary and secondary products. The term "secondary" is used here to indicate that they are not directly formed from the coal but rather by thermal decomposition of other primary products.

### 4.2.4 Occidental Flash Pyrolysis Process

A schematic of the Occidental flash pyrolysis process is given in Figure 4 [1]. In this process, hot recycle char provides the heat for the flash pyrolysis of pulverized coal in an entrained flow reactor at a temperature not to exceed 760°C. The process operates with a short residence time, thereby increasing the coal throughput and also increasing the production of liquid products. As in other processes, cyclones remove fine char particles from the pyrolysis overhead before quenching in the two-stage collector system [1]. The first stage consists of quenching at approximately 99°C to remove the majority of heavier hydrocar-

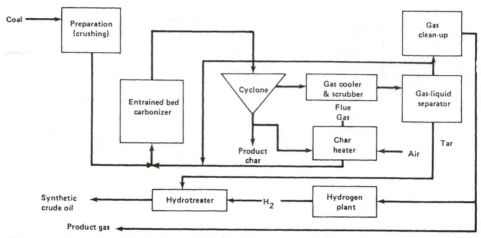

**Figure 4** Schematic of the Occidental flash pyrolysis process. (Source: References 1 and 46.)

bons; the second stage is for quenching at approximately 25°C to cause water and light oils to be removed.

### 4.2.5 Clean Coke Process

A schematic of the clean coke process is shown in Figure 5 [1]. The process involves feeding oxidized clean coal into a fluidized bed reactor at temperatures up to 800°C, whereupon the coal reacts to produce tar, gas, and low-sulfur char. Alternatively, the coal can be processed by noncatalytic hydrogenation at 455–480°C and pressures of up to 5000 psi of hydrogen. The liquid products from both the carbonization and hydrogenation stages are combined for further processing to yield synthetic liquid fuels.

### 4.2.6 Coalcon Process

A schematic of the Coalcon process is shown in Figure 6 [1]. This process uses a dry, noncatalytic fluidized bed of coal particles suspended in hydrogen gas. Hot, oxygen-free flue gas is used to heat the coal to approximately 325°C and also to carry the coal to a feed hopper. A fractionator is employed to subdivide the overhead stream into (1) gases ($H_2$, CO, $CO_2$, and $CH_4$), (2) light oil, (3) heavy oil, and (4) water. Most of the char is removed from the bottom of the reactor, quenched with water, and cooled [1]. The char can then be used as a feed to a Koppers-Totzek gasifier and reacted with oxygen and steam to produce hydrogen for the process [1].

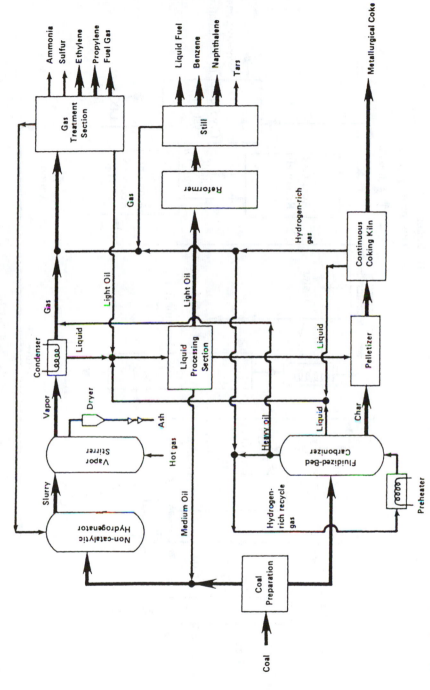

**Figure 5** Schematic representation of the clean coke process. (Source: Reference 1.)

173

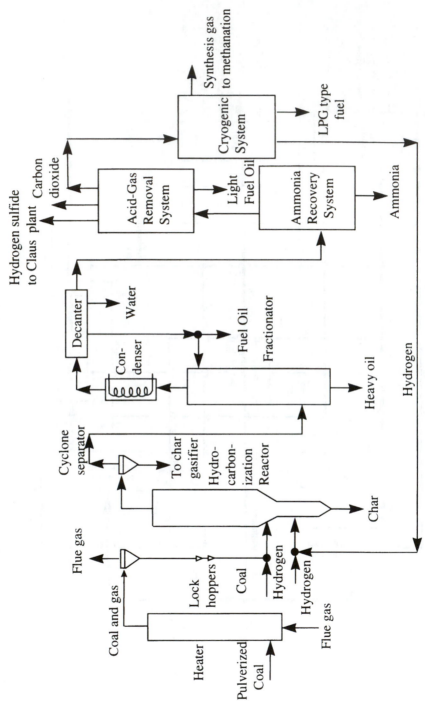

**Figure 6**  Schematic representation of the Coalcon process. (Source: Reference 1.)

174

## 4.3 DIRECT LIQUEFACTION OF COAL

Direct liquefaction of coal is defined as hydroliquefaction, to distinguish it from pyrolysis and coprocessing. Direct liquefaction may be categorized as single stage or two stages. In two-stage processes, the coal is first hydrogenated in a "liquid-phase" stage that transforms it into a deashed, liquid product, and then in a second "vapor-phase" hydrogenation stage the liquid products are catalytically converted to clean, light distillate fuels. Direct liquefaction has a long history and various processes have been successfully operated on large scales. Processes that are ready for demonstration or commercialization include H-Coal, SRC-I, SRC-II, EDS, ITSL, CC-ITSL, and CTSL. In this section, several significant direct coal liquefaction processes are reviewed.

### 4.3.1 Bergius-IG Hydroliquefaction Process

The Bergius process was operated successfully in Germany before and during World War II and was a two-stage process [2]. This process is currently not in operation but contributed greatly to the development of catalytic coal liquefaction technology. The process involves the catalytic conversion of coal (slurried with heavy oil) in the presence of hydrogen and an iron oxide catalyst at 450–500°C and 3000–10,000 psi. The products were usually separated into light oils, middle distillates, and residuum. These oils except residuum were catalytically cracked to motor fuels and light hydrocarbons in a vapor-phase hydrogenation stage, which is the second stage of the process. Some argue that the severe conditions used in the original process might have been due to the fact that German coals are much more difficult to liquefy than U.S. coals. The residence time for catalytic conversion was about 80–85 minutes, which was quite long, and hydrogen consumption was also quite significant, approximately 11% by mass of the daf coal.

### 4.3.2 H-Coal Process

The H-Coal process is a direct catalytic coal liquefaction process developed in 1963 by Hydrocarbon Research, Inc. (HRI). The process development proceeded through several stages from conceptual stages, to bench scale (25 lb/day), to process development unit (PDU) (3 tons/day), and to a pilot plant in Catlettsburg, Kentucky (200–600 tons/day). This pilot plant project was a $300 million project funded by the U.S. DOE, the Commonwealth of Kentucky, EPRI, Mobil, AMOCO, CONOCO, Ruhrkohle, Ashland Oil, SUN Oil, Shell, and ARCO.

A schematic of the H-Coal process is shown in Figure 7 [1, 3]. Pulverized coal, recycle liquids, hydrogen, and catalyst are brought together in the ebullated bed reactor to convert coal into hydrocarbon liquids and gaseous products. The catalyst particles are 0.8- to 1.5-mm-diameter extrudates and pulverized coal is

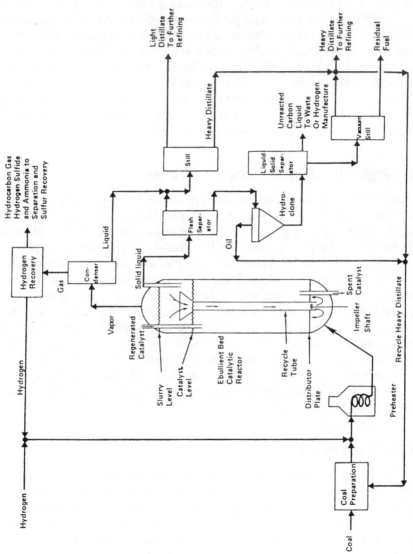

**Figure 7** Schematic representation of the H-Coal process. (Source: Reference 1.)

**Table 4 Product compositions from the H-Coal process**

| | Illinois | | Wyodak synthetic crude |
|---|---|---|---|
| | Synthetic crude | Low-sulfur fuel oil | |
| Product (wt. %) | | | |
| C$_1$–C$_3$ hydrocarbons | 10.7 | 5.4 | 10.2 |
| C$_4$–200°C distillate | 17.2 | 12.1 | 26.1 |
| 200–340°C distillate | 28.2 | 19.3 | 19.8 |
| 340–525°C distillate | 18.6 | 17.3 | 6.5 |
| 525°C + residual oil | 10.2 | 29.5 | 11.1 |
| Unreacted ash-free coal | 5.2 | 6.8 | 9.8 |
| Gases | 15.0 | 12.8 | 22.7 |
| Total (100.0 + H$_2$ reacted) | 104.9 | 103.2 | 106.2 |
| Conversion (%) | 94.8 | 93.2 | 90.2 |
| Hydrogen consumption (scf/ton) | 18,600 | 12,200 | 23,600 |

*Source:* Reference 1.

−60 mesh. Coal slurried with recycle oil is pumped to a pressure of up to 3000 psi and introduced into the bottom of the ebullated bed reactor. The process conditions, 345–370°C, may be altered appropriately to produce different product slates [1]. Table 4 shows product compositions from the H-Coal process [1].

Advantages and disadvantages of the H-Coal process are summarized in Table 5. Like all single-stage processes, H-Coal is best suited for volatile bituminous coal.

### 4.3.3 Solvent Refined Coal (SRC-I)

In 1962 the Spencer Chemical Co. began to develop a process that was later taken up by Gulf, which in 1967 designed a 50 ton/day SRC pilot plant at Fort Lewis, Washington [3]. The plant was operated in the SRC-I mode from 1974 until late 1976. In 1972 Southern Services Co. (SSC) and Edison Electric Insti-

**Table 5 Advantages and shortcomings of H-Coal**

| Advantages | Disadvantages |
|---|---|
| 1. Coal dissolution and upgrading to distillates accomplished in one reactor. | 1. High reaction temperature results in high gas yields (12–15%). |
| 2. Products have a high H/C ratio and low heteroatom content. | 2. Hydrogen consumption is relatively high. |
| 3. High throughput of coal due to fast reaction rates. | 3. Product contains considerable vacuum gas oil (345–525°C, bp), which is difficult to upgrade by standard refinery processes. |
| 4. Ash removed by vacuum distillation, followed by gasification of vacuum tower bottoms to generate hydrogen. | 4. Due to vacuum gas oil, it has utility solely as a boiler fuel. |

tute (EEI) designed and constructed a 6 ton/day SRC-I pilot plant at Wilsonville, Alabama.

The principal objective of the original SRC-I process was a solid boiler fuel with a melting point of about 150°C and a heating value of 16,000 Btu/lb. In the interest of enhancing commercial viability, the product slate was expanded to include liquids that were products of a coker/calciner, an expanded-bed hydrocracker, and a naphtha hydrotreater [3].

SRC-I is a thermal liquefaction process in which solvent, coal, and hydrogen are reacted in a "dissolver" reactor to produce a nondistillable resid, which upon deashing can be used as a clean boiler fuel. Reaction conditions are slightly less severe than for H-Coal. The absence of a catalyst diminishes the hydrogenation rates and the resid has an H/C ratio about the same as that of the coal feed. Again, this process is also ideally suited for bituminous coals, especially those containing high concentrations of pyrite. The pyrite is considered to be the liquefaction "catalyst." A schematic of the SRC-I process is shown in Figure 8 [1]. Advantages and disadvantages of SRC-I are given in Table 6.

Nondistillable SRC-I resid products cannot be deashed by vacuum distillation. Extraction-type separation processes were developed specifically for this process [3]. Typical of these is Kerr-McGee's critical solvent deashing (CSD).

### 4.3.4 Exxon Donor Solvent (EDS) Process

A schematic of the EDS process is shown in Figure 9 [1].

The EDS process utilizes a noncatalytic hydroprocessing step for the liquefaction of coal to produce liquid hydrocarbons. Its salient feature is the hydrogenation of the recycle solvent, which is used as a donor of hydrogen to the slurried coal in a high-pressure reactor. This process is also considered to be a single-stage process, because both coal dissolution and resid upgrading take place in one thermal reactor. The recycle solvent, however, is catalytically hydrogenated in a separate fixed-bed reactor [3]. Reaction conditions are similar to those of SRC-I and H-Coal.

EDS solvent must be well hydrogenated to be an effective hydrogen donor. The recycle solvent "donates" hydrogen to effect rapid hydrogenation of primary liquefaction products; thermal hydrogenation and cracking follow to produce distillates [3]. The product quality is slightly inferior to that of H-Coal, because of the absence of a hydrotreating catalyst. Distillate yields are also lower than for the H-Coal process. Overall, process economics are about equal despite the less expensive thermal reactor and the simple solids removal process.

### 4.3.5 SRC-II Process

The SRC-II process uses direct hydrogenation of coal in a reactor at high pressure and temperature to produce liquid hydrocarbon products instead of the solid

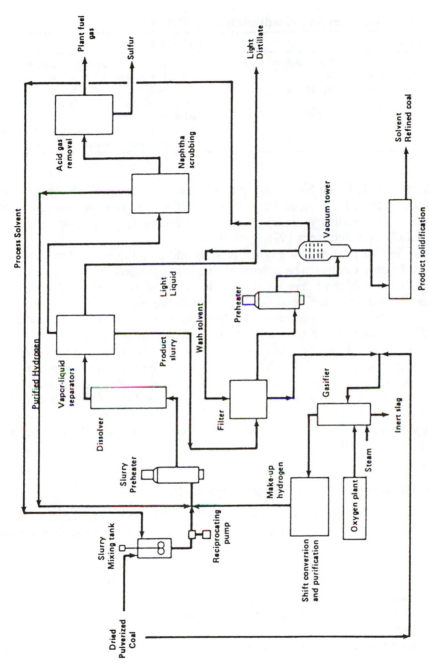

**Figure 8** Schematic representation of the SRC-I process. (Source: Reference 1.)

179

**Table 6 Advantages and disadvantages of SRC-I**

| Advantages | Disadvantages |
| --- | --- |
| 1. A good boiler fuel with high heating value. | 1. Distillate solvent was of poor quality. |
| 2. Reaction conditions less severe. | 2. Solvent is frequently incorporated into the |
| 3. Noncatalytic and easy to operate. | resid product. |
| | 3. Because of 2, solvent balance cannot be achieved. |
| | 4. Nondistillable SRC-I resid cannot be recovered by vacuum distillation. |

products of SRC-I. The 50 ton/day pilot plant at Fort Lewis, Washington, which operated in the SRC-I mode from 1974 to 1976, was modified to run in the SRC-II mode, producing liquid products for testing [3]. The pilot plant was successfully operated from 1978 until 1981.

The SRC-II process is a thermal process and uses the mineral matter in the coal as the only catalyst. The mineral matter concentration in the reactor is kept high by recycle of the heavy oil slurry. The recycled use of mineral matter and the more severe reaction conditions distinguish the SRC-II operation from the SRC-I process and also account for the lighter products. The net product is −540°C distillate, which is recovered by vacuum distillation. The vacuum bottoms including ash are sent to gasification to generate process hydrogen. The SRC-II process is limited to coals that contain "catalytic mineral matter" and therefore excludes all lower rank coals and some bituminous coals. Pulverized coal of particle size smaller than 0.125 in. is used, and the solvent-to-coal ratio is 2.0 for SRC-II, compared with 1.5 for SRC-I. The liquid product quality is inferior to that of the H-coal process. A schematic of the SRC-II process is shown in Figure 10.

### 4.3.6 Nonintegrated Two-Stage Liquefaction (NTSL)

Even though single-stage processes like EDS, SRC-I, SRC-II, and H-Coal are technologically sound, their process economics suffer for the following reasons:

1. The reaction severity is high, with temperatures of 430–460°C and liquid residence times of 20–60 minutes. These severe operating conditions were considered necessary to achieve coal conversions of over 90% (to THF or quinoline solubles).
2. Distillate yields are low, only about 50% of maf bituminous coals and even lower from subbituminous coal.
3. Hydrogen efficiency is low because of high yields of hydrocarbon gases.
4. The costs associated with the SRC-I process or the like are too high to produce a boiler fuel.

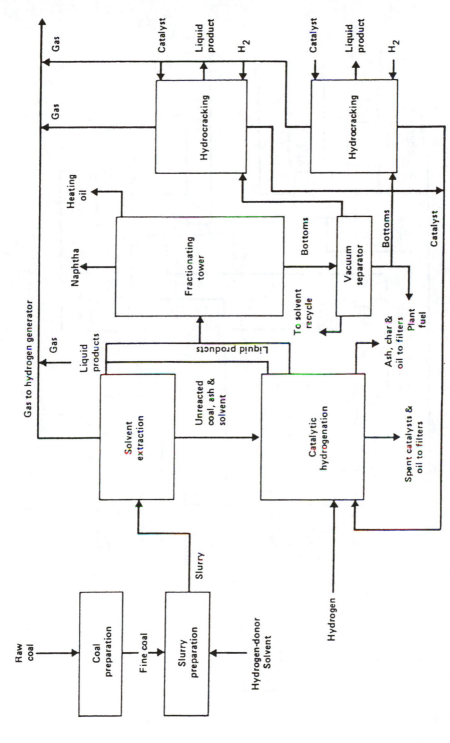

**Figure 9** Schematic representation of the Exxon Donor Solvent process. (Source: Reference 1.)

181

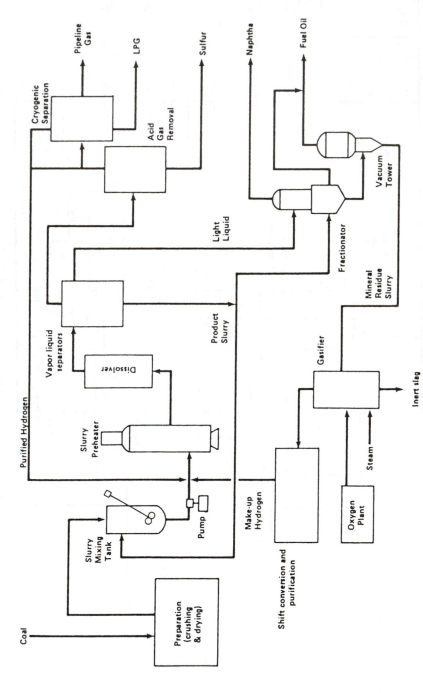

**Figure 10** Schematic representation of the SRC-II process. (Source: Reference 1.)

182

For these reasons, a coal liquefaction process is best applied to make higher value-added products, such as transportation fuels [3]. To produce higher value-added products from the SRC-I process, the resid must first be hydrocracked to distillate liquids. Efforts made by Mobil and Chevron on fixed bed hydrocracking were not entirely successful, because of plugging of the fixed bed ashes and rapid deactivation of the catalyst by coking.

The SRC-I resid was successfully hydrotreated by LC-Fining, a variation of ebullated-bed technology developed by Cities Services R&D [4]. As a result, hydrocracking was added to the SRC-I process to form nonintegrated two-stage liquefaction (NTSL). This unique name was given because the hydrocracking did not contribute solvent to the SRC-I part. The NTSL process was a combination of two separate processes, coal liquefaction and resid upgrading. A block diagram of NTSL operation is shown in Figure 11. Even with the addition of

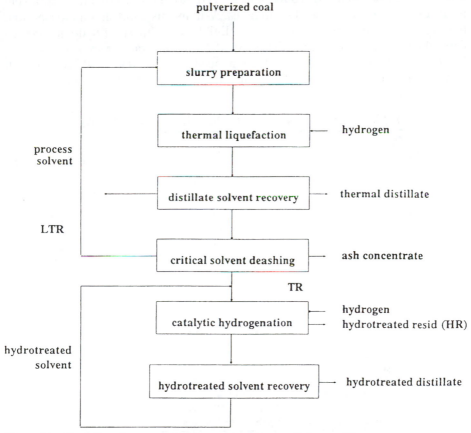

**Figure 11**   Block flow diagram of NTSL operation. (Source: Reference 3.)

the hydrocracking section, NTSL was an inefficient process with the shortcomings listed earlier. The SRC-I product is an unreactive feed to hydrocracking, requiring high temperature (over 430°C) and low space velocity for complete conversion to distillates. In order to keep the temperature and reactor size at reasonable levels, resid conversion was held below 80%. NTSL operation data for the Wilsonville facility are presented in Table 7. Yields were higher than for H-Coal, but hydrogen consumption was still high because of the extensive thermal hydrogenation in the SRC-I dissolver, which was renamed the thermal liquefaction unit (TLU) [3]. NTSL was short-lived, and a newer "integrated" approach was later developed.

## 4.3.7 Thermal Integrated Two-Stage Liquefaction (ITSL)

Thermal coal dissolution studies by Consol, Mobil, and Wilsonville in the late 1970s had shown that coal conversion to THF solubles is essentially complete in an extremely short time, 1–5 minutes. Within this short dissolution period, hydrogenation from the gas phase is negligible and almost all hydrogen comes from the solvent in the liquid phase [3]. If hydrogen transfer from the solvent is insufficient to satisfy the liquefaction needs, the product will have a high concentration of toluene insolubles, causing precipitation or plugging in the re-

**Table 7 NTSL at Wilsonville Facility (Illinois No. 6)**

| Operating conditions | |
|---|---|
| Run | 241CD |
| Configuration | NTSL |
| Catalyst | Armak |
| Thermal stage | |
|     Average reactor temperature, °F | 805 |
|     Coal space velocity, lb/hr ft$^3$ > 700°F | 20 |
|     Pressure, psig | 2170 |
| Catalytic stage | |
|     Average reactor temperature, °F | 780 |
|     Space velocity, lb feed/hr lb catalyst | 1.7 |
|     Catalyst age, lb resid/lb catalyst | 260–387 |

| Yields, weight percent maf coal | |
|---|---|
| C$_1$–C$_3$ gas | 7 |
| C$_4$+ distillate | 40 |
| Resid | 23 |
| Hydrogen consumption | 4.2 |
| Hydrogen efficiency, lb C$_4$+ Distillate/lb H$_2$ Consumed | 9.5 |
| Distillate selectivity, lb C$_1$–C$_3$/lb C$_4$+ Distillate | 0.18 |
| Energy content of feed coal rejected to ash concentrate, % | 20 |

*Source:* Reference 3.

actor or in downstream equipment. With a well-hydrogenated solvent, however, short-contact-time (SCT) liquefaction is the preferred thermal dissolution procedure because it eliminates the inefficient thermal hydrogenation inherent in SRC-I. Cities Services R&D successfully hydrocracked the SRC-I resid by LC-fining at relatively low temperatures of 400–420°C. Gas yield was low and hydrogen efficiency was high. A combination of this process with SCT makes good sense. The low-temperature LC-Fining provides the liquefaction solvent to the SCT first stage, so the two stages become integrated. This combination has the potential to liquefy coal to distillate products in a more efficient process than any of the single-stage processes [3].

**Lummus ITSL (1980–1984).** A combination of SCT and LC-Fining was made by Lummus in the ITSL process [5]. A flow diagram of the Lummus ITSL process is given in Figure 12. Coal is slurried with recycled solvent from LC-Fining and is converted to quinoline solubles in the SCT reactor. The resid is hydrocracked to distillates in the LC-Fining stage, where recycle solvent is also generated. The ash is removed by the Lummus Antisolvent Deashing (ASDA) process, which is similar to the deasphalting operation with petroleum. The net liquid product is either −340°C or −450°C distillate. The recycle solvent is hydrogenated +340°C atmospheric bottoms. It is the recycle of these full-range bottoms, including resid, that couples the two reaction stages and results in high yields of all distillate product [3].

Some of the features of the Lummus ITSL are summarized below:

1. The SCT reactor is actually the preheater for the dissolver in the SRC-I process, thus eliminating a long-residence-time, high-pressure thermal dissolution reactor.
2. Coal conversion in the SCT reactor was 92% of maf coal for bituminous coals and 90% for subbituminous coals.
3. Molecular hydrogen gas consumption was essentially zero, and the hydrogen transferred from the solvent was equivalent to 1.2–2.0% of the coal weight. Gaseous hydrocarbon yield was reduced to 1% for bituminous coal and to 5–6% for subbituminous coal.
4. The SCT resid was more reactive to hydrocracking than SRC-I resid.
5. The LC-Fining second reactor as a hydrotreater (HTR) accomplishes two principal tasks: (a) to make essentially all of the distillate product and (b) to generate recycle solvent capable of supplying the hydrogen required by the SCT reactor.
6. All distillate product was produced as a result of full recycle of unconverted resid to the first stage.
7. A second-stage HTR temperature of 400°C provides sufficient hydrogenation and cracking activity to accomplish both tasks.

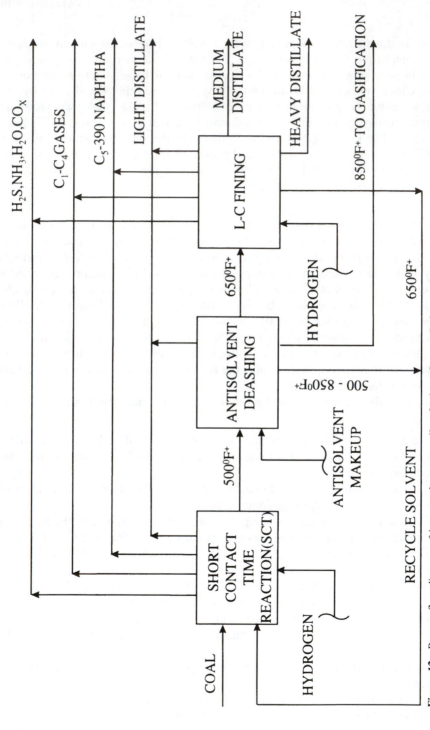

**Figure 12**  Process flow diagram of integrated two-stage liquefaction. (Source: Reference 3.)

8. Catalyst deactivation was much slower than other processes operated at higher temperatures.
9. The SCT resid was more reactive, not only for conversion to distillate but also for heteroatom removal. Product quality surpassed any achieved by the preceding processes. Chevron successfully refined the ITSL products for specification transportation fuels.
10. The ash was removed by antisolvent deashing (ASDA), which used a process-derived naphtha as antisolvent to precipitate the heaviest components of the resid and the solids.
11. The ASDA had the advantage of low pressure (70–100 psi) and low temperature (260–282°C).

Typical product yields for the Lummus ITSL process are given in Table 8 [3]. The product quality of the Lummus ITSL distillates is shown in Table 9 [3].

**Wilsonville ITSL (1982–1985).** The Advanced Coal Liquefaction R&D Facility at Wilsonville, Alabama, sponsored by the U.S. DOE, the Electric Power Research Institute (EPRI), and AMOCO, was operated by Catalytic Inc. under the management of Southern Company Services.

The Wilsonville facility began operations as a 6 ton/day single-stage plant for SRC-I in 1974. In 1978 a Kerr-McGee Critical Solvent Deashing (CSD) unit replaced the filtration equipment that had been used for solids removal from the SRC product. In 1981 an H-oil ebullated-bed hydrotreater was installed for upgrading the recycle solvent and product. In 1985 a second ebullated-bed reactor was added in the hydrotreater area to allow operation with close-coupled reac-

**Table 8  Lummus ITSL product yields**

| | lb/100 lb maf coal | |
| --- | --- | --- |
| | Illinois No. 6 | Wyodak |
| $H_2S$, $H_2O$, $NH_3$, $CO_x$ | 15.08 | 23.08 |
| $C_1$–$C_4$ | 4.16 | 7.30 |
| Total gas | 19.24 | 30.38 |
| $C_5$/390°F | 6.92 | 1.25 |
| 390/500°F | 11.46 | 8.49 |
| 500/650°F | 17.26 | 22.46 |
| 650/850°F | 23.87 | 21.36 |
| Total distillate product | 59.51 | 53.56 |
| Organics rejected with ash | 26.09 | 20.22 |
| Grand total | 104.84 | 104.16 |
| Chemical hydrogen consumption | 4.84 | 4.16 |
| Hydrogen efficiency, lb dist./lb $H_2$ | 12.28 | 12.86 |
| Distillate yield, bbl/ton maf | 3.52 | 3.08 |

*Source:* Reference 3.

**Table 9 Lummus ITSL distillate product quality (Illinois No. 6)**

|  | °API | C | H | O | N | S | HHV (Btu/lb) |
|---|---|---|---|---|---|---|---|
| Naphtha |  |  |  |  |  |  |  |
| NTSL | 36.8 | 86.79 | 11.15 | 1.72 | 0.18 | 0.16 | 19,411 |
| ITSL | 45.4 | 86.01 | 13.16 | 0.62 | 0.12 | 0.09 | 20,628 |
| Light distillate (390–500°F) |  |  |  |  |  |  |  |
| NTSL | 15.5 | 88.62 | 9.51 | 1.50 | 0.28 | 0.09 | 18,673 |
| ITSL | 22.9 | 87.75 | 11.31 | 0.73 | 0.13 | 0.08 | 19,724 |
| Medium distillate (500–650°F) |  |  |  |  |  |  |  |
| NTSL | 7.5 | 90.69 | 8.76 | 0.27 | 0.25 | 0.03 | 18,604 |
| ITSL | 12.9 | 89.29 | 10.26 | 0.28 | 0.12 | 0.05 | 19,331 |
| Heavy distillate (650–850°F) |  |  |  |  |  |  |  |
| NTSL | −1.5 | 91.47 | 7.72 | 0.26 | 0.50 | 0.05 | 18,074 |
| ITSL | 1.8 | 90.77 | 8.47 | 0.45 | 0.23 | 0.08 | 18,424 |

*Source:* Reference 3.

tors. The integrated two-stage liquefaction (ITSL) configuration used at Wilsonville for bituminous runs is shown schematically in Figure 13. A distillate yield of 54% of maf coal was confirmed, as shown in Table 10.

Lummus enhanced the ITSL process by increasing the distillate yield by placing the deasher after the second stage, with no detrimental effect of ashy feed on catalyst activity. This enhanced process is called Reconfigured Two-Stage Liquefaction (RITSL), as shown in Figure 14. The enhancements were experimentally confirmed at Wilsonville.

With the deasher placed after the second-stage reactor and the two stages operating at about the same pressure, the two reactors were close coupled to minimize holding time between the reactors and eliminate pressure letdown and repressurizing between stages [3]. This enhancement was called close-coupled ITSL (or CC-ITSL).

### 4.3.8 Catalytic Two-Stage Liquefaction (CTSL)

Since 1985, all PDU programs in the United States have used two catalyst stages. The two-stage liquefaction was found much more effective than the single stage.

**HRI's CTSL process.** In 1982 Hydrocarbon Research, Inc. (HRI) initiated the development of a catalytic two-stage concept, overcoming the drawbacks of H-coal, which is inherently a high-temperature catalytic process. The first-stage temperature was lowered to 400°C to more closely balance hydrogenation and cracking rates and to allow the recycle solvent to be hydrogenated in situ to facilitate hydrogen transfer to coal dissolution. The second stage was operated at higher temperatures (435–440°C) to promote resid hydrocracking and generate an aromatic solvent, which is then hydrogenated in the first stage [3]. The lower first-stage temperature provides better overall management of hydrogen con-

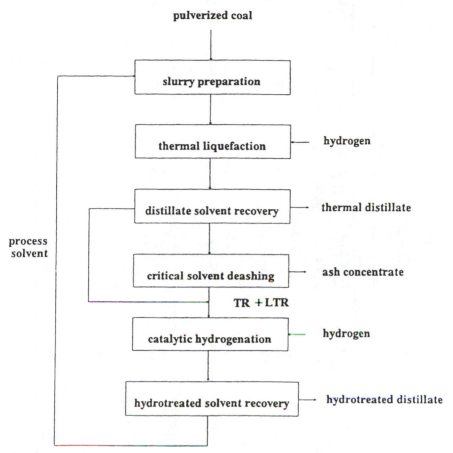

**Figure 13** Block flow diagram of integrated two-stage liquefaction (ITSL).

sumption and reduced hydrocarbon gas yields [3, 6]. A schematic of this process is shown in Figure 15 [3].

The HRI's CTSL had three major changes in comparison with the H-Coal process. The first was the two-stage processing; the second was incorporation of a pressure filter to reduce resid concentration in the reject stream (filter cake) below the 45–50% in the vacuum tower bottoms of the H-Coal process; the third was in the catalyst. The H-Coal process used a cobalt-molybdenum (CoMo)-on-alumina catalyst, American Cyanamid 1442 B, that had been effective in hydrocracking petroleum resids. In coal liquefaction, hydrogenation must occur first, followed by thermal cracking of hydroaromatics, whereas in petroleum applications the contrary is true. The H-Coal catalyst was found unsuitable due to its porosity distribution, which was designed for smaller molecules. For CSTL, the

## Table 10 ITSL at Wilsonville facility (Illinois No. 6)

Operating conditions

| | 241CD NTSL Armak | 7242BC ITSL Shell 324M | 243JK/244B ITSL Shell 324M | 247D RITSL Shell 324M | 250D CC-ITSL Amocat IC | 250G(a) CC-ITSL Amocat IC |
|---|---|---|---|---|---|---|
| **Run** | | | | | | |
| **Configuration** | | | | | | |
| **Catalyst** | | | | | | |
| **Thermal stage** | | | | | | |
| Average reactor temperature, °F | 805 | 860 | 810 | 810 | 824 | 829 |
| Coal space velocity, lb/hr ft$^3$ > 700°F | 20 | 43 | 28 | 27 | 20 | 20 |
| Pressure, psig | 2170 | 2400 | 1500–2400 | 2400 | 2500 | 2500 |
| **Catalytic stage** | | | | | | |
| Average reactor temperature, °F | 780 | 720 | 720 | 711 | 750 | 750 |
| Space velocity, lb feed/hr lb catalyst | 1.7 | 1.0 | 1.0 | 0.9 | 2.08 | 2.23 |
| Catalyst age, lb resid/lb catalyst | 260–387 | 278–441 | 380–850 | 446–671 | 697–786 | 346–439 |
| | | | | | | |
| **Yields, weight percent maf coal** | | | | | | |
| C$_1$–C$_3$ gas | 7 | 4 | 6 | 6 | 7 | 8 |
| C$_4$+ distillate | 40 | 54 | 59 | 62 | 64 | 63 |
| Resid | 23 | 8 | 6 | 3 | 2 | 5 |
| Hydrogen consumption | 4.2 | 4.9 | 5.1 | 6.1 | 6.1 | 6.4 |
| Hydrogen efficiency, lb C$_4$+ distillate/lb H$_2$ consumed | 9.5 | 11 | 11.5 | 10.2 | 10.5 | 9.8 |
| Distillate selectivity, lb C$_1$–C$_3$/lb C$_4$+ distillate | 0.18 | 0.07 | 0.10 | 0.10 | 0.11 | 0.12 |
| Energy content of feed coal reject to ash concentrate, % | 20 | 24 | 20–23 | 22 | 23 | 16 |

*Source:* Reference 3.

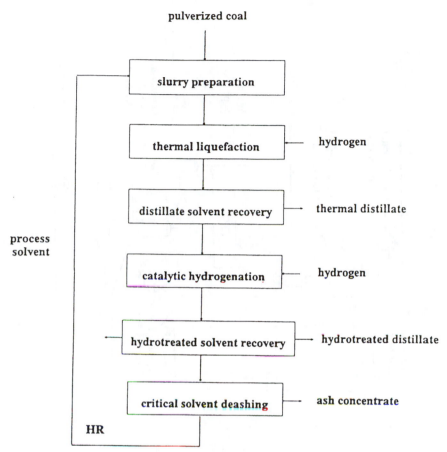

pulverized coal

slurry preparation

thermal liquefaction — hydrogen

distillate solvent recovery — thermal distillate

process
solvent

catalytic hydrogenation — hydrogen

hydrotreated solvent recovery — hydrotreated distillate

critical solvent deashing — ash concentrate

HR

**Figure 14** Block flow diagram of RITSL operation. (Source: Reference 3.)

H-Coal catalyst was replaced by a nickel-molybdenum (NiMo) catalyst of a bimodal pore distribution with larger micropores (115–125Å) as opposed to 60–70Å for H-Coal catalyst. The nickel promoter is also more active for hydrogenation than cobalt. Table 11 shows a comparison of H-Coal and HRI CTSL.

As shown, the two catalytic reaction stages produce a liquid with low heteroatom concentrations and a high H/C ratio, thus making the product closer to petroleum than any coal liquids made by earlier processes.

**Wilsonville CTSL.** A second ebullated-bed reactor was added at the Wilsonville Advanced Coal Liquefaction Facility in 1985. Since then, the plant has been operated in the CTSL mode. As in ITSL, Wilsonville preferred to have most of the thermal cracking take place in the first reactor and solvent hydrogenation in the second reactor [3]. Therefore, the first reactor was at a higher temperature

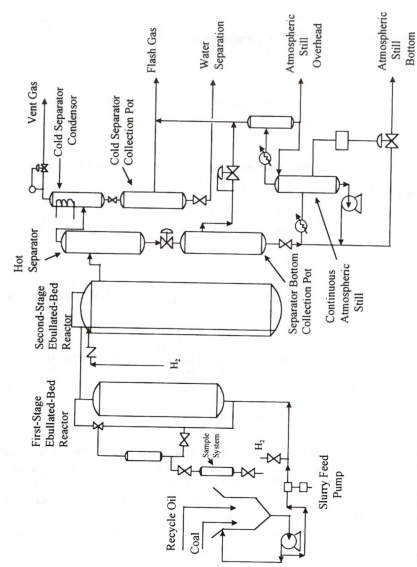

**Figure 15** Schematic of HRI CTSL. (Source: Reference 3.)

**Table 11 CTSL demonstration run comparison with H-Coal (Illinois No. 6 coal)**

| Process | H-Coal PDU-5 | CTSL run (227-20) | CTSL run (227-47) |
|---|---|---|---|
| $C_1$–$C_3$ | 11.3 | 6.6 | 8.6 |
| $C_4$–390°F | 22.3 | 18.2 | 19.7 |
| 390–650°F | 20.5 | 32.6 | 36.0 |
| 650–975°F | 8.2 | 16.4 | 22.2[1] |
| 975°F+ oil | 20.8 | 12.6 | 2.7[1] |
| Hydrogen consumption | 6.1 | 6.3 | 7.3 |
| Coal conversion, wt. % maf | 93.7 | 94.8 | 96.8 |
| 975°F+ conversion, wt. % maf | 72.9 | 82.2 | 94.1[1] |
| $C_4$–975°F, wt. % maf | 51.0 | 67.2 | 77.9[1,2] |
| Hydrogen efficiency | 8.4 | 10.7 | 10.7 |
| $C_4$+ distillate product quality | | | |
|    EP, °F | 975.0 | 975.0 | 750.0 |
|    °API | 26.4 | 23.5 | 27.6 |
|    % hydrogen | 10.63 | 11.19 | 11.73 |
|    % nitrogen | 0.49 | 0.33 | 0.25 |
|    % sulfur | 0.02 | 0.05 | 0.01 |
|    bbl/ton | 3.3 | 4.1 | 5.0 |

*Source:* Reference 3.

(426–438°C) and the second reactor was kept at a lower temperature, 404–424°C. A flow diagram of Wilsonville CTSL is shown in Figure 16 [3]. Run data of Wilsonville CTSL are summarized in Table 12. Distillate yields of up to 78% and organic rejection reduced to 8–15% were achieved at Wilsonville operating at over 4 tons of coal per day.

### 4.3.9 Evolution of Liquefaction Technology

An excellent review by COLIRN panel assessment [3] was published by the U.S. Department of Energy. Substantial technological innovations and enhancements have been realized for the past several decades, especially in the process configurations and in the catalysts. Table 13 summarizes the history of process development improvements in the form of yields and distillate quality [3, 7].

Distillates yields have increased from 41 to 78%, resulting in equivalent liquid yields of about 5 bbl/ton of maf bituminous coal. The distillate quality was comparable to or better than that of No. 2 fuel oil with a good hydrogen content and low heteroatom content.

## 4.4 INDIRECT LIQUEFACTION OF COAL

The indirect liquefaction of coal involves the production of synthesis gas mixture from coal as a *first* stage and the subsequent catalytic production of hydrocarbon

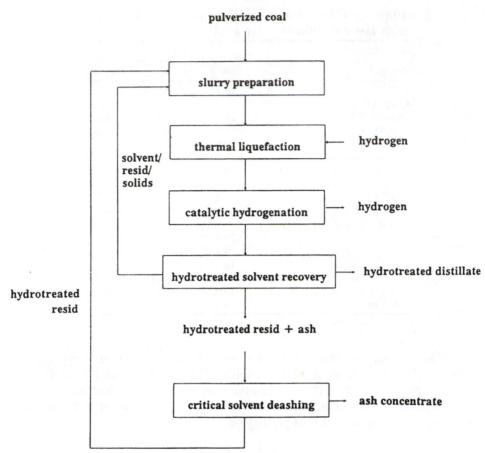

**Figure 16** Block flow diagram of CSTL operation with solids recycle at Wilsonville. (Source: Reference 3.)

fuels and oxygenates from the synthesis gas as a *second* stage. Indirect liquefaction can be classified into two principal areas:

1. Conversion of syngas to light hydrocarbon fuels via Fischer-Tropsch synthesis
2. Conversion of syngas to oxygenates such as methanol, higher alcohols, dimethyl ether, and other ethers

An in-depth review of indirect liquefaction may be found in the 1987 DOE-sponsored COGARN study report entitled Coal Gasification: Direct Application and Synthesis of Chemicals and Fuels, A Research Needs Assessment [8].

**Table 12 CTSL at Wilsonville Facility**

| Operating conditions | | | |
|---|---|---|---|
| Run no. | 253A | 254G | 251-IIIB |
| Coal | Illinois No. 6 | Ohio 6[a] | Wyodak |
| Catalyst | Shell 317 | Shell 317 | Shell 324 |
| First stage | | | |
|   Average reactor temperature, °F | 810 | 811 | 826 |
|   Inlet hydrogen partial pressure, psi | 2040 | 2170 | 2510 |
|   Feed space velocity, lb/hr/lb catalyst | 4.8 | 4.3 | 3.5 |
|   Pressure, psig | 2600 | 2730 | 2600 |
|   Catalyst age, lb resid/lb catalyst | 150–350 | 1003–1124 | 760–1040 |
| Second stage | | | |
|   Average reactor temperature, °F | 760 | 790 | 719 |
|   Space velocity, lb feed/hr lb catalyst | 4.3 | 4.2 | 2.3 |
|   Catalyst age (lb resid/lb catalyst) | 100–250 | 1166–1334 | 371–510 |
| Yield, weight percent maf coal | | | |
|   $C_1$–$C_3$ gas | 6 | 8 | 11 |
|   $C_4$+ distillate | 70 | 78 | 60 |
|   Resid | −1 | −1 | +2 |
|   Hydrogen consumption | 6.8 | 6.9 | 7.7 |
| Hydrogen efficiency, | | | |
|   lb $C_4$+ distillate/$H_2$ consumed | 10.3 | 11.3 | 7.8 |
| Distillate selectivity, | | | |
|   lb $C_1$–$C_3$/lb $C_4$+ distillate | 0.08 | 0.11 | 0.18 |
| Energy content of feed coal rejected to | | | |
|   ash concentrate, % | 20 | 10 | 15 |

[a]Approximately 6% ash.
*Source:* Reference 3.

## 4.4.1 Fischer-Tropsch Synthesis for Liquid Hydrocarbon Fuels

The Fischer-Tropsch process is currently the only commercially operating process for indirect liquefaction of coal. The SASOL plant in South Africa has been in operation since 1956. A generalized flow sheet for the SASOL plant is shown in Figure 17.

**Reaction mechanism.** The Fischer-Tropsch synthesis (FTS) follows a simple polymerization reaction mechanism, the monomer being a $C_1$ species derived from CO. This polymerization reaction follows a molecular weight distribution described mathematically by Anderson [9], and Schulz [10] and Flory [11]. Recognizing these independent groups' work, the description of the FTS product distribution is usually referred to as the Anderson-Schulz-Flory (ASF) distribution, which is generally accepted and constantly used. The ASF distribution equation is written as

$$\log(w_n/n) = n \log x + \log[(1 - x)^2/x]$$

**Table 13 History of process development and performance for bituminous coal liquefaction**

| Process | Configuration | Distillate (wt. % maf coal) | Yield (bbl/t maf coal) | Distillate qty. (gravity °API) | Nonhydrocarbon (wt. %) | | |
|---|---|---|---|---|---|---|---|
| | | | | | S | O | N |
| SRC II (1982) | One-stage noncatalytic | 41 | 2.4 | 12.3 | 0.33 | 2.33 | 1.0 |
| H-Coal (1982) | One-stage catalytic | 52 | 3.3 | 20.2[a] | 0.20 | 1.0 | 0.50 |
| Wilsonville (1985), RITSL | Integrated two-stage, thermal catalytic | 62 | 3.8 | 20.2[b] | 0.23 | 1.9 | 0.25 |
| Wilsonville (1986), CTSL | Integrated close-coupled two-stage catalytic-catalytic | 70 | 4.5 | 26.8[b] | 0.11 | <1 | 0.16 |
| Wilsonville (1987), CTSL | Integrated close-coupled two-stage low-ash coal | 78 | 5.0 | [c] | [c] | [c] | [c] |
| HRI, CTSL (1987) | Catalytic-catalytic | 78 | 5.0 | 27.6 | 0.01 | — | 0.25 |

[a] Light product distribution, with over 30% of product in gasoline boiling range; less than heavy turbine fuel.
[b] Higher boiling point distribution, with 20% of product in gasoline fraction and over 40% turbine fuel range.
[c] API and elemental analysis data unavailable at this time.
*Source:* Reference 3.

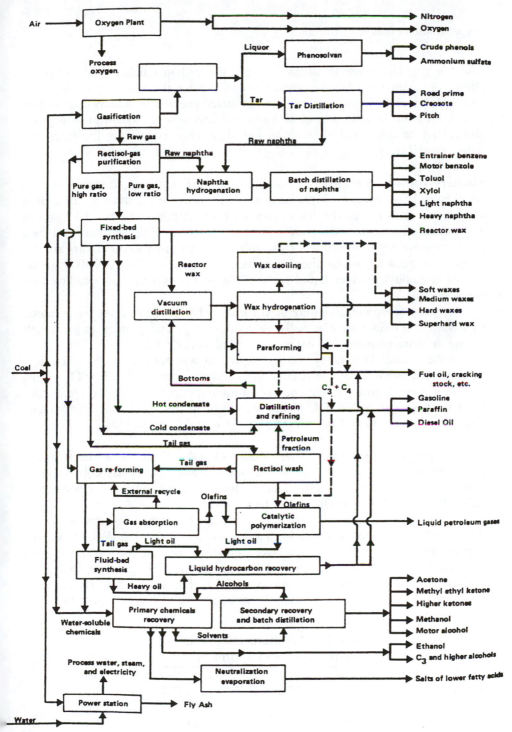

**Figure 17** Generalized flowsheet for the Sasol plant. (From J. C. Hoogendorn and J. M. Salomon, *Br. Chem. Eng.*, 2:238, 1957.

where $w_n$, $n$, and $x$ are the mass fraction, the carbon number, and the probability of chain growth. This equation can predict the maximum selectivity attainable by an F-T synthesis with an optimized process and catalyst, as shown in Table 14. From a linear plot of $\log(w_n/n)$ versus $n$, the chain growth probability can be computed either from the slope ($\log x$) or from the intercept, $\log((1 - x)^2/x)$.

**Fischer-Tropsch catalysis.** According to the ASF equation, catalysts with a small value of $x$ produce a high fraction of methane. On the other hand, a large $x$ value indicates the production of heavier hydrocarbons. The latest FTSs aim at producing high-molecular-weight products and very little methane and then cracking these high-molecular-weight substances to yield lower hydrocarbons. There have been numerous attempts to surpass or exceed the ASF distribution so that one could produce liquid fuels in yields that exceed those predicted by the ASF equation.

Inexpensive iron catalysts are used for the FTS. These catalyst are prepared by fusing iron oxides such as millscale oxides. In practice, either an alkali salt or one or more nonreducible oxides are added to the catalyst.

In the Fischer-Tropsch synthesis, removal of wax formed by the reaction is crucially important, as it can disable the catalytic activity. The SASOL plants furnish a major portion of the country's requirements for fuels and chemicals. Data on the existing SASOL plants are given in Table 15. An approximate distribution of products from SASOL-2 is given in Table 16 [12].

**Fischer-Tropsch processes other than SASOL.** There have been many publications and patents on other F-T synthesis catalysts, mainly cobalt, ruthenium, nickel, rhodium, and molybdenum. However, none of these has been commercially verified by SASOL or by other efforts.

Modern gasifiers, as discussed in Chapter 3, produce syngas with low (0.6–0.7) $H_2/CO$ ratios. Iron is known as a good water gas shift (WGS) catalyst; neither cobalt nor ruthenium is active. In the absence of the water gas shift reaction, the oxygen in CO is rejected as water so that a syngas with an $H_2/CO$ ratio of 2 is needed to produce olefins or alcohols. For paraffins, an $H_2/CO$ ratio larger than 2 is required. When water is formed in the F-T synthesis, it can react

**Table 14 Maximum attainable selectivities by F-T synthesis**

| Product | Max. selectivity (wt. %) |
|---|---|
| Methane | 100 |
| Ethylene | 30 |
| Light olefins ($C_2$–$C_4$) | 50 |
| Gasoline ($C_5$–$C_{11}$) | 48 |

**Table 15 SASOL plants**

| Plant | Start date | Coal (t/d) | Liquids (bbl/d) | Cost ($ billion) |
|-------|-----------|------------|-----------------|------------------|
| SASOL-1 | 1935 | 6,600 | 6,000 | — |
| SASOL-2 | 1981 | 30,000 | 40,000 | 2.9 |
| SASOL-3 | 1982 | 30,000 | 40,000 | 3.8 |

with CO to form more $H_2$ by the WGS reaction, so that syngas with a low $H_2/$CO ratio can still be used with these catalysts.

Most current efforts involve slurry F-T reactors. Mobil in the 1980s studied upgrading a total vaporous F-T reactor effluent over ZSM-5 catalyst. Shell, in 1985, announced its SMDS (Shell Middle-Distillate Synthesis) process for the production of kerosene and gas oil from natural gas [13]. This two-stage process involves the production of long-chain hydrocarbon waxes and subsequent hydroconversion and fractionation into naphtha, kerosene, and gas oil. UOP characterized F-T wax and its potential for upgrading [14]. Dow has developed molybdenum catalysts with a sulfur tolerance up to about 20 ppm. The catalyst system is selective for the synthesis of $C_2$–$C_4$ hydrocarbons, especially when promoted with 0.5–4.0 wt. % potassium [12].

Using a precipitated iron catalyst, the slurry F-T reactor operating with a finely divided catalyst suspended in an oil reactor medium has been shown to yield high single-pass syngas conversion with low $H_2/CO$ ratios [12–15]. A great many studies of three-phase slurry reactors have been published.

## 4.4.2 Conversion of Syngas to Methanol

The expanded production of methyl *tert*-butyl ether (MTBE), ethyl *tert*-butyl ether (ETBE), and *tert*-amyl methyl ether (TAME) as oxygenates has increased the global demand for methanol. The synthesis of methanol from syngas is a

**Table 16 Product distribution of SASOL-2**

| Product | Tons/year |
|---------|-----------|
| Motor fuels | 1,650,000 |
| Ethylene | 204,000 |
| Chemicals | 94,000 |
| Tar products | 204,000 |
| Ammonia (as N) | 110,000 |
| Sulfur | 99,000 |
| Total salable products | 2,361,000 |

*Source:* Reference 12.

well-established technology. Syngas can be produced from natural gas or coal. Synthesis gas is produced via gasification of coal and discussed in Chapter 3. The profitability of a methanol plant is determined on a case-by-case basis to account for location-specific factors such as energy expenditure, environmental impact, and capital cost. Methanol plants exist where there are large reserves of competitively priced natural gas or coal. The advent of methanol synthesis has given a boost to the value of natural gas. Conventional steam reforming produces hydrogen-rich syngas at low pressure. However, this process is well suited to the addition of carbon dioxide, which utilizes the excess hydrogen and hence increases the methanol productivity.

The global methanol demand increased about 8%/year from 1991 to 1995. The global methanol production capacity expanded by about 5.1 million metric tons or 23% in the same time period. Leading the growth is increased methanol demand for MTBE and formaldehyde production. The world methanol supply-demand balance is shown in Table 17.

All industrially produced methanol is made by the catalytic conversion of synthesis gas containing carbon monoxide, carbon dioxide, and hydrogen as the main components. Methanol productivity can be enhanced by synthesis gas enrichment with additional carbon dioxide to a certain limit [16]. However, a $CO_2$-rich environment increases catalyst deactivation, shortens catalyst lifetime, and produces water, which adversely affects the stability of the catalyst matrix and results in crystallite growth via hydrothermal synthesis phenomena [16]. Thus,

**Table 17  World methanol supply-demand balance**

| Demand | Forecast[a] | | | | |
| --- | --- | --- | --- | --- | --- |
| | 1991 | 1992 | 1993 | 1994 | 1995 |
| Formaldehyde | 7154 | 7242 | 7327 | 7512 | 7667 |
| DMT | 644 | 653 | 658 | 665 | 670 |
| Acetic acid | 1324 | 1407 | 1495 | 1517 | 1732 |
| MTBE | 3282 | 4075 | 4670 | 6616 | 8082 |
| Methyl methacrylate | 479 | 479 | 577 | 590 | 615 |
| Gasoline fuels | 541 | 517 | 490 | 428 | 480 |
| Solvents | 730 | 749 | 776 | 796 | 813 |
| Others | 3931 | 4027 | 4149 | 4372 | 4228 |
| Nontabulated countries | 205 | 210 | 215 | 220 | 225 |
| Total demand | 18290 | 19359 | 20357 | 22716 | 24500 |
| Nameplate capacity | 22681 | 23637 | 24172 | 27062 | 27822 |
| Capacity at 90% | 20413 | 21273 | 21755 | 24356 | 25040 |
| Percent utilization at nameplate | 80.60 | 81.90 | 84.20 | 83.90 | 88.10 |
| Percent utilization at 90% nameplate | 89.60 | 91 | 93.60 | 93.30 | 97.80 |

[a]All quantities are in 1000 mty.

Source: *Oil and Gas Journal*, p. 48, May 29, 1993.

a special catalyst has been designed to operate under high-$CO_2$ conditions. This catalyst's crystallites are located on energetic stable sites that lower the tendency to migrate. This stability also minimizes the influence of water formed on the catalyst matrix, which is only slightly affected. This catalyst preserves its higher activity due to a lower deactivation rate over long-term operations [16]. The basic reactions involved in methanol synthesis are:

$$CO_2 + 3H_2 = CH_3OH + H_2O \qquad \Delta H^{\circ}_{298} = -52.81 \text{ kJ/mol} \qquad (1)$$

$$CO + 2H_2 = CH_3OH \qquad \Delta H^{\circ}_{298} = -94.08 \text{ kJ/mol} \qquad (2)$$

$$CO_2 + H_2 = CO + H_2O \qquad \Delta H^{\circ}_{298} = 41.27 \text{ kJ/mol} \qquad (3)$$

Of the three reactions, only two are stoichiometrically independent. The chemical mechanism of the methanol synthesis over the $Cu/ZnO/Al_2O_3$ catalyst has been somewhat controversial [14, 17]. However, more evidence points toward the theory that methanol is produced predominantly via the $CO_2$ hydrogenation and the forward water gas shift reaction [14, 17].

$$CO_2 + 3H_2 = CH_3OH + H_2O$$

$$CO + H_2O = CO_2 + H_2$$

**Different reactor systems for methanol synthesis.** Conventionally, methanol is produced in a two-phase system—the reactants ($CO$, $CO_2$, and $H_2$) and products ($CH_3OH$ and $H_2O$) forming the gas phase and the catalyst being the solid phase. Two reactor types were most popular: an adiabatic reactor containing a continuous bed of catalyst with quenching by cold gas injections (the ICI system) and a multitubular reactor with an internal heat exchanger (the Lurgi system). Both systems are operated at temperatures from 483 to 553 K and low pressures, around 5–7 MPa, using $Cu/ZnO/Al_2O_3$ catalysts. A more detailed account of the various reactor systems utilized for methanol synthesis is given below:

- Fixed bed reactor
- Slurry reactor
- Trickle bed reactor
- Gas-solid-solid trickle flow reactor (GSSTFR)
- Reactor system with interstage product removal (RSIPR)

The reactor configuration most prevalent for methanol synthesis from syngas was the gas-phase fixed bed reactor. This refers to two-phase systems in which the reacting gas flows through a bed of catalyst particles. ICI introduced its low-pressure methanol process in low-tonnage plants. This process typically operates at temperatures between 220 and 280°C and pressures between 5 and 10 MPa. The exothermic nature of the methanol synthesis reaction makes temperature control difficult. The reactor operates adiabatically and the temperature rise in the bed is known to be high. Multibed quench reactors are necessary to counter this problem [16].

In 1975 Chem Systems developed the liquid-phase methanol synthesis process that was later studied by Air Products and Chemicals, Inc., and the University of Akron [16]. The process, called the LPMeOH process, uses a liquid entrained reactor configuration, a special type of slurry reactor. In slurry reactors, synthesis gas containing $CO$, $CO_2$, and $H_2$ passes upward into the reactor co-currently with the slurry, which absorbs the heat liberated during the reaction. The slurry is separated from the vapor and recirculated to the bottom of the reactor via a heat exchanger, where cooling occurs by steam generation. The reactor effluent gases are cooled to condense the products and any inert hydrocarbon liquid that may be vaporized. Methanol and the inert hydrocarbon liquid are immiscible and are separated by a decanter. The methanol stream produced is suitable for use directly or can be sent to a distillation unit to produce chemical-grade product. Due to the excellent reaction temperature control, high conversions of syngas per pass can be achieved, which reduces the recycle gas flow and compression requirement. A small catalyst particle size can be used, which prevents diffusional limitations. The near-isothermal temperature of the system allows favorable conditions to obtain the desired reaction kinetics. However, the slurry reactor has an upper limit for catalyst loading and hence operates at low space velocities. It has low conversion per pass due to the high extent of back-mixing and poses problems in separating the catalyst from the slurry due to catalyst attrition and agglomeration [18].

Methanol synthesis has also been investigated in a trickle bed reactor, and its performance has been compared with that of conventional reactors. In a trickle bed reactor, syngas and mineral oil flow concurrently over a fixed bed of catalyst. Trickle bed reactors tolerate high catalyst loadings and hence can operate at high space velocities. They also operate at nearly plug flow conditions, leading to higher conversions per pass. The fixed bed causes no catalyst attrition, which permits the use of costly catalysts. The trickle bed also consists of a liquid mineral oil phase, which acts as a heat sink to absorb the heat generated by the exothermic reaction. Thus, the trickle bed reactor incorporates the advantages of the gas-phase fixed bed reactor and the slurry reactor. However, small catalyst particle size causes a pressure drop in trickle bed reactors and large catalyst particle size causes mass transfer limitations due to intraparticle diffusion effects. Methanol synthesis reactions were carried out over a Cu/$ZnO$/$Al_2O_3$ catalyst in the temperature range 498–523 K and at a pressure of 5.2 MPa. The $H_2/(CO + CO_2)$ ratio varied at 0.5, 1, and 2, and the $CO/CO_2$ ratio was maintained at 9.0. Methanol productivities as high as 39.5 mol/(h kg of catalyst) were obtained, which were higher than the maximum methanol production rate of gas-phase methanol synthesis of 30 mol/(h kg of catalyst). Studies also showed higher methanol production rates in trickle bed reactors compared with slurry reactors at identical operating conditions. This is attributed to better gas-liquid-solid contact in trickle beds, combined with its close proximity to plug flow conditions as opposed to high extent of back-mixing in the slurry reactors [19].

Two novel converter systems were developed for the manufacture of methanol from synthesis gas: the gas-solid-solid trickle flow reactor (GSSTFR) and the reactor system with interstage product removal (RSIPR). In the GSSTFR system, a solid adsorbent that trickles countercurrently over the catalyst bed removes the product formed at the catalyst surface. Thus, the equilibrium limitation is alleviated, allowing high reactant conversions up to 100% in a single pass. The ratio of the adsorbent flow rate to the reactant feed flow rate determines the driving force for the forward reaction. As compared to the Lurgi low-pressure process, the amount of catalyst used in this reactor system is reduced by 40%, the raw materials consumption is reduced by 10%, and the consumption of cooling water decreases by 50%.

In the RSIPR system, a series of adiabatic or isothermal fixed bed reactors are used and the product is selectively removed in absorbers between the reactor stages. Here the methanol is removed selectively by absorption into a liquid such as tetraethylene glycol dimethyl ether (TEGDME) solution. This absorption takes place at the temperature level of the reactor inlet in order to achieve high energy efficiency. The total solvent/methanol ratio is around 10.0. The specific raw material consumption of the RSIPR is similar to or lower than that of the conventional system. The costs of the converter and the catalyst are reduced because higher specific reaction rates are achieved. The recycle loop is eliminated completely. Thus, the investments are significantly lower than in the case of the Lurgi low-pressure process. The catalyst amount decreases by 40%, and accordingly a smaller reactor volume is needed [20].

Methanol synthesis using fluidized bed technology exhibits better energy utilization, smaller reactor size, and higher per-pass conversion. In fixed bed reactors, the methanol productivity is limited by chemical equilibrium due to the reversible nature of the methanol synthesis reaction. In a fluidized bed reactor, the catalytic hydrogenation of carbon dioxide and carbon monoxide occurs in the dense phase, which contains the catalyst particles. The bubble phase is devoid of catalyst particles. The concentration gradients between the two phases induce the diffusion of methanol and water from the dense phase to the bubble phase and that of carbon dioxide in the opposite direction. Removal of methanol from the immediate vicinity of the catalyst particles enhances the rate of reaction and reactant conversion. The gas-solid trickle flow reactor (GSTF) and the gas-solid-solid trickle flow reactor (GSSTFR) have the advantage of shifting equilibrium in a favorable direction by removal of reaction products. However, these reactor types have the disadvantage of the presence of pore diffusion limitations. In the case of fluidized bed reactors, the problem of pore diffusion resistances is largely eliminated because of the small catalyst particle sizes used. Industrial fixed-bed pellets are of the order of 6–12 mm in diameter, whereas fluidizable catalyst particles can be smaller than 100 μm. Preliminary theoretical investigation shows that the fluidized bed reactor setup resulted in a CO conversion of about 72%, which is higher than the 63% attained in a fixed bed reactor setup. The methanol productivity is higher by 30% in the case of a fluidized bed reactor setup [21, 22].

**Different catalysts used in methanol synthesis.** In 1923 BASF found that methanol was the main product of carbon monoxide hydrogenation when mixed catalysts containing ZnO and $Cr_2O_3$ were used. In realizing this catalytic synthesis reaction, it was also found that the presence of iron or iron alloys might be crucially detrimental to the process, because they might react with CO to yield iron pentacarbonyl, $Fe(CO)_5$, which is a poison to the catalyst. This catalytic system was used until the early 1970s and is also known as Zn-based or ZnO-based catalysis. This catalyst is active at higher temperatures, 350–400°C, and the process is operated under high pressure, 250–350 atm, and called high-pressure methanol synthesis. The methanol synthesis over $ZnO/Cr_2O_3$ proceeds via CO hydrogenation.

In 1963 ICI announced an innovative process system using $Cu/ZnO/Al_2O_3$ catalysts, later called low-pressure synthesis. The major constituents of this catalytic system are Cu (reduced form of CuO) and ZnO on an $Al_2O_3$ support. The pressure and temperature conditions required are much lower than those for the high-pressure process. Typical operating conditions are 220–270°C and 50–100 atm [16, 23]. This catalyst has been found to be susceptible to sulfur and carbonyl poisoning, sintering, fouling, crystallite growth, and thermal aging. Nearly all of commercial methanol syntheses are carried out by low-pressure processes.

In 1975 Chem Systems developed the liquid-phase methanol (LPMeOH) process, which is based on the low pressure synthesis technology, except for the fact that the new process is carried out in an inert oil phase. The catalytic system used is $Cu/ZnO/Al_2O_3$ that is modified for slurry operation, that is, attrition resistant, finely powdered, and leaching resistant. S3.85 and S3.86 catalysts of BASF and EPJ-19 and EPJ-25 catalysts of United Catalysts Inc. have been developed for this process [16, 23, 24].

A slurry-phase concurrent synthesis of methanol using a potassium methoxide/copper chromite mixed catalyst has been developed. This process operates under relatively mild conditions such as temperatures between 100–180°C and pressures between 30 and 65 atm. The reaction pathway involves homogeneous carbonylation of methanol to methyl formate followed by heterogeneous hydrogenolysis of methyl formate to two molecules of methanol, the net result being the reaction of hydrogen with carbon monoxide to give methanol via methyl formate as shown below:

$$CH_3OH + CO = HCOOCH_3$$

$$HCOOCH_3 + 2H_2 = 2CH_3OH$$

$$2H_2 + CO = CH_3OH$$

The carbonylation of methanol to methyl formate is homogeneously catalyzed by alkali alkoxides like potassium methoxide, whereas the hydrogenolysis of methyl formate to methanol is carried out using a copper chromite catalyst. Thus, this concurrent synthesis involves two different catalysts in a single reactor. In this concurrent synthesis, methanol acts as a reactant and a solvent. The slurry-phase operation facilitates efficient heat dissipation, resulting in isothermal operation. The products are methanol and methyl formate with traces of water and dimethyl ether from the dehydration of methanol. Conversion per pass up to 90% and selectivity to methanol up to 94–99% have been achieved. This multistep single-stage methanol synthesis compares favorably with the more prevalent direct synthesis in terms of good heat transfer rates and high per-pass conversions [25]. This catalyst system has not been commercialized.

Studies of the kinetics of methanol synthesis were carried out to confirm the fact that methanol can be formed from either carbon monoxide or carbon dioxide [23]. Various syngas compositions were reacted over two catalysts, $Cu/ZnO$ and $Cu/ZnO/Al_2O_3$, at temperatures between 200 and 275°C and a pressure of 2.86 MPa. These studies were carried out in a differential packed bed reactor constructed of copper tubing. It was noted that when both $CO$ and $CO_2$ were present in the syngas, the rates from either reactant were additive, with a further contribution to methanol arising from conversion of $CO$ to $CO_2$ via the water gas shift reaction, utilizing water formed in the methanol synthesis. At temperatures above 225°C, formation of methanol from $CO_2$ decreased over the $Cu/ZnO$ catalyst, whereas a temperature greater than 250°C was necessary to affect the formation of methanol from $CO_2$ over the $CuO/ZnO/Al_2O_3$ catalyst [26].

Studies carried out for methanol synthesis using $CeCu_2$-derived catalysts show that methanol is synthesized by hydrogenation of $CO$ and not $CO_2$. They also show that the active catalyst surface is extensively covered with a hydrogen-deficient methanol precursor under steady-state conditions and that the cerium oxide surface or its interface with the copper crystallites is intimately involved in the synthesis process. The hydrogenation of this precursor of methanol itself is rate limiting. Transient increases in the exit concentrations of methanol were achieved by displacing methanol from the catalyst surface using pulses of $CO$, $CO_2$, and $O_2$ into $H_2$-containing feed gas streams. Experiments were carried out in a single-pass, fixed bed microreactor that consisted of a $\frac{1}{4}$-in. outer diameter stainless steel tube contained within an aluminum-bronze heater block. The CeCu2 alloy used as catalyst was prepared by high-vacuum, electron-beam melting [27]. This process has not been commercialized.

A new kind of copper catalyst was prepared and optimized for methanol synthesis. The catalyst was prepared using a definite $La_2Zr_2O_7$ support. In the presence of carbon dioxide and water, $Cu/La_2O_3$ catalyst is quickly deactivated by the formation of lanthanum carbonates and hydroxycarbonates. However, the formation of these species can be prevented by incorporating $La_2O_3$ in a stable $LaZr_2O_7$ pyrochlore. After optimization, up to 600 g methanol per gram of catalyst per hour can be produced over the $Cu/La_2Zr_2O_7$ catalysts. Experiments were carried out in a continuous-flow borosilicate glass reactor at standard conditions: 0.5 g catalyst, syngas fed at a flow rate of 4 L $g_{cat}^{-1}$ $hr^{-1}$. Varying feed compositions were used at 230, 250, 270, and 300°C and a pressure of 6 MPa. Copper supported on a well-defined $La_2Zr_2O_7$ stable pyrochlore was shown to be an active catalyst for methanol synthesis even in the presence of carbon dioxide. The deactivation of the catalysts is caused by a high carbon dioxide content. This is mainly due to the formation of lanthanum carbonates and hydroxycarbonates and their spreading onto the metal particles, in that way decreasing the catalytic activity. This carbon dioxide deactivating effect, which is detrimental to methanol synthesis, can be avoided in the presence of $ZrO_2$ or by promotion by ZnO [17]. This catalytic system is still in an investigational stage.

Another study investigated the hydrogenation of variable amounts of carbon dioxide and carbon monoxide in a synthesis gas mixture, using Raney copper, zinc oxide–promoted Raney copper, and a commercial coprecipitated Cu-ZnO-$Al_2O_3$ methanol synthesis catalyst. Similar patterns of carbon monoxide and carbon dioxide conversion at various $CO_2/CO$ ratios were observed for all the catalysts. The activity of the zinc-free Raney copper catalyst indicated that copper is capable of catalyzing the methanol synthesis reaction. However, zinc oxide–promoted catalysts show higher methanol activity than the Raney copper catalyst. This suggested that ZnO plays a role in enhancing the rate-limiting step by interaction with carbon dioxide and carbon monoxide molecules after initial adsorption on the copper surface [28]. The results are not necessarily in agreement with earlier findings by other investigators [16].

**Methanol synthesis process technology.** The technology for low-pressure methanol synthesis is quite well established. A number of companies now offer complete plants for the manufacture of methanol, starting from natural gas, naphtha, or coal as the feedstock. The catalytic systems are more or less standard and contain copper, zinc oxide, and alumina (or chromium oxide). The differences in these designs may be found mainly in the reactor configurations, use of different feedstocks, energy integrations, arrangements of process units, recycle scheme, design of shift converter, and so forth. Some of these major plant designs are discussed.

*The ICI low-pressure methanol synthesis process.* The first ICI (Imperial Chemical Industries, Ltd.) unit was linked to a steam-naphtha reformer, the naphtha feed to which was first thoroughly desulfurized using zinc oxide and cobalt catalysts. A schematic flow diagram of the first ICI [29] unit rated for 300 tons of refined methanol per day is shown in Figure 18. The plant used a syngas feed that contained hydrogen, carbon monoxide, carbon dioxide, and methane. A shift converter was used to adjust the carbon monoxide/carbon dioxide ratio. The feed was then compressed to 50 atm in a centrifugal compressor and fed into a quench-type converter, which was operated at 270°C. The product stream was then cooled, and the methanol was condensed out. A purge gas stream was recycled to a reformer to convert the accumulated methanol into synthesis gas.

As the synthesis reaction proceeded with a decreasing number of moles, low-pressure operation meant lower methanol concentration in the effluent stream and therefore higher recycle rates. However, high operating pressures would involve bulkier equipment and sturdier compressors, and a balance has to be achieved between energy costs and capital costs.

The design of the converter is very critical; it should enable uniform gas distribution, because the channeling of gases can cause local overheating and catalyst deactivation through sintering. Reactor thermal stability and temperature runaway are of utmost importance in designing the converter. A provision must be made to warm the converter during start-up. Special attention must also be paid to ensure that the catalyst can be easily loaded and unloaded from the reactor to reduce downtime. It would also be a good safety measure, as the reduced catalyst is pyrophoric.

The crude methanol that was produced by the low-pressure process was reported to contain water, dimethyl ether, esters, ketones, iron carbonyl, and higher alcohols. However, the impurities, although similar to those formed in the high-pressure process, were much less in quantity.

*ICI Katalco low-pressure methanol process.* The ICI Katalco low-pressure methanol process is divided into three main sections: synthesis gas preparation, methanol synthesis, and methanol purification. The principal feedstocks are natural gas, naphtha, heavier oil fractions, and coal. Synthesis gas is formed by the steam reforming of natural gas. The reformer effluent includes a mixture of hydrogen, carbon oxides, steam, and residual methane at 880°C and 20 bar. The methanol synthesis converter contains a copper-based catalyst and operates in the range 240 to 270°C. Loop operating pressure is in the range of 80 to 100 bar. The methanol synthesis reaction is limited by chemical equilibrium and the unreacted gases are recycled to the converter. Crude methanol from the separator contains water and low levels of by-products, which are removed using a two-column distillation system. The first column removes light ends such as ethers, esters, acetone, and lower hydrocarbons. The second removes water, higher alcohols, and higher hydrocarbons. Blends of gasoline with relatively low levels of methanol have been tested extensively worldwide. Blends of gasoline and methanol, plus high-molecular-weight alcohols, which are added to assist in dissolving the methanol, have been used extensively. Methanol-gasoline blends are subject to phase separation due to the high Reid vapor pressure, brought

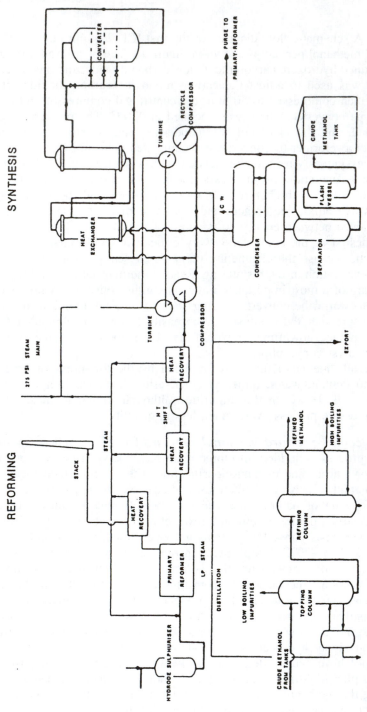

**Figure 18** Schematic of the first ICI methanol synthesis process. (Source: Reference 29.)

about by the inadvertent addition of small amounts of water. Concern over possible deleterious effects of methanol on engines and fuel systems in automobiles not designed for methanol use has also been a negative factor for methanol acceptance. A schematic of the process flowchart is given in Figure 19.

*The Lurgi low-pressure methanol synthesis process.* In the process offered by the Lurgi Corporation [30] for the synthesis of methanol, the converter or synthesis reactor is operated at temperatures ranging from 250 to 260°C. The operating pressure is between 50 and 60 bar. The design of the converter is different from that of ICI. In the ICI design, the catalyst forms a bed in which gas is introduced at various locations along the length of the bed in order to get a uniform temperature distribution. The Lurgi design envisages a shell-and-tube reactor where the tubes are packed with catalyst and the heat of reaction is removed by circulating water on the shell side. Essentially, the reactor also plays a second role, that of a high-pressure steam generator.

The integrated process also accepts gaseous hydrocarbons, such as methane, as well as liquid hydrocarbons, like naphtha, as feedstock. The synthesis gas is

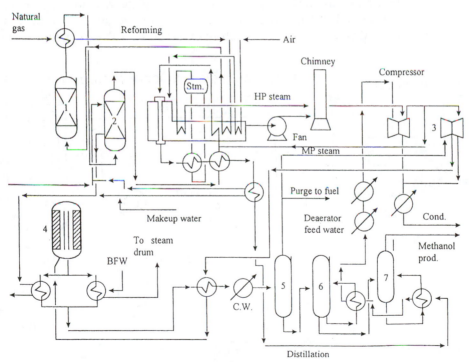

**Figure 19** Schematic of the ICI Katalco low-pressure methanol process. (Source: References 16 and 32.)

generated either by the steam reforming route or by the partial oxidation route. The steam reformer is typically operated at 850 to 860°C, and the previously desulfurized naphtha is contacted with steam to produce hydrogen and carbon oxides. The syngas is compressed in a centrifugal compressor to 50 to 80 bar and fed to the reactor.

In the second route, the heavy residues are fed into a furnace along with oxygen and steam. The feedstock is partially oxidized at 1400 to 1450°C. Because the operating pressure of the partial oxidizer is 55 to 60 bar, there is no need for syngas compression. However, the gas needs to be cleaned to remove hydrogen sulfide and free carbon, and a shift conversion is necessary to adjust the hydrogen/carbon monoxide ratio. A schematic flow diagram is given in Figure 20.

***The MGC low-pressure process.*** A schematic flow diagram of the process developed by the Mitsubishi Gas Chemical Company [30] is given in Figure 21. This process also uses copper-based methanol synthesis catalyst and is operated at temperatures ranging from 200 to 280°C over a pressure range of 50 to 150 atm. The temperature in the catalyst bed is kept under control by using a quench-type converter design, as well as by recovering some of the reaction heat in an intermediate-stage boiler. The process uses hydrocarbon feedstock. The feed is desulfurized and then fed into a steam reformer at 500°C. The exit stream from

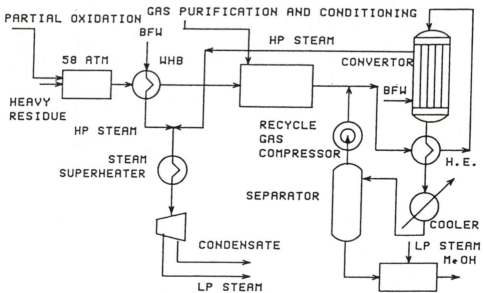

**Figure 20**   Flow sheet of the Lurgi low-pressure methanol synthesis process. (Source: Reference 16.)

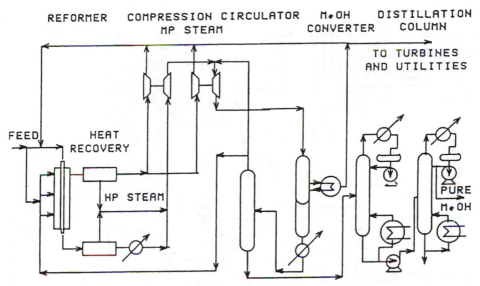

REFORMER    COMPRESSION CIRCULATOR    M•OH    DISTILLATION
            HP STEAM                   CONVERTER   COLUMN

**Figure 21** Flow diagram of the Mitsubishi gas chemical (MGC) methanol synthesis process. (Source: Reference 30.)

the reformer contains hydrogen, carbon monoxide, and carbon dioxide and is at a temperature of 800 to 850°C. The gases are then compressed in a centrifugal compressor and mixed with the recycled stream before being fed into the converter.

*Nissui-Topsøe methanol synthesis process.* A schematic flow diagram [31] of the process offered by Haldor Topsøe of Denmark and Nihon Suiso Kogyo Company of Japan is given in Figure 22. The process was based on a copper-zinc-chrome catalyst that was active at 230 to 280°C and at a pressure range of 100 to 200 atm.

*Japan Gas-Chemical Company process.* The Japan Gas-Chemical Company (JGC) also came up with an intermediate-pressure process that was operated at 150 atm. A flow diagram of the process [31] is given in Figure 23.

*Haldor Topsøe A/S low pressure process.* Figure 24 shows a schematic of Haldor Topsøe A/S process [32], starting from natural gas or associated gas feedstocks using two-step reforming. In the flowchart are the following sections: (1) desulfurizer, (2) process steam generation unit, (3) primary reformer, (4) oxygen-blown secondary reformer, (5) superheated high-pressure steam generator, (6) distillation section, (7) single-stage syngas compressor, (8) synthesis loop, and (9) recirculator compressor for recycle gas. The synthesis reactors are provided

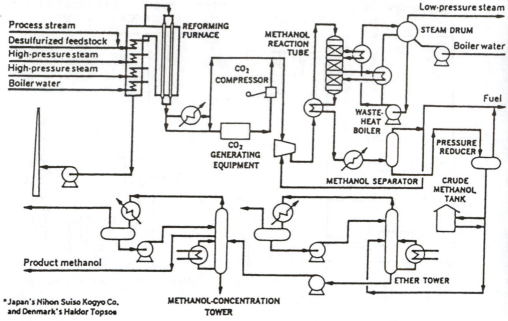

**Figure 22** Flow sheet of the Nissui-Topsøe methanol synthesis process. (Source: Reference 31.)

with heat exchangers between the reactors. Reaction heat is used to heat saturater water. Effluent from the last reactor is cooled by preheating the feed to the first reactor. The distillation section consists of three columns. The total energy consumption for this process is about 29.7 GJ/metric ton including energy requirements for oxygen production.

***M. W. Kellogg low-pressure process.*** Figure 25 shows a schematic of Kellogg's methanol synthesis process [33], starting from hydrocarbon feedstocks using a high-pressure steam reforming process. The process uses BASF low-pressure synthesis catalyst. Desulfurized natural gas is converted to syngas via a single-stage steam reformer. Thermal efficiency is obtained by various schemes of energy integrations including heat exchange between reformer effluent and steam as well as operation of intercoolers and reactors. The reactor type is adiabatic.

***Chem Systems/APCI liquid phase methanol (LPMeOH) process.*** In 1975 Chem Systems developed a novel engineering concept in which methanol is synthesized from syngas in an inert liquid medium. The catalyst used in this processs is essentially the same as that for the conventional low-pressure process.

Air Products and Chemicals, Inc., has operated a process development unit (PDU) at LaPorte, Texas. The process has outstanding merits, with excellent

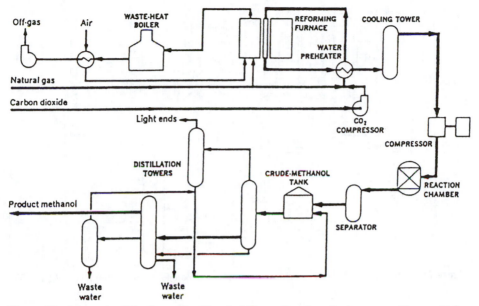

**Figure 23** Schematic of the Japan Gas-Chemical Co. methanol synthesis process. (Source: Reference 31.)

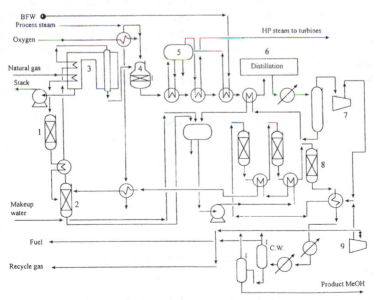

**Figure 24** Schematic of the Haldor Topsøe A/S process for methanol. (Source: References 32 and 33.)

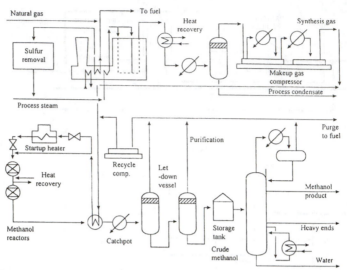

**Figure 25** Schematic of Kellogg's methanol synthesis technology. (Source: References 32 and 33.)

heat transfer, reactor thermal stability, high once-through conversion, and use of CO-rich syngas. The process is ideal for conversion of coal-derived syngas into methanol and can be used as an off-peak process for coal power plants in an IGCC mode. However, the excellent merits are somewhat offset by poorer mass transfer due to the presence of an inert oil medium [16]. The LaPorte PDU was commissioned by APCI under the sponsorship of the U.S. DOE and EPRI in March 1984. Various process enhancements have been made and the process is commercially ready [16]. A schematic of the LaPorte PDU system for the liquid-phase methanol synthesis process is shown in Figure 26 [3].

### 4.4.3 Conversion of Methanol to Gasoline

Methanol itself can be used as a transportation fuel just as liquefied petroleum gas (LPG) and ethanol. However, direct use of methanol as a motor fuel would require nontrivial engine modifications and substantial changes in the lubrication system. This is why the conversion of methanol to gasoline is quite appealing [22, 34].

The Mobil Research and Development Corporation developed the methanol-to-gasoline (MTG) process. The process technology is based on the catalytic reactions using the zeolites of the ZSM-5 class [35]. MTG reactions can be written as

$$n\text{CH}_3\text{OH} \longrightarrow (-\text{CH}_2-)_n + n\text{H}_2\text{O}$$

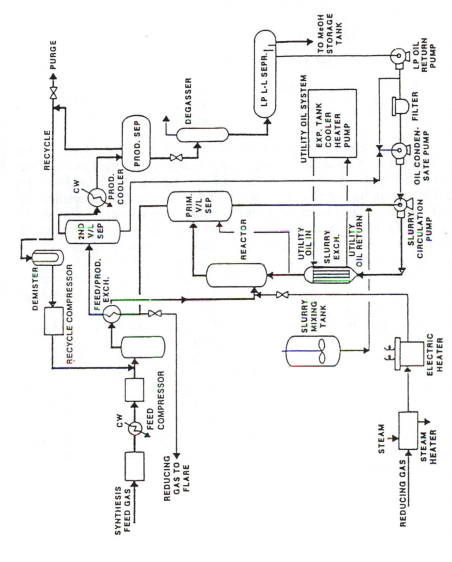

**Figure 26** Simplified process flow sheet for the LaPorte PDU for the LPMeOH process. (Source: Reference 3.)

The detailed reaction path is described in reference 36. The following simplified steps describe the overall reaction path:

$$2CH_3OH \longrightarrow (CH_3)_2O + H_2O$$

$$CH_3OH\ (CH_3)_2O \longrightarrow light\ olefins + H_2O$$

$$light\ olefins \longrightarrow heavy\ olefins$$

$$heavy\ olefins \longrightarrow paraffins$$
$$aromatics$$
$$naphthenes$$

Due to the shape-selective pore structure of the ZSM-5 class catalysts, the hydrocarbons fall predominantly in the gasoline boiling range. The product distributions are influenced by the temperature, pressure, space velocity, reactor type, and Si/Al ratio of the catalyst [37]. Paraffins are dominated by isoparaffins, and aromatics are dominated by highly methyl-substituted aromatics. $C_9+$ aromatics are dominated by symmetrically methylated isomers reflecting the shape-selective nature of the catalyst. The $C_{10}$ aromatics are mostly durene (1,2,4,5-tetramethylbenzene), which has an excellent octane number, but the freezing point is very high at 79°C. Too high a durene content in the gasoline may impair automobile driving characteristics especially in cold weather, due to its tendency to crystallize at low temperature [8]. Mobil's test found no drivability loss at $-18°C$ using a gasoline containing 4 wt. % durene [8]. Mobil also developed a heavy gasoline treating (HGT) process to convert durene into other high-quality gasoline components by isomerization and alkylation [37]. The MTG reactions are exothermic and go through the dimethyl ether intermediate route.

Basically three types of chemical reactors were developed for the MTG process: (1) adiabatic fixed bed, (2) fluidized bed, and (3) direct heat exchange. The first two were developed by Mobil and the last by Lurgi.

The adiabatic fixed bed concept uses a two-stage approach: a first-stage methanol reactor and a second-stage methanol conversion to hydrocarbons. The first commercial plant of 14,500 bbl/day gasoline capacity was constructed in New Zealand. The plant had been running successfully since a 1985 start-up until its recent shutdown. The synthesis gas is generated via steam reforming of natural gas obtained from the offshore Manifield. An HGT plant was also run successfully in New Zealand and reduced the durene content to 2 wt. %. The successful operation of MTG in New Zealand was a very important milestone in human history, as it made possible the chemical synthesis of gasoline from unlikely fossil fuel sources like natural gas and coal. Petroleum crude is no longer the sole source for gasoline.

A fluidized bed MTG concept was concurrently developed by Mobil. The process research went through several stages involving bench-scale fixed fluidized bed, 4 bpd, 100 bpd cold-flow models, and a 100 bpd semiwork plant. Table 18 shows typical MTG process conditions and product yields [8].

During the MTG development, Mobil researchers found that the hydrocarbon product distribution can be shifted to light olefins by increasing the space velocity, decreasing the MeOH partial pressure, or increasing the reaction temperature [38]. Typical yields [8] from 4 bpd operation were: $C_1-C_3$ paraffins, 4 wt. %; $C_4$ paraffins, 4 wt. %; $C_2-C_4$ olefins, 56 wt. %; $C_5+$ gasoline, 35 wt. %. Using olefins from the MTO or FT processes, diesel and gasoline can be made via a process converting olefins to diesel and gasoline (OTG). Using acid catalysts, catalytic polymerization is a standard process and is being used at SASOL to convert $C_3-C_4$ olefins into (G + D). Mobil developed an MOGD (Mobil Olefins to Gasoline and Diesel) process using their commercial zeolite catalyst [39, 40].

## Table 18  Typical process conditions and product yields for the MTG process

| Conditions | Fixed bed reactor | Fluid bed reactor |
|---|---|---|
| MeOH/water charge, w/w | 83/17 | 83/17 |
| Dehydration reactor inlet $T$, °C | 316 | — |
| Dehydration reactor outlet $T$, °C | 404 | — |
| Conversion reactor inlet $T$, °C | 360 | 413 |
| Conversion reactor outlet $T$, °C | 415 | 413 |
| $p$, kPa | 2170 | 275 |
| Recycle ratio, mol/mol charge | 9.0 | — |
| Space velocity, WHSV | 2.0 | 1.0 |
| Yields (wt % of MeOH charged) | | |
|   MeOH + dimethyl ether | 0.0 | 0.2 |
|   HCs | 43.4 | 43.5 |
|   Water | 56.0 | 56.0 |
|   CO, $CO_2$ | 0.4 | 0.1 |
|   Coke, other | 0.2 | 0.2 |
| | 100.0 | 100.0 |
| Hydrocarbon product (wt. %) | | |
|   Light gas | 1.4 | 5.6 |
|   Propane | 5.5 | 5.9 |
|   Propylene | 0.2 | 5.0 |
|   Isobutane | 8.6 | 14.5 |
|   $n$-Butane | 3.3 | 1.7 |
|   Butenes | 1.1 | 7.3 |
|   $C_5$ + gasoline | 79.9 | 60.0 |
| | 100.0 | 100.0 |
| Gasoline (including alkylate), RVP-62kPa (9 psi) | 85.0 | 88.0 |
| LPG | 13.6 | 6.4 |
| Fuel gas | 1.4 | 5.6 |
| | 100.0 | 100.0 |
| Gasoline octane (R+O) | 93 | 97 |

*Source:* Reference 8.

An innovative process enhancement has been made by Lee and co-workers under the sponsorship of the Electric Power Research Institute [41]. Their process, called a DTG (DTH, DTO) process is based on the conversion of dimethyl ether to hydrocarbon over a ZSM-5 type catalyst [41, 42]. Their process is based on the novel, economical, single-stage synthesis of dimethyl ether (DME) from syngas. When DME is produced in a single stage, the intermediate methanol formation is no longer limited by chemical equilibrium, which increases the reactor productivity substantially. This is especially true for the synthesis of methanol in the liquid phase. Furthermore, by feeding DME directly to the ZSM-5 reactor instead of methanol, the stoichiometric conversion and hydrocarbon selectivity increase substantially. The difference between MTG and DTG, therefore, is in the placement of the methanol dehydration reaction step (i.e., DME formation reaction). In the MTG, methanol-to-DME conversion takes place in the gasoline reactor, whereas methanol-to-DME conversion, in the DTG, takes place in the syngas reactor. The process has not yet been tested on a large scale.

## 4.5 COAL-OIL COPROCESSING

Coprocessing is defined as the simultaneous reaction treatment of coal and petroleum resid, or crude oil, with hydrogen to produce distillable liquids. More strictly speaking, this technology should be classified under direct liquefaction as a variation. Petroleum liquids have been often used as a liquefaction solvent, mainly for start-up or whenever coal-derived liquids were unavailable. However, serious consideration has been given to the process possibilities of hydrocracking petroleum resid while liquefying coal in the same reactor. In this sense, coprocessing has an ultimate objective of cobeneficiation.

An early coprocessing patent was granted to UOP, Inc., in 1972 for a process whereby coal is solvent extracted with petroleum [43]. Another early patent on coprocessing was issued to Hydrocarbon Research, Inc. (HRI) in 1977 for the single-stage ebullated-bed COIL process based on the HRI H-Oil/H-Coal technology [44]. Consol R & D tested the use of a South Texas heavy oil for coal hydroextraction but found that, even after hydrogenation, the petroleum made a very poor liquefaction solvent [8]. The Canada Centre for Mineral and Energy Technology (CANMET) developed the CANMET hydrocracking process for petroleum resids. They found that small additions of coal (<5 wt. %) to the petroleum feedstock significantly improved distillate product yields. A 5000 bpd plant using this process was started up in 1985 by Petro-Canada near Montreal, Quebec [45].

In summary, coprocessing has several potential economic and technological advantages relative to coal liquefaction or hydroprocessing of heavy petroleum residua. Synergisms and cobeneficiating effects can be obtained, especially in the areas of (1) replacement of recycle oil, (2) sharing hydrogen between hydrogen-rich and hydrogen-deficient materials, (3) aromaticity of the product, and (4) demetalation and catalyst life. For the current technology, temperatures of 400 to 440°C, 2000 psig hydrogen pressure, and alumina-supported cobalt, molybdenum, nickel, or disposable iron catalysts are frequently used. Various efforts in developing more selective catalysts are being carried out.

# REFERENCES

1. Speight, J. G. *The Chemistry and Technology of Coal.* New York: Marcel Dekker, 1983.
2. Probstein, R. F., and R. E. Hicks. *Synthetic Fuels.* New York: McGraw-Hill, 1982.
3. Schindler, H. D. Coal Liquefaction—A Research & Development Needs Assessment. COLIRN Panel Assessment, DOE/ER-0400,UC-108, Final Report, Vol. II., March 1989.
4. Potts, J. D., K. E. Hastings, R. S. Chillingworth, and K. Unger. Expanded-Bed Hydroprocessing of Solvent Refined Coal (SRC) Extract. Interim Technical Report FE-2038-42, February 1980.
5. Schindler, H. D., J. M. Chen, and J. D. Potts. Integrated Two-Stage Liquefaction. Final Tech. Report, DOE Contract DE-AC22-79 ET14804, June 1993.
6. Comolli, A. G., and J. B. McLean. The low-severity catalytic liquefaction of Illinois No. 6 and Wyodak coals. Pittsburgh Coal Conference, Pittsburgh, September 16–20, 1985.
7. Weber, W., and N. Stewart. *EPRI Monthly Review,* January 1987.
8. U.S. DOE Working Group on Research Needs for Advanced Coal Gasification Techniques (CO-GARN) (S.S. Penner, Chairman). Coal Gasification: Direct Application and Synthesis of Chemicals and Fuels. DOE Contract No. DE-AC01-85 ER30076, DOE Report DE/ER-0326, June 1987.
9. Anderson, R. B. *The Fischer-Tropsch Synthesis.* Orlando, FL: Academic Press, 1984.
10. Schulz, G. V. *Z. Phys. Chem.* 32:27, 1936.
11. Flory, P. J. *J. Am. Chem. Soc.* 58:1877, 1936.
12. Wender, I. Review of indirect liquefaction. In Coal Liquefaction, USDOE Contract DE-AC01-87-ER 30110, ed. by Schindler (chairman of COLIRN), Final Report, March 1989.
13. van der Burgt, M. J. *The Shell Middle Distillate Synthesis Process.* Fifth Synfuels Worldwide Symposium, Washington, DC, New York: McGraw-Hill, 1985.
14. Humbach, J., and N. W. Schoonver. Proc. Ind. Liquefaction Contractor's Meeting, Pittsburgh, PETC, pp. 29–38, 1985.
15. Koelbel, H., and M. Ralek. *Catal. Rev. Sci. Eng.* 21:225, 1980.
16. Lee, S. *Methanol Synthesis Technology.* Boca Raton, FL: CRC Press, 1990.
17. Andriamasinoro, D., R. Kieffer, and A. Kiennemann. Preparation of Stabilized copper-rare earth oxide catalysts for the synthesis of methanol from syngas, *App. Catal. A: General* 106:201–212, 1993.
18. Oxturk, S., and Y. Shah. Comparison of Gas and Liquid Methanol Synthesis Process, *The Chemical Engineering Journal* 37:177–192, 1988.
19. Tjandra, S., R. Anthony, and A. Akgerman. Low $H_2/CO$ ratio synthesis gas conversion to methanol in a trickle bed reactor. *Industrial and Engineering Chemistry* 32:2602–2607, 1993.
20. Westerterp, K., T. Bodewes, M. Vrijland, and M. Kuczynski. Two new methanol converters. *Hydrocarbon Process.* November:69, 1988.
21. Wagialla, K., and S. Elnashaie. Fluidized-bed reactor for methanol synthesis. A theoretical investigation. *Ind. Eng. Res.* 30:2298–2308, 1991.
22. Fox, J. M. *Catal. Rev. Sci. Eng.* 35(2):169–212, 1993.
23. Cybulski, A. *Catal. Rev. Sci. Eng.* 36(4):557–615, 1994.
24. Lee, S. Research Support for Liquid Phase Methanol Synthesis Process Development. EPRI Report AP-4429, pp. 1–312, Palo Alto, CA, February 1986.

25. Pelekar, V., H. Jung, I. Tierney, and I. Wenda. Slurry phase synthesis of methanol with a potassium methoxide/copper chromite catalytic system. *App. Catal. A: General* 102:13–34, 1993.
26. Chanchlanii, K., R. Hudgins, and P. Silveston. Methanol Synthesis from $H_2$, CO, and $CO_2$ over Cu/ZnO Catalysts, *J. of Catalysis* 136:59–75, 1992.
27. Walker, A., R. Lambert, and R. Nix. Methanol Synthesis over Catalysts Derived from $CeCu_2$, *J. of Catalysis* 138:694–713, 1992.
28. Sizek, G., H. Curry-Hyde, and M. Wainwright. Methanol synthesis over copper ZnO promoted copper surfaces, *App. Catal. A: General* 115:15–28, 1994.
29. Rogerson, P. L. Imperical Chemical Industries' Low Pressure Methanol Plant, *Chem. Eng. Prog. Symp. Ser.* 66(98):28, 1970.
30. Hydrocarbon Processing, November 1979.
31. Prescott, J. T. *Chem. Eng.* April 5:60, 1971.
32. *Hydrocarbon Process.*, March 1993.
33. *Hydrocarbon Process.*, March 1995.
34. Mills, G. A. *Fuel* 73(8):1243–1279, 1994.
35. Chang, C. D. *Catal. Rev. Sci. Eng.* 25:1, 1983.
36. Chang, C. D., and A. J. Silvestri. *J. Catal.* 47:249, 1977.
37. Chang, C. D., and A. J. Silvestri. The MTG process: Origin and evolution. Presented at the 21st State-of-the-Art ACS Symposium on Methanol as a Raw Material for Fuels and Chemicals, Marco Island, FL, June 15–18, 1986.
38. Socha, R. F., C. T. W. Chu, and A. A. Avidan. An Overview of Methanol-to-Olefins Research at Mobil: from Conception to Demonstration Plant.
39. Tabak, S. A., A. A. Avidan, and F. J. Krambeck. MTO-MOGD Process, in reference 85.
40. Tabak, S. A., and F. J. Krambeck. *Hydrocarbon Process.* 64:72, September 1985.
41. Lee, S., M. R. Gogate, K. L. Fullerton, and C. J. Kulik. U.S. Patent 5,459,166, October 17, 1995.
42. Lee, S., M. R. Gogate, and C. J. Kulik. *Fuel Sci. Tech. Int.* 13(8):1039–1058, 1995.
43. Gatsis, J. G. U. S. Patent 3,705,092, December 5, 1972.
44. Chervenak, M. C., and E. S. Johanson. U.S. Patent 4,054,504, October 18, 1977.
45. Kelly, J. F., and S. A. Fouda. CANMET coprocessing: An extension of coal liquefaction and heavy oil hydrocaracking technology. Presented at the DOE Direct Liquefaction Contractors' Review Meeting, Albuquerque, NM, October 1984.
46. Speight, J. G. *The Chemistry and Technology of Coal* (Rev. Ed.). New York: Marcel Dekker, 1994.

## PROBLEMS

1. Define the liquefaction of coal from a molecular standpoint.
2. Discuss the advantages and disadvantages of indirect liquefaction in comparison to direct liquefaction.
3. Discuss the advantages of catalytic liquefaction processes.
4. Discuss the advantages of two-stage liquefaction processes.
5. How many of the following chemical reactions are stoichiometrically independent?

$$CO_2 + 3H_2 = CH_3OH + H_2O$$

$$CO + 2H_2 = CH_3OH$$

$$CO + H_2O = CO_2 + H_2O$$

6. The methanol synthesis reaction is one of the least favored reactions from the thermodynamic standpoint at high temperatures. What does this statement mean?

7. What catalytic systems are used for high-pressure methanol syntheses?

8. What are the optimal $CO_2$ concentrations for the vapor-phase synthesis of methanol?

9. What is the ideal $H_2/CO$ ratio (by mole) as a feed for the vapor-phase synthesis of methanol?

10. How is the conventional $CuO/ZnO/Al_2O_3$ catalyst prepared?

11. What are the principal reasons for methanol catalyst crystallite size growth in the liquid phase?

12. In the absence of CO, the methanol synthesis reaction proceeds very slowly. Explain why.

13. In the absence of $CO_2$, the methanol synthesis reaction proceeds very slowly and the catalyst activity decreases very fast. Explain why.

14. Discuss the processing difficulties involved in the Fischer-Tropsch synthesis.

15. Why is wax formation troublesome in the Fischer-Tropsch synthesis?

16. When does the temperature runaway reaction take place? How do we know it is happening?

17. As a rule in industry, $E/RT = 20$ is used for most catalytic reactions that are not limited by the rate of diffusion. Is the low-pressure synthesis of methanol in the liquid phase a diffusion-limited process in general? The activation energy for methanol synthesis in the liquid phase is approximately 18,000 cal/mole.

18. Methanol is not used as oxygenated fuel in the United States. Why is this the case?

19. Do the MTG and DTG processes have identical production distributions?

20. Explain the following reactor types:
    (a) Ebullated-bed reactor
    (b) Trickle-bed reactor
    (c) Liquid-entrained reactor
    (d) Mechanically agitated slurry reactor

21. List the following processes in the order of hydrogen efficiency, from top to bottom: SRC-1, SRC-II, H-Coal, NTSL, CTSL, ITSL, EDS.

22. What are the ideal syngas compositions for the following processes?
    (a) Lurgi's low-pressure methanol
    (b) ChemSystems' liquid-phase methanol
    (c) SASOL process

23. Assess the advantages and disadvantages of the ASDA process (antisolvent deashing).

24. List in the order of Reid vapor pressure the following chemicals: methanol, ethanol, MTBE, ETBE, and TAME.
25. What are the principal reasons for catalyst deactivation in the Fischer-Tropsch synthesis?
26. What are the major causes for catalyst deactivation in the liquid-phase methanol process?

# INTEGRATED GASIFICATION COMBINED CYCLE TECHNOLOGY

## 5.1 INTRODUCTION

Over the past three decades significant efforts have been made to develop cleaner and more efficient technology for power generation. The limited reserves of favored fuel sources such as oil and natural gas and the associated geopolitical constraints have compelled the scientific community to explore alternative resources. Options such as nuclear energy for power generation pose a severe environmental hazard, as witnessed in the Chernobyl accident. Utilization of unconventional resources such as solar, wind, and wave energy and of renewable resources such as biomass, wood, and waste is not viable because of the lack of necessary technological developments for commercial applications. This motivated a second look at coal, with its abundant reserves all over the world, which promises a long-term solution for the world's energy requirements.

Coal gasification technology received a big thrust with the concept of combined cycle power generation. The integration of coal gasification with combined cycle for power generation (IGCC) had the inherent characteristics of gas cleanup and waste minimization, which made this system environmentally preferable. Commercial scale demonstration of the Cool Water plant and other studies have shown that the greenhouse gas and particulate emission from an IGCC plant is drastically lower than the recommended federal New Source Performance Standard (NSPS) levels. IGCC also offers a phased construction and repowering option, which allows multiple-fuel flexibility and the necessary economic viability. Advances in IGCC technology continue to improve efficiency and further reduce emissions, making this the technology of the twenty-first century.

## 5.1.1 Alternative Clean Coal Technologies

In recent years, extensive discussions have been held in public forums and policy meetings on global warming, $CO_2$ generation, acid rain, more stringent New Source Performance Standards (NSPS), and environmental compliance, as well as on the role of fossil-fired power plant effluents in all of the above. It is evident that electric utilities are concerned with these deliberations and their effects on utility use of different coal feedstocks. Especially vulnerable are utilities that use eastern high-sulfur coal. Because of current and future legislation and anticipated stricter emission standards, it is obvious that systems with higher thermal efficiency and lower impact on the environment be the norm of the future. The advanced systems of interest are in various stages of design, development, and demonstration, and most employ gas turbines for electricity generation.

**Pulverized-coal-fired boilers with advanced steam cycles.** Extensive studies have been conducted to develop pulverized coal (PC) boilers with advanced steam cycle conditions to improve the thermal efficiency of these systems. The staged development of these systems has the ultimate goal of achieving a steam pressure and temperature of 34.5 MPa (5000 psig) and 650°C (1200°F), respectively, from the current values of 540°C (1000°F) and 16.5–24 MPa (2400–3500 psig) [1].

**Fluidized-bed-combustion cogeneration (FBC).** This approach employs the technology of coal-fired cogeneration in the form of a combined cycle atmospheric or moderately pressurized fluidized bed in which an air heater heats the air in a gas turbine cycle for the cogeneration of electricity and useful thermal energy. A schematic diagram of typical bubbling and circulating fluid bed systems is shown in Figures 1 and 2 [2–4].

**Magnetohydrodynamic (MHD) topping cycles.** Magnetohydrodynamic (MHD) generation of power is based on the direct conversion of fuel energy into electrical energy by having a heated, electrically conducting fluid flow through a magnetic field. In a coal-fired open-cycle MHD system, the working fluid is utilized on a once-through basis and consists of multiphase fossil fuel combustion products. Figure 3 shows a schematic of the open-cycle MHD system. The overall plant consists of a topping cycle, in which the electrical energy is directly produced by the MHD generator, and a steam-bottoming plant, where additional heat energy and the seed material are recovered [3, 5, 6].

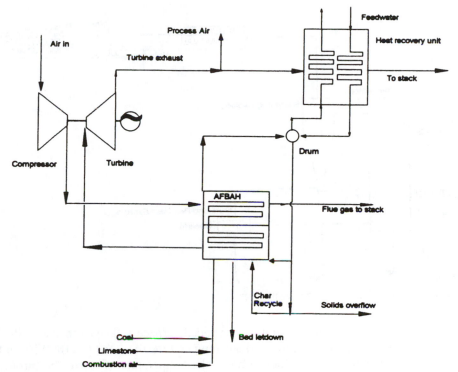

**Figure 1** Bubbling-fluidized-bed combustion.

**High-performance power systems (HIPPS).** The high-performance power system has the ultimate goal of producing electricity from coal with an overall thermal efficiency of 47% or higher, in comparison with ~35% for current systems, and to reduce cost of electricity by 10% compared with current systems. In essence, the HIPPS system emphasizes the use of advanced technologies that enable coal to be used as efficiently as natural gas. HIPPS processes include the use of a high-temperature advanced furnace that integrates combustion, heat transfer, and emission control in a single machine. HIPPS has the following environmental targets:

Reducing the $NO_x$ and $SO_2$ emissions to one-fourth or less of current New Source Performance Standards (NSPS).
Reduce $CO_2$ emissions by 25 to 30% of current systems.
Eliminate waste from the desulfurization process by using and generating usable by-products.

To meet these environmental performance goals, advanced furnace designs as well as advanced flue gas cleanup technologies will have to be developed [7].

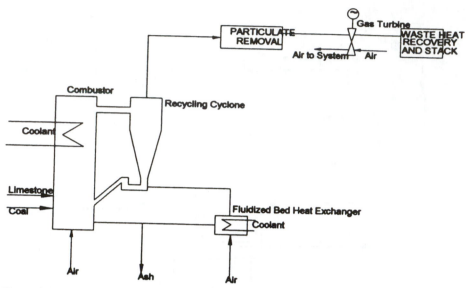

**Figure 2**  Circulating-fluidized-bed combustion.

The pulverized coal high-temperature advanced furnace (HITAF) in the HIPPS concept will heat air to an intermediate temperature of ~980°C (1800°F) and will burn supplemental clean fuel to boost the temperature of air to the turbine inlet to 1260°C (2300°F) or higher. Use of supplemental fuel can be reduced as HITAF technology evolves to permit air to be heated to higher temperatures in the furnace [3].

**Low-emissions boiler system (LEBS).** This system would utilize staged combustion to develop a power plant that will drastically reduce the $SO_2$ and $NO_x$ emissions to one-third and particulate emissions to one-half of current NSPS [3]. The LEBS would also improve efficiency by 1 to 2% and reduce the generation of wastes by production of usable by-products. The LEBS is an extension of current boiler technology with the first stage of combustion having fuel-rich or air-lean conditions. It aims at utilizing advanced combustion techniques that control $NO_x$ without using expensive flue gas treatment. The overall aim of the LEBS program is to use near-term technology for the development of a commercial-scale, low-emission boiler by the year 2003 [7].

**Integrated gasification combined cycle (IGCC).** The conventional IGCC plant is shown in Figure 4 [1]. The gasifier receives coal from the coal handling and preparation equipment and oxidant from an air treatment plant and gasifies the coal through a partial oxidation-reduction process involving coal, oxygen, and steam. The fuel output from the gasifier is in the range of 1000 to 2600°F and

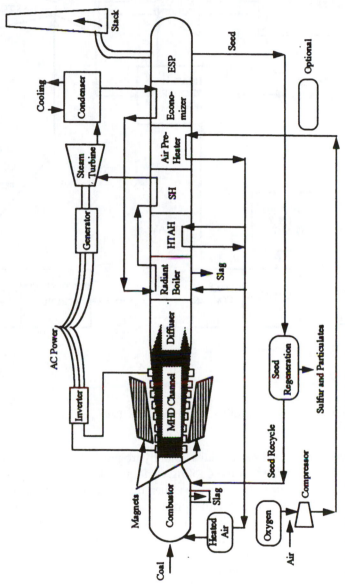

**Figure 3** Magnetohydrodynamic (MHD) topping cycle.

227

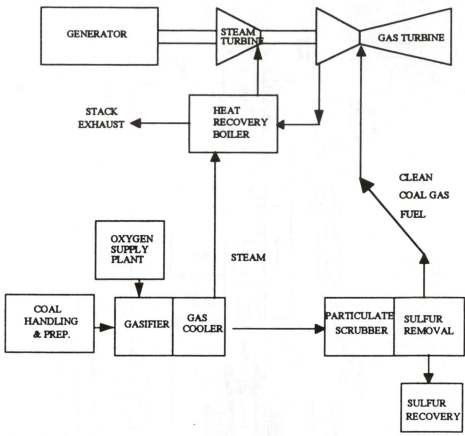

**Figure 4** Schematic of a conventional IGCC plant. (Source: Reference 1.)

is treated for removal of particulates and hydrogen sulfide from sulfur-laden coal. This gas cleanup can involve cooling (cold gas cleanup) or may not involve cooling (hot gas cleanup). A sulfur scrubber removes sulfur from the fuel gas and processes the sulfur to a salable product. The gas cooler produces steam that is utilized by the steam turbine. IGCC has an advantage in cleanup technology compared with a conventional coal steam plant because it is inherently easier to remove hydrogen sulfide from a pressurized fuel stream than to remove sulfur oxides from an atmospheric exhaust stream. Because limestone is not injected for sulfur removal in an IGCC, the solid waste from a gasifier is non-leachable and is a salable by-product. The particulate scrubber system removes particulates, trace alkali metals, and fuel-bound nitrogen converted to ammonia in the gasifier.

The clean coal gas fuel in an oxygen-blown system is approximately 20% of the heating value of pipeline natural gas. Gas turbine fuel handling systems and combustion systems can be modified to accommodate this larger fuel flow. The power island is configured similarly to that of a conventional combined cycle power plant with steam cycle integration between the gasifier and the HRSG. There are many alternatives to the cycle described above, including air-blown gasifiers, integrated air systems, and hot gas cleanup systems.

## 5.2 IGCC TECHNOLOGY OVERVIEW

Integrated gasification combined cycle (IGCC) systems represent a promising new approach to the cleaner and more efficient use of coal for power generation, offering low levels of $SO_2$ and $NO_x$ emissions along with benign solid wastes or by-products and zero or low wastewater discharges. A distinguishing feature of IGCC concepts is the type of fuel gas cleanup strategy employed [8, 9].

Integrated gasification combined cycle (IGCC) systems combine several desired attributes and are becoming an increasingly attractive option among the emerging technologies. First, IGCC systems provide high energy conversion efficiency, with the prospect of even higher efficiencies if higher temperature turbines and hot-gas cleanup systems are employed. Second, very low emission levels for sulfur and nitrogen species have been demonstrated at such facilities as the Cool Water IGCC plant in California. Third, IGCC plants produce flue gas streams with concentrated $CO_2$, as well as high levels of carbon monoxide (CO), which can be easily converted to $CO_2$. Capture of this $CO_2$ before combustion requires the treatment of substantially smaller gas volumes than capture after combustion (the method that would be required with direct coal firing), because the flue gas stream is not yet diluted with atmospheric $N_2$ and excess air. Recovery of $CO_2$ in IGCC systems is potentially less expensive than in conventional combustion systems. Moreover, $CO_2$ recovery can be done in conjunction with hydrogen sulfide ($H_2S$) removal by using several commercial technologies [8].

Typical designs for IGCC systems use cold gas cleanup (CGCU), including low-temperature removal of $SO_2$ and particulates from the coal syngas, sulfur by-product recovery, and syngas moisturization to reduce $NO_x$ formation in the gas turbine. Ongoing research by the U.S. Department of Energy (DOE) and others is focused on developing dry physical and chemical hot gas cleanup (HGCU) techniques to reduce the efficiency penalty associated with syngas cooling. An example of an IGCC system concept with CGCU is shown in Figure 5 [10].

### 5.2.1 Cold Gas Cleanup

The design basis for this system is fluidized bed gasifiers, and the fuel gas cleanup system is representative of the technology employed in the Cool Water

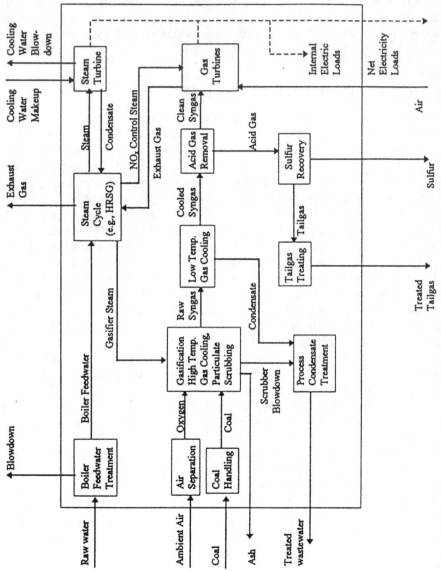

**Figure 5** Schematic of oxygen-blown fluidized bed gasifier IGCC system with cold gas cleanup (conventional design). (*Source:* Reference 9.)

demonstration plant. Coal is partially combusted with oxygen and gasified with steam in a reducing atmosphere to yield a fuel gas containing CO and $H_2$ as key constituents. Oxygen for the gasifier is provided by an air separation plant. Steam is supplied from the plant steam cycle. The hot fuel gas is cooled in a steam generator and then enters a low-temperature cooling section. As part of low-temperature cooling, nearly all of the particulate matter and ammonia in the fuel gas is selectively removed by wet scrubbing. The fuel gas, at a temperature of about 100°F, then enters a Selexol acid removal unit, where $H_2S$, the primary sulfur species in the fuel gas is selectively removed. The clean fuel gas is then combusted in a gas turbine combined cycle system. A portion of the electrical output from the generators must be used to power equipment in the plant, most notably the air separation unit.

In IGCC systems, environmental control is required not just to meet environmental regulations but also for proper plant operation. In particular, contaminants such as sulfur, particulates, and alkali must be removed prior to fuel gas combustion to protect the gas turbine components from erosion, corrosion, and deposition. Because of the close interactions among plant performance, environmental control, and cost, assessment of IGCC technology must be based on integrated analysis of the entire system.

The approach to emissions control in an IGCC plant is fundamentally different from that in a pulverized coal-fired power plant. Emission control strategies focus on the fuel gas, which is pressurized (typically 300–500 psi) and has a substantially lower volumetric flow rate than the flue gas, which flows near atmospheric pressure, of coal combustion power plants. Furthermore, sulfur in the fuel gas is in reduced form (mostly $H_2S$) and can be removed by a variety of commercially available processes. Typically $H_2S$ and COS are removed using a Selexol or similar process, and the concentrated acid gas is then processed for elemental sulfur recovery. Removal of ammonia in the fuel gas in wet scrubbing systems reduces substantially the amount of fuel-bound nitrogen in the fuel gas. In conventional gas turbine combustors, most of the fuel-bound nitrogen is converted to $NO_x$. Thermal $NO_x$ emissions are controlled either by moistening the fuel gas or by gas turbine combustor steam injection to reduce the flame temperature.

In the Cool Water IGCC demonstration plant, which operated from 1984 to 1989, air emissions with low-sulfur coal were reported to be 0.06 lb of $NO_2$/MMBtu, 0.7 lb of $SO_2$/MMBtu, and 0.008 lb of particulate matter (PM)/MMBtu. All three rates are well below federal NSPS levels for conventional coal-fired plants.

In IGCC systems with fuel gas cooling, liquid condensates from the high-temperature fuel gas must be removed and treated prior to discharge. In addition blowdown from the fuel gas wet scrubber must be treated. These wastewater streams are in addition to the steam cycle and cooling water cycle blowdown streams typical of modern thermal power stations.

IGCC solid wastes include gasifier bottom ash and particulate cake from scrubber systems. These solid wastes are suitable for landfilling. Compared to a pulverized coal–fired plant with flue gas desulfurization (FGD), the solid waste burden from an IGCC system is reduced by the production of an elemental sulfur by-product and the lack of spent sorbent waste.

### 5.2.2 Hot Gas Cleanup

Hot gas cleanup systems reduce or eliminate the need for syngas cooling prior to particulate removal and desulfurization. This improves the plant efficiency and reduces or eliminates the need for heat exchangers and process condensate treatment. Thus, HGCU offers a more highly integrated system in which a major wastewater stream is eliminated.

An example of an air-blown fluidized-bed gasifier IGCC system with HGCU is shown in Figure 6. The schematic represents process elements based on design and cost studies prepared for the Gas Research Institute and for DOE. The primary features of this design, compared to the IGCC system with CGCU, are as follows:

1. Elimination of an oxygen plant by using air, instead of oxygen, as the gasifier oxidant
2. In situ gasifier desulfurization with limestone
3. External (e.g., not in the gasifier) desulfurization using a high-temperature removal process
4. High-efficiency cyclones for particulate removal
5. Elimination of heat exchangers for fuel gas cooling at the gasifier exit
6. Elimination of sulfur recovery and tail gas treating
7. Addition of a circulating fluidized bed boiler for sulfation of spent limestone (to produce an environmentally acceptable waste) and conversion of carbon remaining in the gasifier ash

The design basis assumed here includes the use of water quench, rather than heat exchange, for high-temperature syngas cooling from 1850°F at the gasifier outlet to 1100°F prior to gas cleanup. Therefore, there is no knockout drum for process condensate removal. The clean fuel gas is then combusted in a gas turbine modified to fire low-Btu coal gas.

In-bed desulfurization is expected to result in 90% sulfur capture within the gasifier. The external zinc ferrite desulfurization process is expected to reduce the sulfur content of the syngas to 10 ppmv, resulting in low $SO_2$ emissions from the gas turbine combustor. Upon regeneration, the sulfur captured by the zinc ferrite sorbent is evolved in an offgas containing $SO_2$, which is recycled to the gasifier for capture by the calcium-based sorbent.

Thermal $NO_x$ emissions are expected to be quite low for air-blown IGCC. However, the HGCU system assumed here does not remove fuel-bound nitrogen

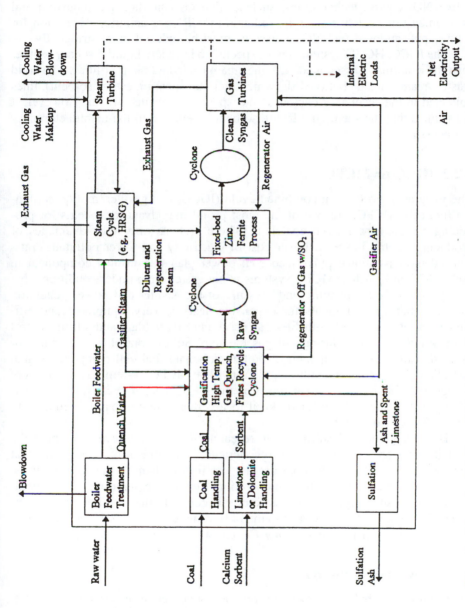

**Figure 6** Schematic of air-blown fluidized-bed gasifier IGCC system with hot gas cleanup (advanced design). (*Source:* Reference 9.)

from the fuel gas. Thus, fuel-bound $NO_x$ emissions may pose a concern. Alternative $NO_x$ control technologies, such as rich or lean staged combustion and postcombustion selective catalytic reduction (SCR), are under consideration for future applications if fuel $NO_x$ emissions must be reduced more stringently.

The IGCC/HGCU system is not expected to have any liquid discharges other than those normally associated with the steam cycle and plant utilities. The solid waste streams include bottom ash (which includes spent limestone sorbent), fines collected in the secondary cyclones, and spent zinc ferrite sorbent. However, it is assumed that the spent zinc ferrite sorbent is returned to the manufacturer for reprocessing.

### 5.2.3 IGCC or MCFC

The integrated gasification combined cycle (IGCC) and gasifier/molten carbonate fuel cell (MCFC) are two of the most promising advanced systems for producing electric power from coal [11]. These systems offer the advantages of producing electric power at lower cost of electricity and lower pollutant emissions than conventional pulverized coal power plants. A critical component in both IGCC and gasifier/MCFC systems is hot gas cleanup under conditions that nearly match the temperature and pressure of the gasifier and power generator. This eliminates requirements for expensive heat recovery equipment and efficiency losses associated with fuel gas scrubbing. Particulate and sulfur control is of primary concern during hot gas cleanup. Sulfur control is important for protection of turbine components, molten carbonate fuel cell components, and compliance with environmental regulations. The MCFC system has much more stringent sulfur tolerances than those of the IGCC system because of the sensitivity of the fuel cell. Sulfur removal to less than 1 ppmv may be required to ensure fuel cell operability.

High-temperature desulfurization research over the past several years in the United States has concentrated on regenerable, mixed-metal oxide sorbents that can reduce the sulfur in the coal gasifier gas to less than 10 ppmv and can be regenerated with air for multicycle operation. Zinc ferrite, an iron and zinc compound with a spinel-type crystal structure, is currently the leading sorbent candidate for application to power systems [43]. Zinc ferrite research was pioneered by Morgantown Energy Technology Center (METC).

### 5.2.4 Advances in the IGCC Field

In response to global concerns regarding energy conservation and environmental emissions, various types of renewable energy have been under development. Often, the popularity of such an energy system is dependent only on its economic feasibility. In general, cogeneration, or combined heat and power (CHP), is an economically feasible energy source with high overall efficiency of energy utilization (60 to 90%) [12]. Cogeneration reduces national fuel consumption and therefore has a beneficial effect on the environment in general. It results in overall reduction in emissions of $CO_2$, $SO_x$, $NO_x$, and particulates [13]. However, its application appraisal is still sensitive to the economic criteria employed.

An energy-environment-economics mathematical model was developed [12] to assess the feasibility of a gas-turbine cogeneration/district heating scheme. The deterioration in the performance of the plant and subsequent maintenance as well as its contribution to global $CO_2$ reduction were investigated. Economic sensitivity analysis and consequent robustness measurements were also performed. Several investment options, which differ with respect to plant numbers and generating capacities, were considered and the optimum system was outlined.

### 5.2.5 Introduction of a Fuel Cell into a Combined Cycle

The need for a new power plant with higher efficiency than a conventional power plant, and therefore better environmental performance, has increased during the past decades. The search for a new concept with an ultrahigh efficiency, environmentally superior, and cost-competitive gas turbine process that can fulfill such demands is taking place all over the world. A high-temperature fuel cell connected to a combined cycle has been thought of as a solution. The concept consists of three major components: a simple gas turbine cycle, a steam cycle, and a solid oxide fuel cell (SOFC). The fuel cell is connected between the compressor and combustion chamber of the gas turbine cycle.

A fuel cell produces both heat and electricity from the chemical reactions within it. The temperature in a high-temperature fuel cell such as the SOFC is very high (1000°C) and should be kept constant. Therefore, both the air and fuel supplies must have similar temperatures. This is achieved by utilizing the heat produced in the fuel cell. However, the heat produced is a function of fuel cell efficiency and the process efficiency decreases with the time of operation. Research has addressed ways to calculate the decrease in efficiency, formulas for solving the problem of the fuel cell as a block and dispelling the lack of clarity concerning fuel cell application in a gas turbine cycle [14].

### 5.2.6 Parallel Compound Dual-Fluid Cycle Gas Turbine Power Plant

The technology of the Parallel Compound Dual Fluid Cycle Gas Turbine (DFC) was invented by Cheng [15]. The technique used waste heat of exhaust gas of a gas turbine to produce superheated steam, which was then injected into the combustion chamber and turbine, mixed with high-temperature gas, and expanded in the same turbine of the gas turbine engine, so the power output of the gas turbine engine was increased significantly. Some important improvements through the use of this technology include greater specific work, higher efficiency, significant reduction in $NO_x$ emissions, greater flexibility in varying the ratio of electricity supply to heat supply in applications of Cogen plants, and easier refitting of existing gas turbine engines. The first commercial steam injection gas turbine plant started working in California in 1985. Since then, this technology has developed rapidly and several power plants of this kind have been sold by companies in the United States, the United Kindgom, and Japan to the global market.

Figure 7 shows a schematic representation of a parallel compound dual-fluid cycle (DFC) gas turbine. It consists of two main parts: the gas turbine and the waste heat boiler (or heat recovery steam generator, HRSG). Preliminary research has been done to study some characteristics and transient performances of DFC power plant SA-02DFC by developing a mathematical model. For modeling the process of steam injection and mixing in the combustion chamber, a simplifying assumption was made. The combustion chamber was conceptually divided into two zones: the combustion zone, where the compressed air and fuel mixed and burned, and the mixing zone, where hot gas and injected steam mixed. The model of gas turbine was similar to the model of a simple cycle gas turbine [16].

### 5.2.7 $CO_2$ Capture in CC and IGCC Power Plants Using a $CO_2$ Gas Turbine

New technological developments should be undertaken for viable capture and disposal of $CO_2$ produced by electricity generation plants. In the framework of $CO_2$ mitigation strategies, a new concept of $CO_2$-based gas turbine using oxyfuel

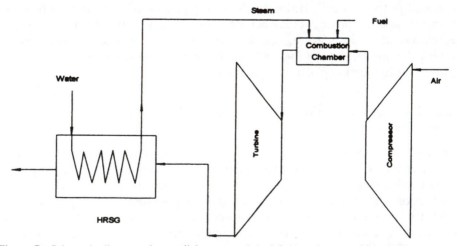

**Figure 7**  Schematic diagram of a parallel compound dual fluid cycle gas turbine (DFC).

and operating in a closed cycle has been presented [17]. To ensure efficient capture, it was proposed to use the $CO_2$ itself, generated by the fuel combustion, as the GT working fluid in a semiclosed cycle. The excess $CO_2$ and water are then easily extracted from the cycle without requiring special costly and energy-consuming separation techniques. The idea was to use the well-known steam-injected (STIG) or evaporative cooling (humid air turbine or HAT) GT cycles in which the nitrogen of air, the working fluid of the regular Brayton cycle is replaced by $CO_2$ and in which the exhaust gases are recirculated to the compressor inlet.

The use of semiclosed cycles running on $CO_2$ rather than on steam or gas is not new, but it never was practiced. This is due to the cost of separating oxygen from air and the need to adapt burners to oxyfuel and to adapt the turbomachinery. Figure 8 gives a schematic representation of a typical steam-injected and evaporative cooling GT cycle. Here, the exhaust gases are recirculated from point B to point A after a cold cleanup, and the heat is introduced by fuel combustion in the GT combustion chamber.

The main ways of mitigating $CO_2$ emissions in power generation systems are the following:

1. $CO_2$ can be extracted from the flue gas using monoethanolamine (MEA), alkanolamine, NaOH, membrane separation, and refrigeration. However, an EPRI-IEA study showed that the power output of a pulverized coal power plant is reduced by about 35% with a loss of efficiency of 12%. In addition, the plant capital cost is multiplied by a factor of up to 2.6 to reduce the $CO_2$ emissions by 90%. For an IGCC plant, this cost is multiplied by about 1.7 while the net power output is reduced by about 12% and the efficiency by 7% [18].
2. The fuel can be reformed (natural gas) or partially combusted to CO and shifted toward $H_2$ and $CO_2$, with subsequent separation. The resulting hydrogen can be burned in any cycle.
3. An open cycle is proposed that uses the Dow process to remove $CO_2$ from the stack gases after the $H_2O$ has been condensed and utilizes the inverted cycle principle [19].
4. Cycles can be designed to run on $CO_2$ as the working fluid, requiring oxygen separation from air and an oxyfuel burner. Pressurized $CO_2$ is available from such a cycle and is easy to recover [17].

The relative merits of these various options using coal as the fuel are largely discussed. However, the first three options using $CO_2$ capture in flue gases are more prevalent because the $CO_2$-based cycles still call for much more technical development.

One of the critical issues related to $CO_2$-based techniques is the requirement for a significant amount of energy for the production of $O_2$ in an air separation unit. The $CO_2$ GT option will be viable only if the cost of an oxygen mass unit

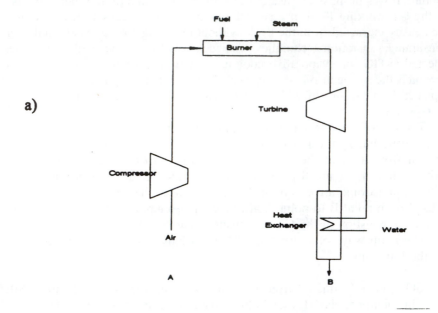

**a)**

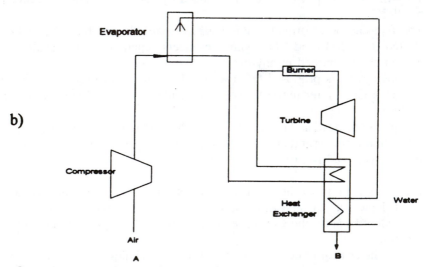

**b)**

**Figure 8** (a) Steam-injected gas turbine cycle (STIG); (b) evaporative cycle.

is lower than the cost of the capture of a $CO_2$ mass unit in flue gases, the costs being dependent on the energy consumption and on the resulting efficiency loss as well as on the electricity price.

## 5.2.8 Development of Advanced Gas Turbine Systems

A Westinghouse-led team, in cooperation with the U.S. Department of Energy's Morgantown Energy Technology Center, is working on phase I of a 10-year, four-phase Advanced Turbine Systems Program to develop the technology required to provide a significant increase in natural gas–fired combined cycle power generation plant efficiency. Environmental performance is to be enhanced, and busbar energy costs are to be 10% less than those of current state-of-the-art turbines.

Intercooling between compressor stages to reduce compression power (which increases shaft power) has long been a feature of centrifugal compressor installations. Thus, inclusion of intercooling by splitting the single axial compressor into a low-pressure (LP) and a high-pressure (HP) section was suggested. Because the HP compressor delivery temperature is greatly decreased with intercooling, additional benefits such as elimination of the rotor air cooler, use of less compressor delivery air, and exhaust gas recuperation can be accrued. Heat exchange between exhaust gas and compressor delivery air significantly reduces the quantity of fuel required to achieve a specified turbine inlet temperature. Because recuperation redirects exhaust energy from the bottoming Rankine cycle into the topping Brayton cycle, as long as the Brayton cycle is more efficient than the Rankine cycle, recuperation should benefit cycle efficiency. With this basis, an intercooled recuperative combined cycle (ICRCC) was analyzed and compared to a base case of a natural gas–fired combined cycle plant. At the same rotor inlet temperature and the same turbine expansion ratio, combined cycle efficiency was found to increase by a very significant 1.5 to 2.0%. In the basic ICRCC configuration, water vapor addition to the Brayton cycle and its effect on efficiency were investigated.

The phase I study showed the feasibility of achieving 60% (LHV) efficiency in a natural gas–fired cycle within a 10-year time frame. The parameters affecting the accomplishment of this goal include cycle innovations, increased firing temperature, reduced cooling air usage, improved component efficiencies, and improved material/coating systems. The resulting advanced turbine system will be environmentally superior, as well as adaptable to coal-fired fuel systems [20].

**System design and optimization of district heat and power plants based on combined cycle plants.** An efficient district heating system can be achieved by adapting a combined cycle plant, simultaneously generating heat and power, to the system. In a conventional district heating system with hot water boilers used as heat suppliers, the boiler load will control the heat supply depending on the heat demand by the consumers. This will result in a varying heat supply to the boiler. For the gas turbine–based combined cycle, however, the gas turbine will be operated continuously on base load for efficiency reasons. This also means a constant heat supply to the heat supplier, which will result in excess heat production during a certain period of the year. For the system design this excess amount of heat has to be taken into consideration in order to obtain optimal overall efficiency. Different solutions, together with their technical and financial consequences, are given [21]. It was found that for an optimal return on investment, when using such a combined cycle district heating plant, the steam system and the gas turbine type have to be selected on the basis of the load duration curve of the consumer grid.

A typical schematic of a district heating system is shown in Figure 9. In an integration of a combined cycle (CC) plant in a district heating (DH) system, the heat can be obtained from two sources:

- The heat available at the exhaust of the steam turbine, which can be used by installing a backpressure steam turbine exhausting into a so-called district heating condenser. Here, the backpressure level depends on the required district heating water supply temperature.
- The heat available at the outlet of the HRSG. The gas outlet temperature of the HRSG is sufficient for generating district heat on the required temperature level by using a so-called district heating coil.

To improve the electrical efficiency, the gas turbine should be operated at a base load independent of the thermal load of the plant. Therefore, the heat produced must be balanced with the heat demand using one of the following options:

1. Cooling down of the excess heat produced in a separate heat exchanger by means of water or air cooling (constant electric efficiency, lower thermal efficiency).
2. Accumulation of the excess heat production in accumulators for later use. Here the combined cycle plant operation hours will be reduced to time only when accumulated heat has been consumed. However, this will provide constant thermal and electrical efficiency.
3. Generation of electric power by using an extraction condensing steam turbine. Here, excess steam not needed for heat production in the district heat condenser will be fed to a low-pressure turbine to generate electric power. In this design, the water outlet temperature of district heat condenser depends on the extraction pressure and can be controlled independently of the district heating water flow by means of the control valve in the crossover pipe to the low-pressure turbine. In this case, the thermal output of a constant gas turbine load can be reduced, resulting in increased electrical power

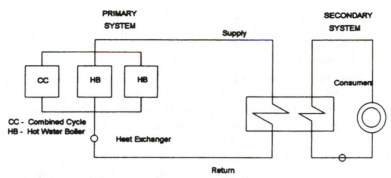

PRIMARY
SYSTEM

Supply

SECONDARY
SYSTEM

CC  HB  HB

Consumer

CC - Combined Cycle
HB - Hot Water Boiler

Heat Exchanger

Return

**Figure 9**  Typical schematic of a district heating system.

output. It also results in a decrease in overall efficiency under part load conditions with the advantage of electrical efficiency increase.

The disadvantage of the system is the need to cool the low-pressure turbine at maximum thermal power output, due to the ventilation losses of the LP turbine. This will reduce the maximum thermal output in comparison with the back-pressure design.
The following conclusions can be drawn from the results of this study.

1. If the heat capacity of the combined cycle plant is large compared with the heat demand of the consumer grid, the condensing design is preferable.
2. If the heating capacity of the combined cycle plant is small compared with the heat demand of the consumer grid, the back-pressure designs are preferable.
3. The back-pressure design with accumulators is not profitable.

**Steam recompression in combined cycles with steam injection.** Combined gas-steam cycles are used worldwide and offer wide efficiency improvement possibilities. Enhanced efficiency can be obtained by carefully designing the heat recovery steam generator (HRSG), which maximizes the utilization of the energy contained in the gas turbine output gases. To minimize the heat transfer energy losses in the HRSG, one suggested method includes a steam recompression bottoming cycle (SRBC). This offers wider applicability to medium-sized and small plants, in addition to performance similar to that of pressure-staged cycles. Low volumetric flow rates in HRSGs, resulting from the use of low-power gas turbines, can affect the performance of conventional dual-pressure bottoming cycles. These are characterized by separated high-pressure and low-pressure steam turbines, the first of which works with partial flow rate. In the SRBC, only one steam turbine is needed, which works on the entire steam flow rate. This advantage is counterbalanced by the presence of the steam compressor, which absorbs a large part of the mechanical work. This puts the SRBC in the middle of a hypothetical performance scale ranging from single-pressure to conventional dual-pressure cycles.

In order to improve gas turbine power plant performance in terms of specific work $W_{sp}$ (combined cycle output power specific to the air mass flow rate at the compressor inlet), injection of steam or water has proved to be an efficient technique. This attractive solution is generally used for peak load operations, as it is counterbalanced by a reduction in efficiency and by the need for makeup water mass flow rates to be treated or to be recovered from the waste gases. The drawback in terms of efficiency can be limited if the water-steam to be injected is supplied by a regenerative process. An energy (second law) analysis of the SRBC, including gas turbine topping cycle, to evaluate the combined cycle overall performances has been presented [22]. It also considers steam injection in the gas turbine, by evaluating feasibility and performance, in terms of energy efficiency and shaft work, of SRBC circuits characterized by different steam extraction points.

**Development of an IGCC flow scheme for commercial refinery residue disposal.** The development of gasification linked with gas turbines for power generation has until recently been confined to the use of coal feedstocks. Refinery residues, which contain much less ash than coal, are gasified for synthesis gas feedstock (e.g., for ammonia and methanol), but when used for power generation these residues are usually burned directly in a boiler to raise steam. There is no universally accepted process flow scheme for the production of electric power from refinery residues using IGCC technology. A study was conducted to this effect for a commercial application of this technology [23]. The study outlines the design philosophy including technological, commercial, and technical considerations. The criteria for selection of design method, gasification and heat recovery techniques, sulfur removal and hydrogen extraction, and options in combined cycle selection are outlined. Based on this, an optimal flow scheme is generated for the commercial application on hand.

**Method for assessing and comparing the potential efficiencies of advanced fossil-fueled combined cycles.** Fuel efficiency can be defined as the efficiency of a combined cycle with a particular fuel-cycle system expressed as a percentage of the efficiency of an optimized clean fueled cycle of similar sophistication, operating under similar conditions and at the same rotor inlet temperature. The fuel efficiency of a clean fueled cycle is thus the base at 100% efficiency. Combined cycles with different arrangements and fuels maintain constant relationships of their efficiencies when compared at equal rotor inlet temperatures. This characteristic has allowed rational quantification and comparison of fuel-cycle systems of combined cycles by fuel efficiency. The designation fuel efficiency is justified because the characteristics of fuels are a major reason for different cycle arrangements. A study conducted used fuel efficiency to compare the performance potential of several combined cycle coal firing systems [24].

The scope of the work was narrowed down to five cycles, with variations, that are either being researched or deserve to be researched.

1. Integrated gasification oxygen blown
   a. Cool water $O_2$ technology
   b. With integrated air separation
   c. With hot gas cleanup
2. Integrated gasification air blown
   a. KRW airblown with gas scrubbing
   b. With hot gas cleanup
3. Pressurized fluid bed
   a. Tidd plant technology
   b. With topping combustion by pyrolysis
4. Externally fired air turbine
   a. Metal air heater
   b. Ceramic air heater
5. Atmospheric fluid bed, metal air heater, pyrolysis topping

From the study it was concluded that the known efficiency of a combined cycle with one fuel system can be used to predict the efficiency of the same basic combined cycle machinery using another fuel system by using the ratio of the fuel efficiencies of the two fuel systems. Furthermore, the fuel efficiency method of evaluating combined cycles allowed cycle efficiency at any rotor inlet temperature to be prorated to any other rotor inlet temperature by the ratio of efficiencies of a clean cycle between the same two temperatures. The study allowed the efficiencies of the different cycles to be predicted and compared over a range of rotor inlet temperatures and presented the impact of limitations on rotor inlet temperature due to fuel, cleanup, heat exchanger, or gas turbine technology.

**New prospects for the use of regeneration in gas turbine cycles.** Excessive complication was one of the main obstacles to the diffusion of the internally regenerative gas cycle, in which compressed air is preheated by the exhaust gases before entering the combustion chamber. The main difficulties were linked to the development of the regenerative gas-air exchanger, to machine adjustment, and to start-up; moreover, as the compression ratio rises the air temperature increases, reaching finally a condition at which regeneration becomes less and less effective to the point of becoming impossible. The marked progress in the field of gas turbines has led to the creation of increasingly complex systems such as the combined bottomer cycle with two or more pressure levels with efficiency of the order of 50%, which is considerably better than the state of the art for Rankine steam cycles. These improvements were achieved by exploiting sensible heat at the turbine exhaust, with the disadvantage of considerable complications in the basic gas cycle requiring major investments and construction time. The use of alternative fluids has been proposed for the bottomer cycle to improve heat recovery [25].

Several modifications of the traditional Joule cycle have also been proposed with the aim of increasing performance without having to add bottomer cycles. These proposals share the common feature of using water as the working fluid in the gas cycle. Two examples are the steam injection cycle (STIG) and the more recent humid air cycle (HAT). All this leads to considerable complications, the need for a suitable water treatment system and high water consumption, which is no longer viable.

In view of economic and R & D impetus in the field of gas turbines for land applications, the technological limits preventing the full development of regenerative gas turbine cycles are no longer as restrictive. Furthermore, by making appropriate modification to the Joule cycle, it is possible to create a regenerative cycle not affected by the same limitation on the maximum temperature on the compressor output or on high compression ratios as placed on traditional regeneration [26].

**Advanced combined cycle alternatives with advanced gas turbines.** New gas turbine designs based on higher turbine entry temperatures (TETs) are on the market. These new machines hold the promise of even higher combined cycle efficiencies for two reasons: when the gas turbine TET is increased, both the gas turbine efficiency and the gas turbine exhaust temperature are increased for the optimum pressure ratio under combined cycle operation. The increase in GT exhaust temperature allows the use of higher superheated steam temperatures, in the vicinity of the classical 540°C, which is common practice in classical steam power plants. A considerable improvement in combined cycle efficiency is obtained under these circumstances, because the Rankine cycle efficiency is highly dependent on the steam superheat temperature.

There is also the possibility that some advanced combined cycle configurations, especially those involving reheat or reheat with supercritical steam conditions, come into force because of this increase in GT exhaust temperatures. Such cycle configurations are usually thought to take better advantage of the amount of heat that is available above the highest pinch point temperature in the HRSG. Moreover, new IGCC power plants often incorporate advanced triple pressure (sometimes even more) reheat systems in order to fulfill the steam requirements on the gasification side. All this results in an important increase in the use of gas-fired combined cycle plants for power generation in many countries.

The relative merits of such advanced combined cycle options have been compared with those of a state-of-the-art two-pressure system, for both a 150-

MW class GT with 1100°C TET and a 200-MW class GT with 1288°C TET. In all, six different combined cycle schemes were considered in the study [27]. These options were:

Dual-pressure nonreheat cycle (2P)
Triple-pressure nonreheat cycle (3P)
Dual-pressure reheat cycle (2PR)
Triple-pressure reheat cycle (3PR)
Dual-pressure reheat supercritical cycle (2PRS)
Triple-pressure reheat supercritical cycle (3PRS)

It was concluded that of all the improvements that can be made in a state-of-the art 2P combined cycle, the introduction of reheat holds the most promise in terms of fuel savings.

## 5.3 CLEAN COAL TECHNOLOGY (CCT) PROGRAM

Congress initiated a program in 1985 to cofund with industry the demonstration of environmentally sound, coal use technologies that attempt to reduce or eliminate the environmental and economic factors influencing the gap between energy reserves and energy use patterns in the United States.

The first four rounds of solicitations represented $2.148 billion in government support and an even greater amount in industry money [28]. The program includes 42 projects, of which five projects are either new, repowering, or cogeneration IGCC demonstrations that range from 55 to 265 MW in size. A synopsis of these projects, representing a total investment of $1.60 billion, is given [29].

After the Chernobyl accident in April 1986, most Western countries have slowed down, suspended, or stopped their nuclear programs [30]. In many cases, natural gas has been regarded as the best alternative to nuclear fuel in the short and medium term, and numerous countries have opted for combined cycle plants. However, dependence on natural gas as the only fuel would lead to a situation similar to that previously experienced with oil. Hence, the need to diversify the fuel sources and to develop a combined cycle technology with high efficiency and multifuel flexibility is felt worldwide (refer to Figure 10).

Given this background and the development of advanced technologies such as IGCC, interest in coals as a fuel source for future base load power plants has reemerged in the United States and the world. The IGCC concept offers several advantages over conventional technology. These include high thermal efficiencies; considerably reduced emission levels of $SO_x$, $NO_x$, and particulates that easily satisfy the federal limits; and lower costs of emission control. Moreover, it offers utility planners the necessary flexibility in matching future load growth.

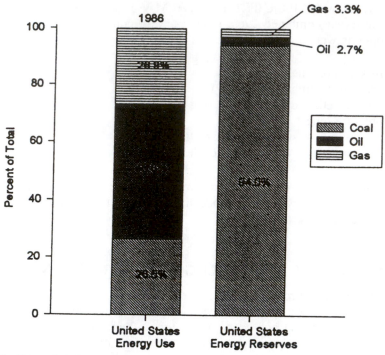

**Figure 10**  Energy imbalance.

IGCC technology is modular in nature and offers the possibility of phased construction. A schematic representation of one of many possible phased-construction scenarios for an IGCC plant is presented in Figure 11 [31].

The simple scheme in Figure 11 is based on two gas turbines. In the first phase, a single gas turbine is installed, to be fired by natural gas and using a peaking unit in a utility system. A second gas turbine is added in the second phase, increasing the capacity of the plant, which continues to be operated as a peaking unit. A steam cycle is added next, increasing the efficiency of the plant and enabling it to operate as a base load combined cycle plant fired by natural gas. In the fourth and the final phase, a coal gasification plant is constructed and the entire facility is operated as a coal-fired IGCC power plant. Detailed engineering analyses are required to determine the performance of such schemes during each phase of operation [31].

The disadvantage in the above strategy, starting with a combined cycle plant and adding gasification units several years later in order to switch from natural gas to coal, is the capital cost involved and also a loss of performances of the resulting plant [30]. A second option, wherein, a combined cycle plant is built with some of its components oversized in order to allow easy integration of the

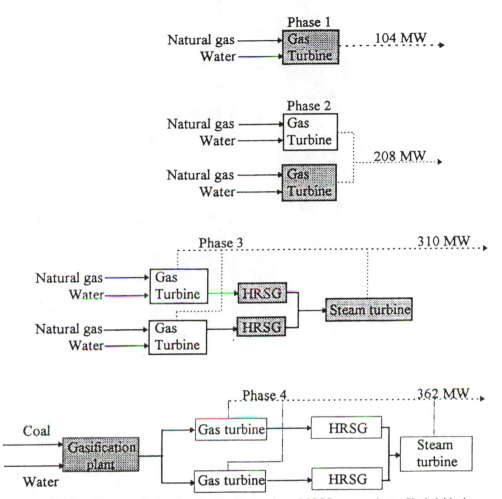

**Figure 11** Possible scenario for the phased construction of IGCC power plants. Shaded blocks represent new capacity; open blocks, existing capacity. (Source: Reference 31.)

future connections from and to the gasification unit erected in the second stage of the construction, is investigated [27]. The performances of the combined cycle plant for various pre-investment options, inside the heat recovery boiler, in the steam turbine, and in the gas turbine are calculated. In particular, the perform-ances of the resulting IGCC plants with four types of gasifiers used, namely, Texaco, Shell, Dow, and British-Gas-Lurgi, are compared. The paper uses the performances of the same plants built in stages without pre-investment and built as genuine IGCC plants as the yardsticks for comparison.

**Table 1 Synopsis of five IGCC projects**

| | | |
|---|---|---|
| Title | Combustion Engineering (CE) IGCC Repowering Project | Tampa Electric IGCC Project |
| Participants | CE: Springfield City Water, Light & Power (CWL&P); Illinois Department of Energy & Natural Resources (ENR) | Tampa Electric Company |
| Location | Lakeside Generating Station, Springfield, IL | Polk County, FL (20 miles south of Lakeland) |
| Gasifier | Pressurized, air-blown, two-stage, entrained flow | Pressurized, oxygen blown, entrained flow |
| Gas cleanup | Cyclones, hot particulate removal and GE moving-bed zinc ferrite desulfurizer | Cyclones and GE moving-bed zinc titanate desulfurizer |
| Combined cycle | 40 MW, 2000°F gas turbine, HRSG, 25-MW repowered steam turbine | 190-MW, 2300°F advanced gas turbine, HRSG, 110-MW steam turbine, 260 MW net |
| Cost | $270 million, 48% DOE funded | $241 million, 50% DOE (part of larger project) |

| | | | |
|---|---|---|---|
| Title | Toms Creek IGCC Demonstration Project | Piñon Pine Power Project | Wabash River Coal Gasification Project |
| Proposer | TAMCO Power Partners | Sierra Pacific Power Co. | Wabash River Coal Gasification Repowering Project Joint Venture |
| Participants | Tampella Power Corp.; Coastal Power Production Co.; Stone & Webster; Institute of Gas Technology | Foster Wheeler USA Corp.; MW Kellogg Co. | PSI Energy, Inc.; Destec Energy, Inc. |
| Location | Coeburn, VA | Tracy Station, Reno, NV | West Terre Haute, IN |
| Gasifier | Tampella U-Gas fluidized-bed with in-bed sulfur capture | KRW fluidized bed with in-bed sulfur capture | DOW oxygen-blown, entrained bed |
| Gas cleanup | Cyclones, zinc titanate desulf, and barrier filter | Cyclones, barrier filter, and zinc ferrite desulf. | Warm particulate; low temperature acid gas removal |
| Combined cycle | 55-MW GE MS 6001 Gas turbine | 80 MW | 265 MW net output utilizing 192-MW advanced gas turbine |
| Project cost | $219 million | $341 million | $592 million |
| DOE request | $109 million[a] | $170 million[a] | $243 million[a] |

[a] Funding amounts reflect those contained in proposals and are subject to negotiation.

## 5.3.1 IGCC for Acid Rain Control [32]

Emissions of sulfur dioxide and nitrogen oxides into the atmosphere increase the acidity of moisture in the air, which is commonly called "acid rain". Man-made sulfur dioxide and nitrogen oxides emissions in the U.S. are estimated to be over 40 million ton/yr. Coal-fired utility power plants account for 55% of all the man-made sulfur dioxide emissions in the United States [33]. Coal-fired utility power plants also account for 27% of all the man-made nitrogen oxides emissions in the United States [33]. Sulfur dioxide emissions appear to acidify freshwater lakes located in acidic soil areas, whereas nitrogen oxides emissions appear to kill trees. It was, therefore, obligatory that the utility industries employ cleaner technology, either using coal or other fuel sources, to mitigate the environmental impacts such as acid rain resulting from its emissions.

Coal gasification is commercially demonstrated and can reduce $SO_2$ and $NO_x$ to significantly lower levels than other technologies. However, to use this technology effectively and economically for acid rain control, it has to be integrated with a combined cycle for power generation. For coal gasification to be effective, the following concepts have to be used together:

- Repowering
- Coprocessing
- Phased construction

Several coal gasification processes are available, each with its own distinctive characteristics, preferred feed coals, and applications. Of these, four types of commercial coal gasification processes have been successfully used worldwide for 40 years.

- Lurgi
- Koppers-Totzek
- Winkler
- Gas Producers

However, all these systems have operating problems associated with handling eastern and midwestern bituminous coals, and none is a serious candidate for acid rain control. The three most important coal gasification technologies for acid rain control application are:

- Texaco
- Dow
- British Gas Corp/Lurgi Slagger

These technologies are ideally suited for using eastern and midwestern U.S. bituminous coal and have been operating in commercial-scale demonstration plants with high efficiency and at over 30 atm pressure.

## 5.3.2 Cool Water IGCC Plant

Texaco's is considered the most prominent system because of the success of the Cool Water demonstration plant [34, 35]. This plant, located near Barstow, California, produces 120 MW of electricity from 1000 ton/d of bituminous coal using IGCC technology. The coal is gasified in a single Texaco gasifier. The raw coal gas is cooled by generating steam and then desulfurized before combusting the 220 Btu/scf gas in a single General Electric Frame 7 combustion turbine. The hot flue gas, at about 1000°F, is cooled by superheating steam produced in the gasifier raw gas coolers and by generating additional steam. This steam is used in a steam turbine (bottoming cycle) to generate additional electricity.

Engineering of this "first of a kind" IGCC project had begun in February 1980 and was completed in April 1984 significantly under budget and ahead of schedule. As of December 1985, the plant produced over 700 million kWh of electricity from over 300,000 tons of coal. The plant had significant environmental performance during its operation. The $SO_2$ emissions were consistently less than 0.034 lb/$10^6$ Btu feed coal and $NO_x$ emissions were less than 0.059 lb/$10^6$ Btu feed coal. These emissions are an order of magnitude lower than the current NSPS standards for new coal-fired utility boilers. The only technology that can match this performance is combustion of natural gas with special modification to reduce $NO_x$.

A large number of advanced coal gasification processes are under development. However, these systems have yet to be demonstrated at the large commercial demonstration scale of operation of Texaco, Dow, and BGC/Lurgi coal gasification processes. Important advanced coal gasification processes include:

| | |
|---|---|
| Shell | CGT |
| PRENFLO (pressurized K-T) | U-Gas |
| VEW | HKV |
| Sumitomo/KHD | CMRC (Japan) |
| HTW | Lurgi (fluidized bed) |
| KRW (Westinghouse) | Kilngas |
| Kloeckner | |

## 5.3.3 Economics of IGCC

The commercial-scale Cool Water plant demonstrated that coal gasification can reduce $SO_2$ and $NO_x$ to significantly lower levels than any other coal-based acid rain control option. A study from EPRI detailed the designs and capital cost estimates for IGCC plants based on the Texaco and the BGC/Lurgi processes, whose designs were based on:

Nominal 500-MW plant size
Green-field southern Illinois site
High-sulfur Illinois No. 6 bituminous coal
95% sulfur removal and low $NO_x$

The Texaco-based IGCC design and cost estimate were produced by Fluor Engineers Inc. and based on constant 1983 dollars [36]. The BGC/Lurgi-based IGCC design and cost estimate were made by R.M. Parsons Company and based on constant 1984 dollars [37]. The newer IGCC plants use hot gas cleanup technology unlike the cold gas cleanup technology used in the Cool Water demonstration plant.

## 5.3.4 Repowering and Coprocessing

Repowering refers to the integration of a new combustion turbine power generation unit with an existing utility steam boiler to create a combined cycle system. This concept provides many advantages:

A combined cycle offers significantly better thermal efficiency than a steam cycle.
The capital cost of combustion turbine power generation is significantly less than that of any other new power generation option.
Combustion turbines can be installed in significantly less time than any other new power generation option.
Repowering increases plant capacity: depending on configuration and desired increase, it can be as much as 70%.

The only way coal gasification can be economical for emission reduction in existing power plants is via repowering to assure high efficiency. Also, repowering increases power output and efficiency, whereas operations such as flue gas treating reduce power output and efficiency. Repowering existing high-sulfur residual oil–fired boilers via coal gasification should be considered because of higher price of residual oil. Coal gasification provides total sulfur removal and produces essentially no $NO_x$, so repowering existing high-sulfur coal-fired boilers with coal gasification should be considered.

## 5.3.5 Phased Construction

Phased or staged construction usually refers to construction of natural gas–fired combustion turbines or combined cycle power units. The units are built one by one depending on the demand. There are a number of advantages of phased construction IGCC over direct coal combustion including:

Gradual capacity increase.
Gradual capital expenditure.
Short time from start of construction to initial electric generation.
Lower economic risk if the coal gasification investment is delayed. The combined cycle can operate on natural gas until the price and supply of gas warrant the change to coal.

Phased construction IGCC should also be attractive as an alternative emissions control for existing high-sulfur coal-fired utilities. Natural gas–fired combustion turbines can be quickly and inexpensively installed to repower existing coal-fired boilers. Coal and natural gas can be coprocessed with the ratio determined by reduction requirements for sulfur dioxide and nitrogen oxides. The short-term price and economics of natural gas make this economically attractive. As natural gas prices increase and the current oversupply of natural gas is consumed, the plant can be converted to coal-derived gas.

### 5.3.6 Integrating Thermal Energy Storage in Power Plants

The National Energy Strategy forecast estimates that 200,000 megawatts-electric (MWe) of new electric-generating capacity must be added in the United States by the year 2010 [38]. Approximately 40% of this new generating capacity will be for peak or intermediate loads, with the rest providing continuous base load power generation. Natural gas-fired combustion turbine technologies, including cogeneration, combined cycle, and integrated gasification combined cycle power plants, are becoming the generating options of choice because of their flexibility, relatively low capital cost, reduced environmental impact, and increasing thermal efficiency. Thermal energy storage (TES) for utility applications includes a range of technologies that can further improve the efficiency, flexibility, and economics of gas turbine options [39].

The storage system selected will depend on the quality and quantity of recoverable thermal energy and on the nature of the thermal load to be supplied from the storage system. The TES systems and technologies that are being considered for utility applications can be categorized by storage temperature. High-temperature storage can be used to store thermal energy at high temperatures (heat storage) from sources like gas turbine exhaust and hot syngas from coal gasifiers [49]. High-temperature TES options such as oil-rock storage, molten nitrate salt storage, and combined molten salt and oil-rock storage are well developed and commercially available. Low-temperature TES technologies store thermal energy at temperatures below ambient temperature (chill storage) and can be used to cool the air entering gas turbines. Examples of chill storage concepts include commercially available options such as diurnal ice storage and more advanced schemes represented by complex compound chemisorption TES systems.

## 5.3.7 Biomass Fueled Cogeneration

Carbon dioxide emissions associated with electricity generation as well as transportation fuel combustion contribute to the greenhouse effect and need to be controlled. The various measures required to decrease $CO_2$ emissions include energy saving, replacing fossil fuel–fired power plants by combined heat and power (CHP), nuclear energy, natural gas, renewable wind, and biomass, as well as reforestation to provide a carbon sink. For biomass, the most attractive new alternatives are IGCC on a large-scale and bio-oil-fueled diesel plants on a small scale parallel to today's commercial fluid bed combustion technology [40].

Another strategy for reducing $CO_2$ emissions from the transportation sector involves methanol synthesis using gases derived from biomass sources. The production of methanol is integrated with processes for the generation of electricity from coal or natural gas. This approach assumes that nearly all the carbon of the biomass may be incorporated in the methanol or may compensate, partially or fully, for the emissions of fossil fuel carbon released in electricity generation. If the quantity of $CO_2$ captured and sequestered equals the quantity of carbon entering the process with the fossil fuel, both the methanol synthesized and the electricity generated are neutral with respect to $CO_2$ emissions in the atmosphere [41].

## 5.3.8 IGCC Coproducing Electricity and Fertilizer (IGCC/F)

The concept of coproduction is based on the use of commercial coal gasification and related process units to produce electricity and chemicals concurrently. The combined cycle unit to produce electricity is sized for the full gasification synthesis gas output. The production of chemicals can be bypassed during periods of peak power demand. Coproduction of fertilizers requires integration of the following additional process units with IGCC:

- Ammonia unit: hydrogen separation or enhancement and catalytic gas-phase reaction of $H_2$ and nitrogen, from the air separation unit, to produce ammonia.
- Urea unit: two-stage reaction of ammonia and carbon dioxide, from acid gas removal, to produce urea.

Figure 12 shows the process concept for a coproduction demonstration project [42]. The coproduction concept utilizes coal gasification to optimize revenues from the sale of the primary coproducts (fertilizer or organic chemicals and electricity) and several by-products. Urea has a higher net value, is easier to

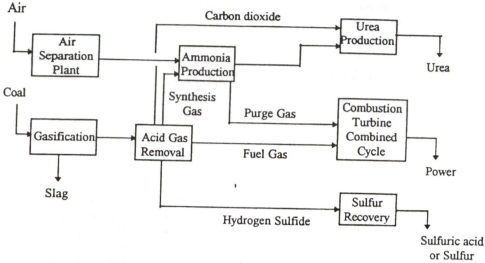

**Figure 12**   Process concept: coproduction demonstration project.

store, and is more readily marketable than anhydrous ammonia. The United States is a net importer of nitrogen fertilizers, and imports are expected to continue to grow because of the lack of economic incentives to build new ammonia or urea plants in the United States. A coproduction project can either provide new capacity to replace future imports or replace existing capacity as old plants are retired.

# REFERENCES

1. MacGregor, P. R., C. E. Maslak, and H. G. Stoll. Integrated gasification combined-cycles: A key generation option for the 1990s. Industrial Power Conference, pp. 45–51, ASME, 1992.
2. Almhem, P., and J. J. Lofe. The 750 Mwe PFBC power plant: Reliable, clean and efficient power. Presented at the Joint ASME/IEEE Power Generation Conference, Kansas City, Kansas, October 17–22, 1993.
3. Natesan, K. Materials performance in advanced combustion systems. *J. Eng. Gas Turbines Power* 116:331–337, April 1994.
4. Rehmat, A., and A. Goyal. Advanced pressurized fluidized-bed staged combustion. Presented at the Joint ASME/IEEE Power Generation Conference, Kansas City, KS, October 17–22, 1993.
5. Chapman, J. N., and N. R. Johanson. Design considerations for a class of 600 MWe MHD steam combined cycle power plants, ASME 93-GT-159, pp 1.983–1.988, 1993.
6. Hart, A. T., K. D. Filius, D. A. Micheletti, and P. V. Cashell. Coal-Fired MHD Test Progress at the Component Development and Integration Facility, ASME 93-GT-159, pp 1.989–1.994, 1993.
7. Klara, J. M., and J. H. Ward. High performance power systems: State-of-the-art configurations. Presented at the International Power Generation Conference, Atlanta, October 18–22, 1992.
8. Doctor, R. D., J. C. Molburg, P. Thimmapuram, G. F. Berry, C. D. Livengood, and R. A. Johnson. Gasification combined cycle: Carbon dioxide recovery, transport, and disposal. *Energy Convers. Mgmt.*, 34(9–11):1113–1120, 1993.

9. Frey, C. H., and E. Rubin. Integration of coal utilization and environmental control in integrated gasification combined cycle systems. *Environ. Sci. Technol.* 26(10):1982–1990, 1992.
10. Frey, C. H., and E. Rubin. Evaluation of advanced coal gasification combined cycle systems under uncertainty. *Ind. Eng. Chem. Res.* 31(5):1299–1307, 1992.
11. Gangwal, S. K., S. M. Harkins, M. C. Woods, S. C. Jain, and S. J. Bossart. Bench-scale testing of high-temperature desulfurization sorbents. *Environ. Prog.* 8(4):    , 1989.
12. Taki, Y., and R. Babus'Haq. Cogeneration/District-Heating Model: Energy-Environment-Economics Approach. IGTI-Vol. 8, ASME COGEN-TURBO, pp. 177–87, 1993.
13. Taki, Y., R. F. Babus'Haq, and S. D. Probert. Combined heat and power as a contributory means of maintaining a green environment. *Applied Energy* 39:(2):83–91, 1991a.
14. Rokni, M. Introduction of a Fuel Cell into a Combined Cycle: A Competitive Choice for Future Cogeneration. IGTI-Vol. 8, ASME COGEN-TURBO, pp. 255–261, 1993.
15. Cheng D. Y. Parallel-compound dual-fluid heat engine. Patent 3978661, September 7, 1976.
16. Qing, W., F. Shikai, S. Yufeng, Y. Jianxiong, W. Guoxue, and Z. Zhengyi. Preliminary Research on Some Characteristics and Transient Performances of a Parallel Compound Dual Fluid Cycle Gas Turbine Powerplant S1A-02DFC. IGTI-Vol. 8, ASME COGEN-TURBO, pp. 281–287, 1993.
17. Mathieu, P., and J. De Ruyck. $CO_2$ Capture in CC and IGCC Power Plants Using a $CO_2$ Gas Turbine. IGTI-Vol. 8, ASME COGEN-TURBO, pp. 77–83, 1993.
18. Booras, G. S., and S. C. Smelser. An engineering and economic evaluation of $CO_2$ removal from fossil-fuel-fired power plants. *Energy* 16(11/12):1991.
19. Rice, I. G. Steam injected gas turbine analysis part III—steam-regenerated heat. ASME IGTI Conference, Cincinnati, 1993.
20. Little, D. A., R. L. Bannister, and B. C. Wiant. Development of Advanced Gas Turbine Systems. IGTI-Vol. 8, ASME COGEN-TURBO, pp. 271–280, 1993.
21. Linnemeijer, M. J. J. System Design and Optimization of District Heat and Power Plants Based on Combined Cycle Plants. IGTI-Vol. 8, ASME COGEN-TURBO, pp. 171–176, 1993.
22. Grimaldi, C. N., and G. Bidini. Steam Recompression in Combined Cycles with Steam Injection. IGTI-Vol. 8, ASME COGEN-TURBO, pp. 397–407, 1993.
23. Griffiths, J., and A. F. Magrath. Development of an IGCC Flowscheme for Commercial Refinery Residue Disposal: Quench—The Thirst for Power. IGTI-Vol. 8, ASME COGEN-TURBO, pp. 127–142, 1993.
24. Foster-Pegg, R. W. A Method for Assessing and Comparing the Potential Efficiencies of Advanced Fossil Fuelled Combined Cycles. IGTI-Vol. 8, ASME COGEN-TURBO, pp. 409–415, 1993.
25. Kalina, A. Combined cycle system with novel bottoming cycle. *ASME J. Eng. Gas Turbines Power* 106:737–742, 1984.
26. Facchini, B. New Prospects for the Use of Regeneration in Gas Turbine Cycles. IGTI-Vol. 8, ASME COGEN-TURBO, pp. 263–69, 1993.
27. Dechamps, P. J., D. Magain, and P. Mathieu. Advanced Combined Cycle Alternatives with Advanced Gas Turbines. IGTI-Vol. 8, ASME COGEN-TURBO, pp. 387–396, 1993.
28. Schmidt, D. K. Integrated gasification combined cycle projects in the clean coal technology program. Presented at the 1991 International Joint Power Generation Conference, San Diego, CA, October 1991.
29. Salvador, L. A., and T. F. Bechtel. "Coal gasification: A key for advanced power systems. Presented at the 1991 EPRI Gasification Power Conference, Palo Alto, CA, October 1991.
30. Dechamps, P. J., and Ph. Mathieu. Phased construction of IGCC plants for fuel flexibility. Paper ASME 92-GT-144, presented at the ASME Turbo-Expo 92, Koln, June 1992.
31. Erbes, M. R., J. N. Phillips, M. S. Johnson, Paffenbarger, M. Gluckman, and R. H. Eustis. Off-design performance of power plants: An integrated gasification combined-cycle example. *Science*, 237:379–383, 1987.
32. Simbeck, D. R., and R. L. Dickenson. Integrated gasification combined cycle for acid rain control. *Chem. Eng. Prog.* October: 28–33, 1986.

33. DOE. *Acid Rain Information Book*, 2nd ed. U.S. Dept. of Energy, DOE/EP-0018/1, Washington, DC, May, 1983.
34. Holt, N. A., et al. The cool water project—preliminary operating results. AIChE meeting, Houston, March 26, 1985.
35. McCarthy, C. B., and W. N. Clark. Integrated gasification/combined cycle (IGCC) electric power production—A rapidly emerging energy alternative. Symposium on Coal Gasification and Synthetic Fuels for Power Generation, San Francisco, April 15, 1985.
36. Matchak, T. A., et al. Cost and Performance for Commercial Applications of Texaco-Based Gasification–Combined Cycle Plants. EPRI, AP-3486, Palo Alto, CA, April 1984.
37. dela Mora, J. A., et al. Evaluation of the British Gas Corporation/Lurgi Slagger Gasifier in Gasification–Combined-Cycle Power Generation. EPRI, AP-3980, Palo Alto, CA, March 1985.
38. Stuntz, L. G. The national energy strategy: DOE's view from the top. *Electricity J.* 3(8):18–23.
39. Somasundaram, S., M. K. Drost, Z. I. Antoniak, and D. R. Brown. Integrating thermal energy storage in power plants. *Mech. Eng.* September: 84–90, 1993.
40. Sipila, K., Y. Solantausta, and E. Kurkela. Long-term cogeneration and biomass strategies for reducing $CO_2$ emissions in Finland. *Energy Convers. Mgmt.* 34(9–11):1051–1058, 1993.
41. Walsh, J. H. The synthesis of atmospherically-neutral methanol integrated with the generation of electricity in process equipped for the capture and sequestering of carbon dioxide. *Energy Convers. Mgmt.* 34(9–11):1031–1049, 1993.
42. Bradshaw, D. T., T. L. Wright, and R. Weatherington. An integrated coal gasification combined cycle coproducing electricity and fertilizer (IGCC/F). Proceedings of the American Power Conference, pp. 1205–1207, 1994.
43. Ayala, R. E., and D. W. Marsh. Characterization and long-range reactivity of zinc ferrite in high-temperature desulfurization processes. *Ind. Eng. Chem. Res.* 30:55–60, 1991.
44. Dechamps, P. J., and Ph. Mathieu. Phased construction of IGCC plants with pre-investment in the CC plant. Paper ASME 93-GT-159, presented at the ASME International Gas Turbine and Aeroengine Congress and Exposition, Cincinnati, May 1993.
45. Diwekar, U. M., C. H. Frey, and E. S. Rubin. Synthesizing optimal flowsheets: Applications to IGCC system environmental control. *Ind. Eng. Chem. Res.* 31(8):1927–1936, 1992.
46. Rath, L. K., and R. C. Bedick. Research and Development Efforts at the Department of Energy (DOE) Supporting Integrated Gasification Combined Cycle (IGCC) Demonstrations. IGTI-Vol. 7, ASME COGEN-TURBO, ASME, pp. 87–94, 1992.
47. Smith, D. J., ed. Repowering could be preferable to new construction. *Power Eng.* September: 38–40, 1988.
48. Smith, D. J., ed. Commercialization of IGCC technology looks promising. *Power Eng.* February: 30–32, 1992.
49. Somasundaram S., M. K. Drost, Z. I. Antoniak, and D. R. Brown. Thermal energy storage for power plant applications. *Energy Eng.* 87(3):34–48, 1990.

## PROBLEMS

1. What do the following abbreviations stand for?
   (a) IGCC
   (b) FGD
   (c) NSPS
   (d) CC
   (e) FBC
   (f) MHD

(g) HIPPS
(h) LEBS
(i) CGCU
(j) HGCU
(k) MCFC
(l) SCR
(m) CHP
(n) DFC
(o) SOFC
(p) HRSG
(q) STIG
(r) HAT
(s) MEA
(t) ICRCC
(u) SRBC
(v) TET
(w) CCT
(x) IGCC/F

2. List the advantages of the IGCC technology.
3. With adoption of the IGCC technology, what kind of results are expected (i.e., increase, decrease, or remain unchanged)?
    (a) efficiency
    (b) emission of particulates
    (c) emission of greenhouse chemicals
    (d) multiple-fuel flexibility
    (e) capital cost of plant
4. What are included in "$NO_x$"?
5. What are greenhouse gases?
6. List process industries where $CO_2$ can be used as raw material.
7. List $SO_2$ control technologies.
8. List $NO_x$ control technologies.
9. Explain briefly "fuel switching".
10. Describe the fluidized-bed-combustion cogenation process.
11. List the advantages of high-performance power systems.
12. List the technological advantages of low emissions boiler system.
13. What is the discerning difference between the hot gas cleanup and cold gas cleanup?
14. Discuss the process technology employed in the cool water demonstration plant.
15. Discuss the Selexol process.
16. Is thermal $NO_x$ emission expected to be higher or lower for low temperature burner? Why?
17. What are the advantages of using zinc ferrite as a sorbent?

18. What is a fuel cell?
19. How does the heat recovery steam generator operate?
20. Explain the Rankine cycle and the Brayton cycle.
21. Discuss the energy integration nature of electric power generation using an extraction condensing steam turbine.
22. What is the cause of "acid rain"?
23. How does the IGCC technology contribute to the solution of the acid rain problem?
24. For coal gasification, explain the following concepts: repowering, coprocessing, and phased construction.
25. List various coal gasification processes that are candidates for the IGCC technology.
26. What additional process units are required for coproduction of fertilizers with the IGCC?
27. What are the advantages of urea over anhydrous ammonia?

# COAL SLURRY FUEL

## 6.1 INTRODUCTION

Coal slurry fuels consist of finely ground coal dispersed into one or more liquids such as water, oil, or methanol. Slurry fuels have the advantages of convenient handling (similar to heavy fuel oil) and a high energy density as shown in Table 1 [1, 2]. Slurries have been investigated for replacement of oil in boilers and furnaces, internal combustion engines, and recently for cofiring of coal fines in utility boilers. Coal slurry is used around the world in countries such as the United States, Russia, Japan, China, and Italy.

Coal slurry fuels have been investigated since the nineteenth century, but economic issues have kept it from becoming a usable energy source. Typically, interest in coal slurry develops when oil availability is in doubt, such as periods during both world wars and again in the energy crises of 1973 and 1979 [3]. Much of the work during these time periods was on coal-oil fuels, which could quickly and easily replace oil in furnaces and boilers. However, research since 1980 has centered on coal-water fuels (CWFs) for complete replacement of oil in industrial steam boilers, utility boilers, blast furnaces, process kilns, and diesel engines [4].

The initial development of coal slurry was performed over a hundred years ago by Smith and Munsell [5]. By World War I a full-scale slurry test was

## Table 1 Fuel energy densities [2]

| Fuel | Density (lb/gal) | Btu/lb | Btu/gal | Btu/ft$^3$ |
|---|---|---|---|---|
| Coal in bulk (7% moisture) | 6.2–9.4 | 12,500 | 76,000–116,500 | 573,000–872,000 |
| Residual oil | 8.2 | 18,263 | 150,000 | 1,122,000 |
| 60% coal/40% water blend | 9.8 | 8,000 | 78,700 | 589,000 |
| 70% coal/30% water blend | 10.2 | 9,373 | 95,600 | 715,000 |

successfully made on the U.S. Navy Scout ship *Gem*. The test revealed problems such as high ash content, visible fluid track left in the ship's wake, and stability problems with the slurry [6].

In the 1930s, the Cunard ship company used coal-oil slurries in both land and sea trials in an attempt to reduce oil imports and develop new markets for coal [7–9]. At the same time, tests were conducted in Japan on coal-in-oil fuels at the National Fuel Laboratory [10]. Similar tests were performed in Germany on a mixture of powdered coal (55 wt. %) and tar oil (45 wt. %) called *Flies-skhloe*. The tests showed that the mixture burned well and had thermal efficiencies of 70–75% [11]. Although the systems worked well, economic limitations limited further development.

Development during World War II consisted of two comprehensive programs at the Bureau of Mines and Kansas State College. The programs explored methods of preparation, flow, stability, and burning processes of coal-oil slurry [6]. After the war development on coal-oil slurries ceased until the 1970s. However, work on coal-water fuels started in the USSR in the 1950s [12, 13], and similar work was conducted in the United States and Germany on storage, pumping, and combustion properties [14].

The oil crisis of 1973 again propelled research and development into coal-oil slurries. A consortium of companies was formed in 1973, led by General Motors, to develop the technology [15]. In 1975 the Department of Energy (DOE) joined in supporting the project, and by 1976 the program had expanded into switching utility gas and oil boilers to coal-oil mixtures. The initial comprehensive investigations were completed in 1977, and the DOE transferred GM projects to places like New England Power and Service Co. (NEPSCO) and Pittsburgh Energy Technology Center (PETC) [6].

Research since 1980 has centered on coal-water fuels for the complete replacement of oil [4]. The economic incentive for replacing oil with coal-oil slurries disappeared as oil prices dropped. This has spurred the development of coal-water slurries for complete replacement of oil or coburning of coal fines in a slurry. A significant amount of development has been accomplished on coal slurry rheology, characterization, atomization, combustion mechanisms, and transport techniques.

## 6.2 COAL SLURRY CHARACTERIZATION

Coal slurry mixtures can be made from a combination of various liquids; the most common are oil, water, and methanol. Detailed descriptions are given below:

1. *Coal-oil mixture (COM)*—a suspension of coal in fuel. Sometimes referred to as coal-oil dispersion (COD).
2. *Coal-oil-water (COW)*—a suspension of coal in fuel oil with less than 10 wt. % water. Oil is the main ingredient.
3. *Coal-water-oil (CWO)*—a suspension of coal in fuel oil with more than 10 wt. % water. Water is the main ingredient.
4. *Coal-water fuel (CWF), coal-water mixtures (CWM), coal-water slurry (CWS), or coal-water slurry fuel (CWSF)*—a suspension of coal in water.
5. *Coal-methanol fuel (CMF)*—a suspension of coal in methanol.
6. *Coal-methanol-water (CMW)*—a suspension of coal in methanol and water.

The CMF and CMW slurries have favorable properties; however, the cost of methanol has eliminated them from further development. COM, once highly investigated, has now been abandoned for economic reasons. CWF has been investigated for complete oil replacement in boilers and furnaces and internal combustion engines, but low oil prices have eliminated or reduced the economic advantage. CWFs developed from waste streams and tailings are being investigated for cofiring in boilers and furnaces [16].

Important slurry characteristics are stability, pumping, atomizability, and combustion characteristics. These properties control the hydrodynamics and rheology of the coal slurry system. A coal slurry must have low viscosity at pumping shear rates ($10-200$ s$^{-1}$) and at atomization shear rates ($5000-30,000$ s$^{-1}$). This allows low pumping requirements and increased boiler and furnace efficiencies related to smaller droplets sizes [17].

In order to understand coal slurry hydrodynamics and rheology, an understanding of dispersed systems is required. Solid-liquid dispersed systems are classified into two systems based on particle size: colloidal and coarse particle. Colloidal dispersed systems consist of particles smaller than 1 μm and coarse particle dispersion systems (suspension) consist of particles larger than 1 μm. In colloidal dispersion systems sedimentation is prevented by Brownian motion (thermal activity). However, suspensions are thermodynamically unstable and tend to precipitate because of the large particle size.

### 6.2.1 Particle Size Distribution

Typically, coal slurry fuels have a particle size distribution (PSD) with 10–80% of the particles smaller than 74 μm (200 mesh). Micronized CWF has a PSD with a mean particle diameter of <15 μm and 98% of the particles are smaller than 44 μm (325 mesh). This type of slurry is typically produced by coal beneficiation systems in the removal of mineral matter.

The sizing of coal is a multistep process consisting of coal crushing, pulverization, and finishing steps. Finishing steps encompass coarse, fine, and ultrafine crushing of coal. Coal crushing reduces the coal size to 20–7.6 cm and 5–3.2 cm depending on the application, and coarse pulverization further reduces the coal size to <3.2 mm. The finishing processes can be done by wet or dry grinding. Grinding reduces the particle size to <1 mm for coarse, <250 μm for fine, and <44 μm for ultrafine grinding. In slurry preparation wet grinding is often used to minimize oxidation of the coal, which is detrimental to many beneficiation processes.

Coal slurries are most economical when they have the maximum amount of coal at the lowest possible viscosity. In order to obtain the highest possible loading, a bimodal or polymodal PSD is utilized as shown in Figure 1 [2]. The smaller (finer) coal particles fit into the interstices of the larger coal particles, forming a more concentrated network of particles. These particles may also act as a lubricant, leading to lower viscosity [18]. A monomodal slurry has a peak solids loading of ~65% at which the viscosity becomes infinite; idealized multimodal systems offer a theoretical loading in excess of 80% as shown in Figure 2 [19]. The lowest viscosity of a coal-water mixture occurs at a fine/coarse ratio of $35 \pm 5{:}65 \pm 5$ regardless of the ratio of the mean diameters [18]. Multimodal systems are commonly used, because they can be generated easily by common grinding schemes. A typical multimodal distribution formulated for minimum viscosity is shown in Figure 3 [19].

Settling of coal particles is a complex phenomenon from a theoretical point of view, involving hydrodynamic and physiochemical forces. The falling movement of fine coal in slurries has a Re << 1, which leads to the use of Stokes' equation for hindered settling rate as shown in Eq. (1) [3].

$$v = \frac{d^2 g (\rho_1 - \rho_2)}{18 \mu} f(\phi) \tag{1}$$

$$f(\phi) = \frac{v}{v_t} \leqslant 1 \tag{2}$$

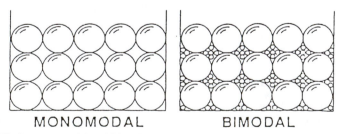

MONOMODAL          BIMODAL

**Figure 1** PSD for monomodal and bimodal distributions [2].

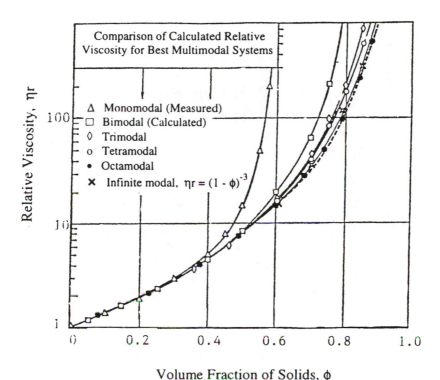

Figure 2  PSD effect on viscosity [19].

where   $v$ = hindered settling rate

$v_t$ = single-particle setting velocity

$\mu$ = viscosity of dispersing medium

$\rho_1$ = density of dispersed medium

$\rho_2$ = density of dispersing medium

$d$ = diameter of dispersed particles

$g$ = acceleration due to gravity

$f(\phi)$ = function of volume fraction of suspended solids

Stokes' law suggests that in order to reduce the sedimentation velocity, the particle diameter should be reduced, the viscosity of the dispersing medium should be increased, and the difference in density between the solid and the liquid phase should be decreased. However, optimal slurry processing demands high loading at low viscosity for transportation and atomization requirements.

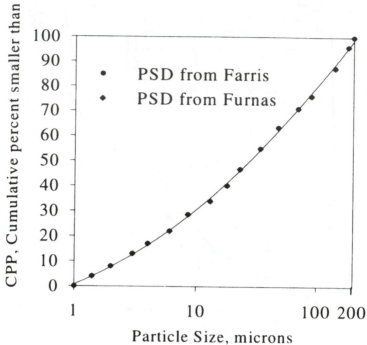

**Figure 3** PSD formulated for minimal viscosity [19].

## 6.2.2 Rheology

Rheology is the study of a system's response to a mechanical perturbation in terms of elastic deformations and viscous flow [20, 21]. In most systems elastic response is associated with solids and viscous response is associated with liquids. A suspension system, such as a coal slurry, exhibits both elastic and viscous responses. These responses in a coal slurry are a function of the type of coal, coal concentration, particle size distribution, properties of the dispersing phase, and additive package [4, 22].

In general, coal slurries exhibit non-Newtonian behavior. However they do exhibit a wide range of responses including Newtonian, dilatant, pseudoplastic (shear thinning), and plastic flow characteristics. Each of these responses is shown graphically in Figure 4, a plot of the shear stress versus the rate of shear [23]. Newtonian is the simplest response and exhibits linear functionality with respect to shear stress and shear rate. Typically, slurries exhibit pseudoplastic or shear thinning behavior, meaning that as the shear stress is increased the shear rate increases at a lower rate. This behavior is typical of a material that has a fragile internal structure that degrades with shearing stresses. The material does not have a yield point, but the apparent viscosity continually decreases with

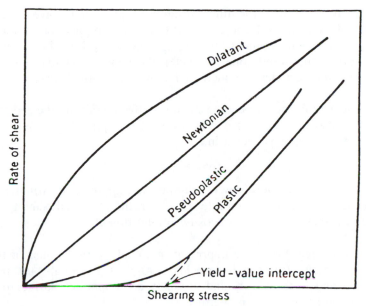

**Figure 4** Rheograms of various flow behaviors [23].

applied stress. An extension of this behavior is thixotropy, which is characterized by shear thinning that requires significant periods of time to reform the internal structure. The period of time can range from minutes to several days. Dilatant behavior is the opposite of shear thinning, as resistance to flow increases with shear stress. In plastic behavior, a sufficient stress, the yield stress, must be applied before flow begins. Once the suspension yields, the shear stress is linear with shearing rate. The important rheological characteristics are yield stress, viscosity, and plasticity (thixotropy) [24]. These values are determined experimentally.

The viscosity has proved to be very difficult to model because it exhibits both elastic and viscous responses. The Einstein equation for viscosity in dilute suspensions is

$$\mu_r = \frac{\mu}{\mu_0} = 1 + 2.5\phi \tag{3}$$

where $\mu_0$ = viscosity of dispersing phase

$\mu_r$ = relative viscosity

$\mu$ = absolute viscosity

$\phi$ = volume fraction of solids

This equation works well for dilute systems. A number of investigators have developed relations of the form $\mu_r = f(\phi)$ to describe higher concentrations that reduce to Einstein's equation at low concentrations [4]. Modeling of these systems, however, utilizes only one parameter, the volume fraction of solids. This assumes that the particles are inert and that interaction between particles is negligible.

The viscosity of the suspending medium affects not only the sedimentation velocity but also the rate of agglomeration. An empirical expression for $f(\phi)$ illustrates this example as shown in

$$f(\phi) = e^{-5.9\phi} \tag{4}$$

This equation illustrates how rapidly an increase in solids volume fraction, $\phi$, can reduce the sedimentation rate. However, the increased stability compromises slurry properties such as viscosity, combustion characteristics, and overall handling [25].

Physicochemical forces are important in a coal slurry, because of the small size of the particles (most coal particles <50 $\mu$m). Although bulk properties such as coal and water density and viscosity are important, surface properties have a large effect on slurry properties.

### 6.2.3 Stability

The stability of a slurry is an important factor in processability. The factors that affect stability are the density, size, concentration, surface properties (relative hydrophilic nature), surface charge (zeta potential), and morphology of coal and density and type of liquid [26]. Stability is classified into three broad categories: sedimentative (static), mechanical (dynamic), and aggregative.

The stability of a slurry against gravity is called sedimentative stability. A statically unstable slurry will settle, but as the system becomes more stable the degree of settling decreases. Static stability in a fluid requires a yield stress in the fluid sufficient to support the largest particle. The stability of a dynamic system is called dynamic stability. Dynamic stability involves the superposition of mechanical stress; some examples are pumping and mixing [27]. The third stability type, aggregative, is a function of interparticle forces.

### 6.2.4 Suspension Types

A suspension can be classified into three broad classifications: aggregatively stable, flocculated, and coagulated as shown in Figure 5 [3]. In an aggregatively stable suspension repulsion, forces do not allow particles to adhere to each other. They tend to settle due to gravity based on size, leading to a highly classified and compact sediment with coarse particles on the bottom and the finest particles on the top.

(a). Agglomeration

(b). Flocculation

(c). Dispersion
(Deflocculation)

**Figure 5**  Illustrations of suspension types [3].

In the second suspension type, flocculated, the particles interact weakly to form porous clusters called flocs. They tend to settle slowly because of increased drag forces from the floc structure. The sediments formed are very loose and occupy a large fraction of the original slurry volume. The slurry is easily brought back to original uniform concentrations with mild agitation.

In the last suspension type, coagulated, the particles interact strongly. The strong attractive interparticle forces promote the formation of compact and tightly bound clusters that are difficult to break even with significant agitation. These unstable slurries have fast settling rates and often display non-Newtonian behavior such as thixotropic (time-dependent), pseudoplastic (shear thinning), or plastic behavior.

## 6.2.5 Interparticle Interactions

Studies have shown that stability in slurries is achieved by promoting networks through weak interparticle interactions [26]. The properties of coal slurries are governed by the nature of the forces between particles. Six important particle-particle interactions may exist in aqueous dispersions [26]. A more comprehensive account of these phenomena can be found in the literature [27, 28, 29].

1. Interaction between electrical double layers (EDLs)
2. Van der Waals attraction
3. Steric interactions
4. Polymer flocculation
5. Hydration- and solvation-induced interactions
6. Hydrophobic interactions

When a substance is brought into contact with an aqueous polar medium, it acquires a surface electrical charge through mechanisms such as ionization, ion adsorption, or ion dissolution. The surface charge influences the distribution of nearby ions in solution; ions of opposite charge are attracted while ions of like charges are repelled. This, coupled with the mixing effects of thermal motion, leads to the formation of the electrical double layer (EDL). The EDL consists of a surface charge with a neutralizing excess counterion and, farther from the surface, coions distributed in a diffuse manner as shown in Figure 6 [26]. The EDL is important because the interaction between charged particles is governed by the overlap of their diffuse double layers. This creates a potential (the Stern potential) at the interface of the Stern plane and the diffuse layer. Unfortunately, direct measurement of the Stern potential is impossible; however, it is possible to measure the zeta potential ($\zeta$), which corresponds to the shear plane adjacent to the Stern plane as shown in Figure 7 [26]. Although the zeta potential may not necessarily give a good indication of the Stern potential, in certain cases an

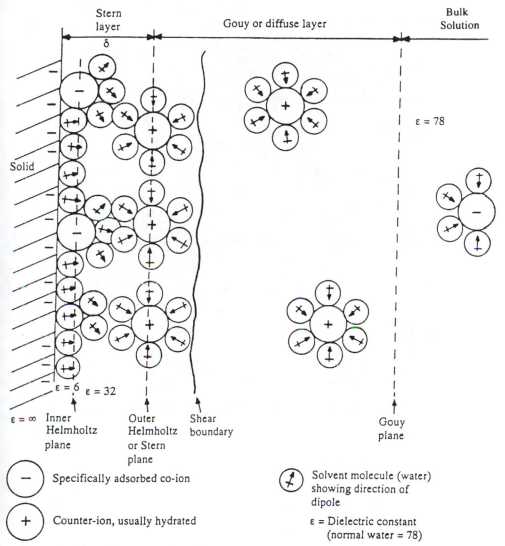

**Figure 6** Schematic representation of the EDL in the vicinity of a liquid-solid interface [26].

expression such as Eq. (5) can be formulated for the repulsive energy ($V_R$) of interaction between particles based on surface roughness, shape, and other factors [30].

$$V_R = 2\pi\varepsilon a\zeta^2 e^{(-\kappa h)}$$ (5)

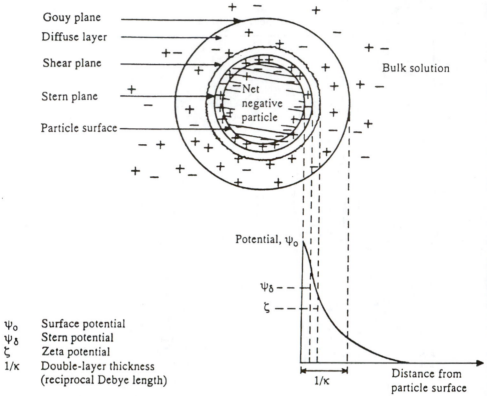

**Figure 7** Distribution of electric potential in the double-layer region surrounding a charged particle [26].

where   $a$ = radius of two particles

   $\zeta$ = zeta potential

   $\kappa$ = inverse Debye length

   $\varepsilon$ = permissivity of the medium

   $h$ = distance between particles

The attractive van der Waals force encourages aggregation between particles when the distance between them is very small. These forces are due to spontaneous electric and magnetic polarizations giving a fluctuating electromagnetic field within the dispersed solids and the aqueous medium separating the particles [26]. Two methods are commonly used for predicting these forces: the Hamaker approach and the more rigorous Lifshitz approach. The Hamaker approach adds up the forces pairwise between the two bodies; the Lifshitz method directly computes the attractive forces based on the electromagnetic properties of the medium. The Hamaker method for identical spheres is shown in Eq. (6).

$$V_A = -\frac{A_{12}a}{12h} \tag{6}$$

where $A_{12}$ is the Hamaker constant

Now it is possible to predict the interactions of the EDL and van der Waals forces using Eq. (7).

$$V_T = V_A + V_R \tag{7}$$

This forms the basis of the DVLO (Derjaguin, Landua, Verwey, Overbeek) theory of colloid stability [24]. The energy interactions for EDL and van der Waals forces are shown in Figure 8 [26]. The subsequent total energy curve allows prediction of aggregation at close distances (primary minimum) and the possibility of a weak and reversible aggregation in the secondary minimum [26]. The form of the curve depends on the size of the particles and surface charge [31]. For example, coarse particles are more likely to be vulnerable to aggregation at the secondary minimum.

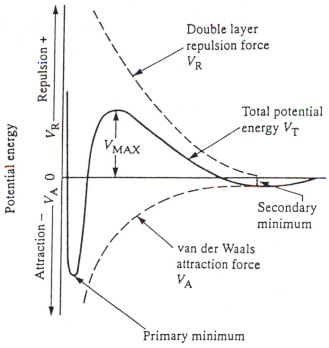

**Figure 8** Particle-particle potential interaction energy [26].

Kinetic effects are important because thermodynamic prediction may not yield sufficient information. Coagulation rates are described in terms of the stability ratio. The stability ratio, $W$, can be thought of as the efficiency of interparticle collisions resulting in coagulations as shown in Eq. (7) for two identical particles brought together by diffusion [26].

$$W = 2 \int_0^\infty \frac{\exp (V_T/kT)}{(2 + s)^2} \, ds \tag{8}$$

where $s = h/a$.

The other types of interparticle interactions (steric interactions, polymer flocculation, hydration- and solvent-induced interactions, and hydrophobic interactions) represent special cases. The most significant of these are steric interactions and polymer flocculations. Steric interactions develop when molecules (usually polymers) are adsorbed onto the particle surface at high coverages. The polymer molecules protrude from the surface of the particle into solution. When particles approach one another, the polymer chains overlap and often dehydrate, increasing the stability of the slurry [32, 33]. Polymer flocculation occurs when the particle surface has low coverage of a high-molecular-weight polymer. The polymers bridge the particles and form flocs [32, 34].

Hydration- and solvation-induced interactions become important when interparticle distance is on the order of a solvent molecule. For aqueous systems these solvation effects manifest themselves in structuring of the water near the interface surface (with the bound water), which interact with hydrated ions from solution. The net effect is increased stability or net repulsion between particles as it becomes necessary for the ions to lose their bound water to allow the approach to continue [26].

Hydrophobic interactions are analogous to hydration and solvation effects, because stability can be enhanced by the attraction between two hydrophobic particles. Therefore hydrophobic particles tend to associate with each other. The hydrophobic forces are greater than the van der Waals forces and have a longer range [26].

## 6.3 COAL-WATER SLURRY

Coal-water slurries initially attracted attention as a replacement for oil in furnaces and boilers and in cofiring of boilers and furnaces in the use of coal fines. CWMs (coal-water mixtures) and coal-water slurry fuels (CWSFs) have received a great deal of attention for use as fuel because of relative ease of handling (similar to that of fuel oil and not explosive like coal dust), storage in tanks, and injection into furnaces and boilers. CWMs typically have extremely high loadings in the range of 60–75 wt. % coal, which leads to high energy densities (concentrations) [72–74]. Possible applications include gas turbines, diesel engines, fluid bed combustors, blast furnaces, and gasification systems [2]. However, coal slurries for cofiring purposes are limited by economics to the use of minimal additives and lower concentrations (50 wt. %) [35].

The physical properties of a coal slurry are extremely important in the processing of the fuel. A slurry must be stable and exhibit low viscosity at the shear rates of pumping and atomization. The flow characteristics of coal depend on (1) physicochemical properties of the coal, (2) the volume fraction $\phi$ of the suspension, (3) the particle size range and distribution, (4) interparticle interactions (affected by the nature of surface groups, pH, electrolytes, and chemical additives), and (5) temperature [36]. Rheological and hydrodynamic behavior of coal-water slurries varies from coal to coal. Each coal has a unique package of PSD, concentration, and additives to reach the desired processability.

A parameter that measures how well a coal will slurry, the slurryability, is the equilibrium moisture content. The equilibrium moisture content is a measure (index) of the hydrophilic nature of the coal. The more hydrophilic a coal is, the more water it will hold and the less likely it is to produce a highly concentrated slurry [37]. The viscosity increases with the hydrophilic nature of the coal. Therefore, a hydrophobic coal can more easily form a slurry of low viscosity at high loadings.

Conventional high-rank coals (black coal), except anthracite, have a hydrophobic nature due to lack of acid groups and form a slurry of ~80 wt. % (dry coal weight basis). Anthracite coals have low reactivity and volatility, leading to poor ignition stability [2]. However, low-rank coals (brown coal) are hydrophilic due to an abundance of oxygen functional groups and form slurries of only 25–20 wt. % (dry coal weight basis) [38]. These slurries have low concentrations but are nonagglomerating and have high reactivities. Coal slurry stability difference from coal to coal is a function of the relative balance of acid and base groups on the coal surface [39].

CWMs are loaded to the highest possible concentration at acceptable viscosities. However, viscosity increases with coal concentration and reducing the viscosity compromises the stability of the slurry [3]. To stabilize coal-water slurries additives such as surfactants and electrolytes are added to enhance the stability of the coal dispersion.

Surfactants are used as dispersants to wet and separate coal particles by reducing the interfacial surface tension of the water. Surfactants are short-chain molecules with a hydrophobic group and a hydrophilic oxide (nonionic) or a charged ionic group (ionic). These molecules attach themselves to the coal particle through adsorption or ionic interaction. Generally dispersants are ionic. Some examples of dispersants are sodium, calcium, and ammonium lignosulfates and the sodium and ammonium salts of naphthalene formaldehyde sulfonates [4].

Ionic surfactants adsorb onto the alkyl groups at hydrophobic sites on the coal particle. This gives the coal particles a negative charge, which affects the

EDL, enhancing the repulsive forces and thus preventing agglomeration [17]. Anionic surfactants decrease the viscosity of the slurry up to a critical loading. At this point the coal adsorption sites are saturated and the remaining surfactant forms micelles in the slurry, leading to an enhanced structure and increased viscosity.

Nonionic surfactants function by two different methods, depending on the nature of the coal. On a hydrophilic coal surface the hydrophobic end of the surfactant is toward the aqueous phase. The water then acts as a lubricating material between coal particles. The second method of attachment is on a hydrophilic coal. The hydrophilic end of the surfactant attaches to the coal, leaving the hydrophobic end into the aqueous medium. This increases the amount of water near the surface of the coal particle, producing a hydration layer or solvation shell. This prevents agglomeration by cushioning coal particles and lowers the viscosity [17].

The ionic strength of water in the CWM is an important parameter in the rheological and hydrodynamic characteristics of a slurry. Because coal is not a uniform substance but a mixture of carbonaceous material and mineral matter, the ionic strength of the water affects the interaction with coal. In a hydrophobic colloidal system dispersed by electrical repulsive forces, the electrolyte concentration (ionic strength) has a considerable effect on the stability against flocculation of particles [17]. The cation concentration causes an increase in the viscosity of the slurry with decreasing pH [40]. Electrolytes strongly affect the degree of particle dispersion and thus rheology in a CWM that uses an anionic dispersant [40]. Addition of electrolyte to a slurry using nonionic dispersants has no effect on the viscosity [40].

In highly concentrated slurries minimal settling is expected, but viscosity-reducing additives increase the settling rate. In order to stabilize the dispersion flocculating agents are added, which produce a gel. Some examples of this are nonionic amphoteric polymers of polyoxyethylene, starches, natural gums, salts, clays, and water-soluble resins [2].

Polymers have been used for drag reduction, or viscosity reduction [41]. Both ionic and nonionic polymer solutions show reduction in viscosity, although the reduction is more pronounced for anionic polymers.

## 6.4 COAL-OIL SLURRY

Coal-oil slurries have been investigated for over 100 years. Typically, interest peaks in times of high prices and shortages of oil. The most recent interest was fueled by the energy crises of 1973 and 1979, when tremendous effort was put forth to find a quick, viable alternative to oil in boilers and furnaces. Since the mid-1980s, however, most slurry investigation has been directed toward CWMs.

In general, coal-oil slurries exhibit non-Newtonian behavior, mostly pseudoplastic except at low coal loadings, where the slurry is Newtonian (provided the oil is Newtonian). The viscoelastic properties of the dispersion depend on coal concentration, PSD, coal type, oil type, and chemical additives. COM rheological properties are highly sensitive to coal concentration. At this critical concentration a dramatic increase in viscosity occurs with incremental changes in concentration.

The standard particle size distribution in a COM is listed in Table 2 [3]. The particle size distribution makes a coal-oil slurry a coarse particle system. COM mixtures are classified as lyophobic because the dispersed particles are not compatible with the dispersion medium. These systems are thermodynamically unstable and separate into two continuous phases.

Ultrafine COMs with 95% particles smaller than 325 mesh (44 μm) and slurry concentrations of 50 wt. % have been investigated. These slurries reduce abrasiveness, have improved combustion characteristics, and do not have additives [4, 42]. However, the grinding costs are higher and coal concentration is limited to 50 wt. %.

Chemical additives for COMs are surfactants and polymers. The surfactants add stability to the mixture by preventing agglomeration and enhancing flocculation. Cationic polymers are the most effective surfactants for stabilization [43], and polymers are used to reduce the drag [44].

In many instances water is added to coal-oil slurries forming COW or CWO, depending on the water concentration. Water, a flocculating agent, is added to increase the stability of the slurry and for cost savings. Water increases the viscosity of the slurry through the formation of aggregates and particle bridging [44], although the combustion properties are retained.

## 6.5 ADVANCED TRANSPORT

The transport of coal slurries can take place in truck and railroad tanks, slurry tankers, and pipelines. Historically, coal water slurries transportation schemes have received the most development. Although all transportation schemes have been investigated, few new processing developments have occurred in truck, railroad, or slurry tankers. In these cases, the slurry stability has been enhanced to minimize settling during transportation or the slurry has been dewatered to

**Table 2 COM particle distribution [3]**

| Percent[a] | Particle size (μm) |
|-----------|--------------------|
| 100 | <200 |
| 80 | <74 |
| 65 | <44 |
| 15 | <10–20 |
| 1 | <1 (colloidal) |

[a]Percent: passing through mesh size.

maximize energy density and cost effectiveness of transportation. Coal water pipeline systems across the world have undergone almost constant development since the 1950's. Coal pipelines can be broken into four different systems: conventional fine coal, conventional coarse coal, stabilized flow, and coal water mixture. These systems differ by the particle size of coal in the slurry, as shown in Table 3 [45].

Conventional fine-coal slurry pipeline systems account for only two systems in the United States. The first was the Consolidation Coal Company's Ohio pipeline, which was built in 1957 and operated for several years until competitive transportation mode, the unit train, became available. The slurry traveled at moderate velocities and had a coal content of 50 wt. %. The second pipeline system built, Black Mesa pipeline, followed the same basic design as the Consolidation pipeline. The pipeline began operation in 1970 and is still operating today. The pipeline operates at 4–6 mph, optimized for minimal erosion and settling, at a 50% coal loading [45, 46].

In conventional coarse-coal slurry transport, run of mine (ROM) coal is transported at high velocities in coal loadings ranging from 35 to 60% depending on the coal particle size (150–50 mm) [45]. In this pipeline coal slurry preparation costs are minimal, but energy requirements are high. This type of system is limited to short distances where other transportation modes, such as rail, barge, truck, or conveyor, can be used.

Stabilized flow coal slurry systems use smaller particles to support the coarse coal particles. The particle distribution for the fine coal is <0.2 mm and for the coarse coal <50 mm [45]. This particle distribution was utilized to enhance the stability of the slurry for long transport distances. Coal concentrations up to 70 wt. % can be used. The coarse coal is easier to dewater than fine coals because of the particle size.

Coal-water mixture pipeline systems are designed for direct use at the final destination site, most likely in utility boilers and furnaces. The coal concentration is between 70 and 75 wt. % with bimodal and polymodal particle size distributions of coal particles smaller than 200 mesh [45]. CWMs use approximately 1% surfactants and dispersants for slurry stability. These pipelines have yet to be investigated over long distances.

Many different coal processing configurations exist for coal use at the end of a coal slurry pipeline. The options include direct use (CWM), dewatering and

**Table 3 Summary of particle sizes in coal-water slurry pipeline systems [45]**

| Coal-water slurry systems | Particle size |
| --- | --- |
| Conventional fine coal | Less than 1 mm |
| Conventional coarse coal | 50–150 mm |
| Coal-water mixtures | −30 and −150 $\mu$m (less than 200 mesh) |
| Stabilized-flow coal | Less than 0.2 mm (fine) and less than 50 mm (coarse) |

conveying to a plant (Figure 9), piping to an offshore platform and dewatering into a ship (Figure 10), dewatering into a railcar or truck, and piping into an offshore buoy system into a ship (CWM) [34–48]. Table 4 is a comparison of coal content in coal slurries for the pipeline systems discussed [45]. Each coal transportation system has merits and decisions have to be made on an individual basis. The transport systems are evaluated in Table 5 [45].

## 6.6 COMBUSTION

Combustion of coal in utility boilers (and furnaces) and internal combustion engines was widely investigated throughout the 1970s and 1980s. The bulk of the combustion research and development has been in the area of CWSFs.

Generally COM combustion occurs in three stages: heating of droplet, combustion of volatile matter, and combustion of char. For coal, char combustion is controlled by the mass transfer process for particles larger than 100 μm; for smaller particles it is limited by kinetics [4]. Carbon conversions are comparable to those of fuel oil.

COM and CWSF differ in burning such that CWSF heat is generated only by the coal; the water is evaporated from the coal particle. There are problems associated with CWSF such as flame stability, incomplete combustion, erosion, and slagging. Flame stability is a function of atomized droplet size, fuel stability, agglomeration properties, burner swirl, burner throat construction, and preheat air temperature [2]. Incomplete combustion is a function of inadequate preheating, delayed ignition (vaporization of water), higher oxygen requirements, and agglomeration. Erosion is related to excess velocities of particle-laden gases. Slagging is a function of inorganic content and slow char particle burnout. In

- Stabilized Flow ( Stabflow ) ( short or long distance )
- Conventional Fine Coal ( CFC ) ( long distance only )
- Conventional Coarse Coal ( very short distance only )

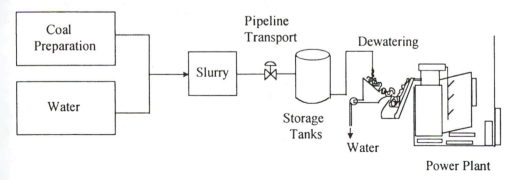

**Figure 9**  Domestic combustion pipeline scheme [45].

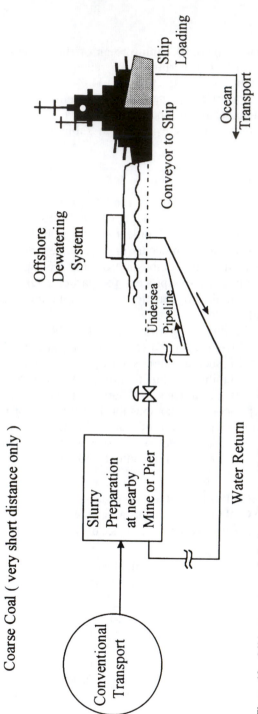

**Figure 10** Offshore coarse coal pipeline scheme with dewatering [45].

**Table 4  Comparison of coal content in coal slurries [45]**

| Coal-water slurry systems | Percentage of coal in mixture |
|---|---|
| Conventional fine coal | 50 (overland and shiploading) |
| | 75 (aboard ship) |
| Conventional coarse coal | 30–60 (overland-short distances only and ship loading) |
| | 90–92 (aboard ship) |
| Coal-water mixtures | 70 (overland) |
| | 85–90 (shiploading and aboard ship) |
| Stabilized-flow coal | 70–75 (overland, shiploading, and aboard ship) |

cases of using CWSF in oil boilers and furnaces and boiler duty is derated, unless modifications are made to the unit.

The combustion mechanism for CWSFs, shown in Figure 11 [46], is similar to that for pulverized coal with the additional stage of water evaporation. The coal slurry droplet is injected into a hot gas stream and quickly dries. The particles in the droplet agglomerate as a result of surface forces and become tightly bound while undergoing plastic deformation during pyrolysis. Ignition occurs and devolatilization produces rotation and fragmentation. The char further fragments during burnout.

Coal combustion is considered slow when compared with the combustion of oil [47]. The flame temperature is lowered because of the energy absorbed in order to vaporize the water. The atomization of the slurry strongly influences the combustion efficiency [48]. Fine atomization increases the rate of evaporation through surface area increase and reduces the size of the agglomerates. Smaller droplets produce more stable flames and greater carbon burnout [2].

**Table 5  Coal-water slurry pipeline system selection [45]**

| Objective | System characteristics Length (miles) | Type | Best pipeline selection |
|---|---|---|---|
| Rapid implementation | <5 | Domestic | Coarse coal |
| | | Export | Coarse coal |
| | 50–100 | Domestic | Conventional fine coal |
| | | Export | Conventional fine coal |
| | >100 | Domestic | Conventional fine coal |
| | | Export | Conventional fine coal |
| Lowest cost and water use | <5 | Domestic | Coarse coal |
| | | Export | Coarse coal |
| | 50–100 | Domestic | Stabilized flow |
| | | Export | Stabilized flow |
| | >100 | Domestic | Stabilized flow |
| | | Export | Stabilized flow |
| Oil displacement | All | Domestic | CWM |
| | | Export | CWM |

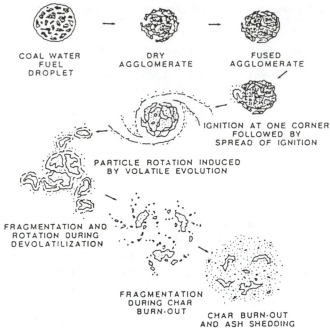

COAL WATER
FUEL
DROPLET

DRY
AGGLOMERATE

FUSED
AGGLOMERATE

IGNITION AT ONE CORNER
FOLLOWED BY
SPREAD OF IGNITION

PARTICLE ROTATION INDUCED
BY VOLATILE EVOLUTION

FRAGMENTATION AND
ROTATION DURING
DEVOLATILIZATION

FRAGMENTATION
DURING CHAR
BURN-OUT

CHAR BURN-OUT
AND ASH SHEDDING

**Figure 11**  CWSF combustion mechanism [46].

Coal-water slurry firing of conventional boilers has been performed over extended periods of time with few operational problems. The boilers were successfully started, reliably and safely operated, and complied with emissions. The combustion efficiency was 95% and thermal efficiency was 79%. However, in order to increase efficiency to 99+% costly modifications must be made and anticipated maintenance schedules are up to six times a year [49, 50].

Until recently, coal has been used only in external combustion engines (Rankine cycles). Attempts at coal utilization date back to the development of the diesel engine during the 1920s on large-bore diesel engines. Problems such as fuel introduction, combustion efficiency, and engine wear could not be resolved and the work was halted [51].

Investigation of CWSFs has brought about renewed interest in coal-fired internal combustion engines. Feasibility studies have shown that slurry engines have thermal efficiencies similar to those of oil engines and the use of new materials has limited erosive effects [52]. The presence of water in the fuel allows control of $NO_x$ emissions and hydrocarbon and CO emissions are low although $CO_2$ emissions are higher than for oil because of the high carbon content of the coal [53, 54]. The particles generated can be handled easily with particulate traps [54].

The difficulties in the utilization of CWSFs in diesel engines are fuel injection (atomization), ignition, erosion, and corrosion [55]. The fuel injection system efficiency depends on slurry properties, combustion chamber layout, fuel compatibility, ignition delays, and erosion. Initial ignition of CWMs is difficult, as the water must first be evaporated from the coal [56]. Operating combustion temperatures are higher, with the optimal range being 1000–1100 K [51, 56]. This is to limit the delay in ignition of coal resulting from the evaporative effects of water and the slower burning of coal [56, 57]. Erosion of parts in the injection system as well as in the cylinders is a major concern for the life of the engine. Engine wear is six times higher in CWM-fired engine than a diesel, but can be reduced significantly with special alloys and redesigned parts [54]. These efforts have shown great promise for operating an internal combustion engine on CWSFs.

## 6.7 ENVIRONMENTAL ISSUES

Environmental issues play an important role in the implementation of coal slurry fuels. The transportation systems must take into account possible leaks and spills encountered in slurry handling. Combustion processes must be thoroughly investigated to find the different combustion mechanisms in coal slurry, specifically the mineral matter and sulfur content of the coal.

In the transportation of coal slurries, spills, leaks, and catastrophic disasters are important factors in safe handling. Transportation of CWMs has been widely investigated over short as well as long distances. Slurries of water are considered nontoxic and nonhazardous. In the United States, when there have been handling accidents in pipeline systems, cleanup of the coal has not been necessary. The coal has been allowed to reenter the ground naturally. CWMs have the advantage that they are not readily combustible in accidents.

In the combustion of slurries, many processes are not able to handle the high sulfur and mineral matter content of coal. Utility boilers and furnaces may exhibit slagging, fouling, and erosion. Other applications such as cement and asphalt kiln and fluid bed combustors are able to easily handle the increases in inorganic material. The amount of ash present that induces fouling in boilers varies from boiler to boiler. Table 6 shows the predicted acceptable ash levels for different types of boilers [2].

An important constituent in the coal is sulfur in both organic and inorganic forms. Sulfur is a precursor for ash formation and can form $SO_x$, a precursor of

**Table 6 Permissible coal ash content for utility boilers [2]**

| Ash content (%) | Boiler type |
| --- | --- |
| 5–7 | Coal designed/coal capable |
| 2–3 | Liberal oil design |
| <0.5 | Compact oil designed |

acid rain, which has regulated emissions. In many instances coal beneficiation techniques are used to clean the coal after grinding. These processes are broken into physical and chemical methods. Physical methods are based on differences in coal and mineral matter density or surface properties. Examples of these processes are gravity separation [58], froth flotation [57], selective agglomeration [59], heavy liquid separation [60], high-gradient magnetic separation [61], microbubble flotation [62], and biological method [63]. Chemical methods have been widely investigated at the laboratory scale but are not commonly used industrially [64]. These have been thoroughly investigated and will not be discussed here in detail.

Other environmentally regulated emissions of coal slurries are $NO_x$, hydrocarbons, and particles. Burning coal slurries reduces hyrocarbon emissions significantly in oil design boilers and furnaces. $NO_x$ emissions are a function of fuel nitrogen content, fuel composition, combustion temperature, and excess air [4]. $NO_x$ emissions are controllable by conventional methods such as low excess air, stage combustion, low-temperature combustion, and flue gas recirculation [2]. The particulate emissions are of the ash content in the coal slurry and can be significantly higher than for oil-burning furnaces. However, conventional particle control techniques are effective at removing the particles [4].

## 6.8 RECENT ADVANCES AND FUTURE

Coal-water fuels have received the most process development recently. Coal slurry facilities have been built in Australia, Canada, China, Italy, Japan, Sweden, and the United States. The most active in development have been China, Japan, and Russia. China has built several slurry production facilities and boiler units. Japan has done considerable research and technological development on slurry processes and has converted several boilers. Russia has built several pipelines, production facilities, and boilers. Work in China and Australia has been fueled by Japan's need for coal and space limitations.

In Australia a group of companies comprising Ube Industries, Nissho Iwai Corp., and Coal & Allied Industries has been working since 1987 on developing the production, transportation, and marketing of CWMs from Australia to Japan [65]. In 1991 the first bilateral trading of coal slurry began between China and Japan [66]. A coal-water slurry plant was built in Rizhao, located in Shandong province, by Yanzhou Coal Mining Bureau, Nisshon Iwai Corp., and JWG Corporation. The plant has a capacity of 250,000 ton/yr. The coal is mined, then transported by train to Rizhao. The coal is processed into a CWM and shipped to an overseas terminal. Once in the relay terminal the CWM is transferred to coastal shipping and then finally to end uses. The manufacture, transportation, and combustion of coal slurry have had minimal technical problems [66].

In the United States the greatest potential use for CWMs may be in the utilization of coal fines for cofiring boilers. Over 40 million tons of coal fines are discarded annually in the United States into slurry ponds, and some 2.3 billion tons of coal fines were estimated to reside in ponds in 1994 [16]. Coal fines ($-100$ mesh) production has increased over the years with the increase in demand for cleaner coal. Increased demand for high-quality coal (often benefi-ciated) has led to rejecting 20–50% of the coal mined as coal fines [67]. The environmental impact of coal fines includes nonproductive use of land, loss of aesthetics, danger of slides, dam failure, significant permitting costs, and possible water pollution [67].

Penelec, New York State Electric and Gas Corporation (NYSEG), Pennsyl-vania Electric Energy Development Authority (PEEDA), Pennsylvania Electric Energy Research Council (PEERC), and the Electric Power Research Institute (EPRI) commissioned a project to investigate the utilization of CWMs developed for coal fines. The project investigated laboratory-scale as well as full-scale cofiring of a 32-MW boiler [68].

CWMs created from coal fines will reduce coal handling problems and elim-inate the need for costly dryers and their environmental hazards. CWMs will eliminate the need for the addition of oil at start-up, which is done to stabilize combustion. Finally, the equipment life of the pulverizer will be extended by reducing the load on the unit. The use of a CWM from coal fines can stabilize the cost of fuel to the boiler, because it is a lower cost fuel and the fuel ratio can be controlled up to 50% [69, 70]. Environmentally, the addition of coal slurry reduces $NO_x$ emissions [70].

In order to be economical the lowest cost, slurries must be developed that lead to a "low-tech" slurry, that is, one without stabilizers, dispersants, or need for further grinding. This slurry can be developed from coal pond fines, the fine coal fraction of existing coal supplies, or in the future from advanced coal tech-nologies to deep-clean fine coal [54]. The slurry has been tested with solids loadings ranging from 54 to 67 wt. % coal and has had excellent handling and storage properties [71].

However, slurry developed from coal fines is highly variable. The slurry ability is a function of PSD, ash level, and extent of oxidation. Slurry preparation can vary from minimal processing to significant processing. Minimal processing arises with coals derived from wet, fine coal that has been cleaned. Significant processing is needed with coal fines that have high ash levels and high oxidation. High oxidation makes the coals more wettable through oxygen bonds, which increase the viscosity and stability of the slurry.

Currently, slurry produced from coal fines at Homer City is being cofired along with powdered coal in the 32-MW boiler No. 14 of the Seward Station. The plant has been operating without disruption. The $NO_x$ production has de-creased by 10–20%; CO, $SO_2$, and particulates have remained essentially the same as for powder coal, but the CO levels are highly variable [71].

Coal slurry development finds itself adjusting once again to process eco-nomics. Since the mid-1980s, COM mixtures, once the most investigated slur-ries, have become uneconomical and development of CWMs has occurred. In places where coal is plentiful but transportation is lacking, such as China and Russia, slurry development continues on production, pipelines, and combustion. In other areas such as Japan, where natural resources and space are limited, the storage and transportation ease that CWM represents is now being exploited. In the United States coal slurry development is now centered on coal utilization. CWM enables the use of higher amounts of coal after beneficiation and use of the environmental liability of discarded coal fines. Coal slurry development continues for internal combustion engines in injection systems, atomizer design, and materials for construction [70]. Coal slurry process development will continue and may be poised to represent a major form of coal energy.

# REFERENCES

1. Choudhury, R. Slurry fuels. *Prog. Energy Combust. Sci.* 18:409–427, 1992.
2. Kesavan, S. Stablilization of coal particle suspensions using coal liquids. M.S. thesis, University of Akron, 1985.
3. Papachristodoulou, G., and O. Trass. Coal slurry fuel technology. *Can. J. Chem. Eng.* 65: 177–201, 1987.
4. Smith, H. R., and H. M. Munsell. Liquid fuels. U. S. Patent 219181, 1879.
5. Lord, N. W., R. P. Ouellette, and O. G. Farah. *Coal-Oil Mixture Technology.* Ann Arbor, MI: Ann Arbor Science, 1982.
6. Manning, A. B., and R. A. A. Taylor. Colloidal fuel. *J. Inst. Fuel.* 9:303, 1936.
7. Manning, A. B., and R. A. A. Taylor. Colloidal fuel. *Trans. Inst. Chem. Eng.* 14:45, 1936.
8. Adams, R. A., F. C. V. Holmes, and A. W. Perrin. Colloidal fuel using cracked oil and high carbon residue. U.K. Patent 396, 432, August 2, 1933.
9. Barkley, J. F., A. B. Hersberger, and L.R. Burdick. *Trans. Am. Soc. Mech. Eng.* 66:185, 1944.
10. Nakabayashi, Y. Outline of COM R&D in Japan. Proc. 1st Int. Symp. on Coal-Oil Mixture Combustion, M78–97, St. Petersburg, FL, May 1978.
11. Basta, N., and M. D'Anastaio. The pulse quickens for coal-slurry projects. *Chem. Eng.*, September 17, 22–25, 1984.
12. Bergman, P. D., L. Kirkland, and T. J. George. Why coal slurries stir worldwide interest. *Coal Mining Process.* 19(10):24–42, 1982.
13. Marnell, P. Direct firing of coal water suspensions, state of the art review. Proc. Coal Techn. '80, Industrial Presentation Inc., Houston, 1980.
14. Brown, A., Jr. Powdered COM Program. Final Report of the General Motors Corporation, October 1977.
15. Miller, B. G. Coal-water slurry fuel utilization in utility and industrial boilers. *Chem. Eng. Prog.* March: 29–38, 1989.
16. Meyer, C. W. Stabilization of coal/fuel oil slurries. Proc. 2nd Int. Symp. on Coal-Oil Mixture Combustion, vol. 2, CONF-791160, Danvers, MA, November 1979.
17. Williams, R. A. Characterization of process dispersions. In *Colloid and Surface Engineering: Applications in the Process Industries,* ed. R. A. Williams. Boston: Butterworth Heinemann, 1992.
18. Barnes, H. A., J. F. Hutton, and K. Walters. *An Introduction to Rheology.* New York: Elsevier, 1989.
19. Hunter, R. J. *Foundations of Colloidal Science,* vol. 1. New York: Oxford University Press, 1987.
20. Shaw, D. J. *Introduction to Colloid and Surface Chemistry,* 4th ed. Butterworth Heinemann, 1991.

COAL SLURRY FUEL **285**

21. Napper, D. H. *Polymeric Stabilization of Colloidal Dispersions.* New York: Academic Press, 1983.
22. Gregory, J. Flocculation of polymers and polyelectrolytes. In *Solid Liquid Dispersions*, ed. T. F. Tadros, pp. 163–181. Orlando, FL: Academic Press, 1987.
23. Evans, D. F., and H. Wennerstrom. *The Colloidal Domain Where Physics, Chemistry, Biology, and Technology Meet.* New York: VCH, 1994.
24. Vossoughi, S., and O. S. Al-Husaini. Rheological characterization of the coal/oil/water slurries and the effect of polymer. Proc. of the 19th Int. Tech. Conf. on Coal Utilization and Fuel Systems, pp. 115–122, Clearwater, FL, March 21–24, 1994.
25. Ross, S., and I. D. Morrison. *Colloidal Systems and Interfaces.* New York: Wiley, 1988.
26. Rowell, R. L. The Cinderella synfuel. *CHEMTECH* April:244–248, 1989.
27. Turian, R. M., M. K. Fakhreddine, K. S. Avramidis, and D.-J. Sung. Yield stress of coal-water mixtures. *Fuel* 72(9):1305–1315, 1993.
28. Roh, N.-S., D.-H. Shin, D.-C. Kim, and J.-D. Kim. Rheological behavior of coal-water mixtures 1. Effects of coal type, loading and particle size. *Fuel* 74(8):1220–1225, 1995.
29. Woskoboenko, F., S. R. Siemon, and D. E. Creasy. Rheology of Victorian brown coal slurries, 1. Raw coal-water. *Fuel* 66:1299–1304, 1987.
30. Roh, N.-S., D.-H. Shin, D.-C. Kim, and J.-D. Kim. Rheological behavior of coal-water mixtures 2. Effect of surfactants and temperature. *Fuel* 74(9):1313–1318, 1995.
31. Kaji, R., Y. Muranaka, H. Miyadera, and Y. Hishinuma. Effect of electrolyte on the rheological properties of coal-water mixtures. *AIChE J.* 33(1):11–18, 1987.
32. Rowell, R. L., S. R. Vasconcellos, R. J. Sala, and R. S. Farinato. Coal-oil mixtures 2. Surfactant effectiveness on coal oil mixture stability with a sedimentation column. *Ind. Eng. Chem. Proc. Des. Dev.* 20:283–288, 1981.
33. Veal, C. J., and D. R. Wall. Coal-oil dispersions—overview. *Fuel* 60:873–882, 1982.
34. Vossoughi, S., and O. S. Al-Husaini. Rheological characterization of the coal/oil/water slurries and the effect of polymer. Proceedings of the 19th International Technical Conference on Coal Utilization and Fuels Systems, pp. 115–122. Clearwater, FL, 1994.
35. Bertram, K. M., and G. M. Kaszynski. A comparison of coal-water slurry pipeline systems. *Energy* 11(11/12):1167–1180, 1986.
36. Brolick, H. J., and J. D. Tennant. Innovative transport modes: Coal slurry pipelines. ASME Fuels and Combustion Division Pub. *FACT* 8:85–91, 1990.
37. Manford, R. K. Coal-water slurry: A status report. *Energy* 11(11/12):1157–1162, 1986.
38. Ng, K. L. Coal unloading system using a slurry system. 281–287. ASME Fuels and Combustion Division Pub. *FACT* 8:281–287, 1990.
39. Hapeman, M. J. Review and update of the coal fired diesel engine. ASME Power Division Pub. PWR 9:47–50, 1990.
40. Likos, W. E., and T. W. Ryan, III. Experiments with coal fuels in a high-temperature diesel engine. *J. Eng. Gas Turbines Power* 110:444–452, 1988.
41. Rao, A. K., C. H. Melcher, R. P. Wilson, Jr., E. N. Balles, F. S. Schaub, and J. A. Kimberly. Operating results of the Cooper-Bessemer JS-1 engine on coal-water slurry. *J. Eng. Gas Turbines Power.* 110:431–436, 1988.
42. Urban, C. M., H. E. Mecredy, T. W. Ryan, III, M. N. Ingalls, and B. T. Jetss. Coal-water slurry operation in an EMD diesel engine. *J. Eng. Gas Turbines Power* 110:437–443, 1988.
43. Hsu, B. D. Progress on the investigation of coal-water slurry fuel combustion in a medium speed diesel engine: Part 1—Ignition studies. *J. Eng. Gas Turbines Power* 110:415–422, 1988.
44. Hsu, B. D. Progress on the investigation of coal-water slurry fuel combustion in a medium speed diesel engine: Part 2—preliminary full load test. *J. Eng. Gas Turbines Power* 110:423–430, 1988.
45. Hsu, B. D., G. L. Leonard, and R. N. Johnson. Progress on the investigation of coal-water slurry fuel combustion in a medium speed diesel engine: Part 3—Accumulator injector performance. *J. Eng. Gas Turbines Power* 110:516–520, 1988.

46. Miller, S. F., J. L. Morrison, and A. W. Scaroni. The formulation and combustion of coal water slurry fuels from impounded coal fines. 19th International Technical Conference on Coal Utilization and Fuel Systems, 643–650, 1994.
47. Schimmeller, B. K., P. S. Jacobsen, and R. E. Hocko. Industrial use of technologies potentially applicable to the cleaning of slurry pond fines. 19th International Technical Conference on Coal Utilization and Fuel Systems, 805–817, 1994.
48. Crippa, E. R. 50,000 HP coal slurry diesel engine. 19th International Technical Conference on Coal Utilization and Fuel Systems, 821–828, 1994.
49. Bradish, T. J., J. J. Battista, and E. A. Zawadzki. Co-firing of water slurry in a 32 MW pulverized coal boiler. 18th International Technical Conference on Coal Utilization and Fuel Systems, 303–313, 1993.
50. Yanagimacho, H., O. Matsumoto, and M. Tsuri. CWM production in China and CWM properties in all stages from production to combustion in the world's first bilateral CWM trade. 18th International Technical Conference on Coal Utilization and Fuel Systems, 327–337, 1993.
51. Morrison, D. K., T. A. Melick, and T. M. Sommer. Utilization of coal water fuels in fire tube boilers. 18th International Technical Conference on Coal Utilization and Fuel Systems, 339–347, 1993.
52. Pisupati, S. V., S. A. Britton, B. G. Miller, and A. W. Scarani. Combustion performance of coal water slurry fuel in an off-the-shelf 15,000 lb steam/h fuel oil designed industrial boiler. 18th International Technical Conference on Coal Utilization and Fuel Systems, 349–360, 1993.
53. Morrison, J. L., B. G. Miller, and A. W. Scaroni. Preparing and handling coal water slurry fuels: Potential problems and solutions. 18th International Technical Conference on Coal Utilization and Fuel Systems, 361–368, 1993.
54. Battista, J. J., and E. A. Zawadzki. Economics of coal water slurry. 18th International Technical Conference on Coal Utilization and Fuel Systems, 455–466, 1993.
55. Ohene, F., D. Luther, and U. Simon. Fundamental investigation of non-Newtonian behavior of coal water slurry on atomization. 18th International Technical Conference on Coal Utilization and Fuel Systems, 607–617, 1993.
56. Kihm, K. D., and S. S. Kim. Investigation of dynamic surface tension of coal water slurry (CWS) fuels for application to atomization characteristics. 18th International Technical Conference on Coal Utilization and Fuel Systems, 637–648, 1993.
57. Vossoughi, S., and O. S. Al-Husaini. Rheological characterization of the coal/oil/water slurries and the effect of polymer. 19th International Technical Conference on Coal Utilization and Fuel Systems, 115–122, 1994.
58. Hamieh, T. Optimization of the interaction energy between particles of coal water suspensions. 19th International Technical Conference on Coal Utilization and Fuel Systems, 103–114, 1994.
59. Tobori, N., T. Ukigia, H. Sugawara, and H. Arai. Optimization of additives for CWM commercial plant with a production rate of 500 thousand tons per year. 19th International Technical Conference on Coal Utilization and Fuel Systems, 123–133, 1994.
60. Addy, S. N., and T. J. Considine. Retrofitting oil-fired boilers to fire coal water slurry: An economic evaluation. 19th International Technical Conference on Coal Utilization and Fuel Systems, 341–352, 1994.
61. Kaneko, S., H. Suganuma, and Y. Kabayashi. Fundamental study on the combustion process of CWM. 19th International Technical Conference on Coal Utilization and Fuel System, 403–414, 1994.
62. Takahashi, Y., and K. Shoji. Development and scale-up of CWM preparation process. 19th International Technical Conference on Coal Utilization and Fuel Systems, 485–495, 1994.
63. Tu, J., K. Cefa, J. Zhou, Q. Yao, H. Fan, X. Cao, Y. Qiu, Z. Huang, X. Wu, and J. Liu. The comparing research on the ignition of the pulverized coal and the coal water slurry. 19th International Technical Conference on Coal Utilization and Fuel Systems, 517–528, 1994.

64. Battista, J. J., T. Bradish, and E. A. Zawadzki. Test results from the co-firing of coal water slurry fuel in a 32 MW pulverized coal boiler. 19th International Technical Conference on Coal Utilization and Fuel Systems, 619–630, 1994.
65. Tu, J., Q. Yao, X. Cao, K. Cen, J. Ren, Z. Huang, J. Liu, X. Wu, and X. Zhao. Studies on thermal radiation ignition of coal water slurry. 18th International Technical Conference on Coal Utilization and Fuel Systems, 659–668, 1993.
66. Hamich, T., and B. Siffert. Physical-chemical properties of coals in aqueous medium. 18th International Technical Conference on Coal Utilization and Fuel Systems, 771–782, 1993.
67. Fullerton, K. L., S. Lee, and S. Kesavan. Laboratory scale dynamic stability testing for coal slurry fuel development. 18th International Technical Conference on Coal Utilization and Fuel Systems, 799–808, 1993.
68. Zang, Z.-X., L. Zhang, X. Fu, and L. Jiang. Additive for coal water slurry made from weak slurrability coal. 18th International Technical Conference on Coal Utilization and Fuel Systems, 821–833, 1993.
69. Hamieh and B. Siffert. Rheological properties of coal water highly concentrated suspensions. 18th International Technical Conference on Coal Utilization and Fuel Systems, Clearwater, FL, 809–820, 1993.
70. Sadler, L. Y., and K. G. Sim. Minimize solid-liquid mixture viscosity by optimizing particle size distribution. *Chem. Eng. Prog.* 87(3):68–71, 1991.
71. Everett, D. H., Basic Principles of Colloid Science. London: Royal Society of Chemistry, 1988.
72. Wenglarz, R. A., and R. G. Fox, Jr. Physical aspects of deposition from coal-water fuels under gas turbine conditions. *J. Eng. Gas Turbines Power* 112:9–14, 1990.
73. Wenglarz, R. A., and R. G. Fox, Jr. Chemical aspects of deposition/corrosion from coal-water fuels under gas turbine conditions. *J. Eng. Gas Turbines Power* 112:1–8, 1990.
74. Dwyer, J. G. Australian coal water mixtures (CWM) plant development at Newcastle, NSW. 19th International Technical Conference on Coal Utilization and Fuel Systems, 35–38, 1994.

# PROBLEMS

1. Coal A has a heating value of 14,500 Btu/lb. If the coal is used to make a coal slurry fuel, estimate the heating value of resultant slurry fuel for the following cases:
   (a) 50:50 coal-water slurry
   (b) 70:30 coal-water slurry
   (c) 50:50 coal-oil slurry (oil heating value of 16,800 Btu/lb)
   (d) 70:30 coal-oil slurry (oil heating value of 16,800 Btu/lb)
2. In preparation of a coal-water slurry, which of the following characteristics of coal are considered positive, negative, or neutral?
   (a) Low ash content
   (b) High moisture content
   (c) Low particle size
   (d) High porosity
   (e) High material density
   (f) High swelling property
   (g) Highly agglomerating

   (h) Viscosity increasing
   (i) High calorific value
   (j) Low sulfur content
3. Discuss the interfacial chemistry of a coal-water slurry.
4. Discuss the force balance in relation to a nonsettling coal-water slurry.
5. Discuss the interfacial agent in coal-water slurry fuel.
6. What are the advantages of low-temperature combustion of coal-water slurry fuel?
7. What are the potential problems in utilizing a coal-water slurry in internal combustion engines?
8. In preparing a coal-water slurry, which rank of coal is considered ideal?
   (a) lignite
   (b) subbituminous
   (c) bituminous
   (d) anthracite
9. Devise a method of measuring the viscosity of coal-water slurry.

# OIL SHALE AND SHALE OIL

## 7.1 PROPERTIES OF OIL SHALE AND SHALE OIL

In order to develop efficient retorting processes as well as to design a cost-effective commercial-scale retort, physical, chemical, and physicochemical properties of oil shales (raw materials) and shale oils (products) must be fully known. However, difficulties exist in measuring various physical and chemical properties of oil shale on a consistent basis. In this section, the properties that are needed in designing a retort as well as in understanding oil shale retorting are discussed. Even though a good deal of literature data is presented in this section, it is not the author's only intention to build a data bank of oil shale properties. Instead, appropriate utilization of data and their measurements are stressed.

### 7.1.1 Physical and Transport Properties

**Fischer assay.** The nominal amount of condensable oil that can be extracted from oil shale is commonly expressed by the Fischer assay of the oil shale. The actual oil content in the oil shale normally exceeds the Fischer assay. Depending on the treatment processes, the oil yield from oil shale often exceeds the Fischer assay (e.g., retorting in a hydrogen-rich environment, supercritical extraction of oil shale). The procedure is modified from a Fischer assay procedure for carbonization of coal at low temperature. A brief description of the Fischer assay procedure is given below.

Take a crushed oil shale sample of 100 g and subject it to a programmed heating schedule in an inert (such as nitrogen) environment. The oil shale is heated from 298 to 772 K linearly over a 50-minute period while being purged with nitrogen. Following the heatup period, the sample is held at 772 K for 20–40 minutes and the oil and gas collected are measured.

The procedure does not recover all of the organic matter in the shale but leaves a char associated with ash. Nevertheless, the Fischer assay is used as a measure of recoverable organic content. If this value is higher than 100 L/ton, it is a rich shale; if less than 30 L/ton, a lean shale.

**Porosity.** Porosity of porous material can be classified into various kinds that result in different kinds of definitions. They include interparticle porosity, intra-particle porosity, internal porosity, porosity by liquid penetration, porosity by saturable volume, porosity by liquid absorption, superficial porosity, total open porosity, bed porosity (void), and packing porosity.

Porosity of the mineral matrix of oil shale cannot be determined by the methods used in determining porosity of petroleum reservoir rocks because the organic matter in the shale is solid and essentially insoluble. However, results of a laboratory study at the Laramie Center [1] have shown that the inorganic particles contain some micropore structure, about 2.36 to 2.66 vol. %. Although the particles have an appreciable surface area, 4.24 to 4.73 $m^2$/g for shale as-saying 29 to 75 gal/ton, it seems to be limited mainly externally rather than by pore structure. Measured porosities of the raw oil shales are shown in Table 1 [2].

**Table 1 Porosities and permeabilities of raw and treated oil shales**

| | Porosity | | | Permeability | |
|---|---|---|---|---|---|
| Fischer assay | Raw | Heated to 815°C | Plane | Raw | Heated to 815°C |
| | | | A[c] | | 0.36[d] |
| 1.0[a] | 9.0 | 11.9[b] | B | | 0.56 |
| | | | A | | 0.21 |
| 6.5 | 5.5 | 12.5 | B | | 0.65 |
| | | | A | | 4.53 |
| 13.5 | 0.5 | 16.4 | B | | 8.02 |
| | | | A | | — |
| 20.0 | <0.03 | 25.0 | B | | — |
| | | | A | | — |
| 40.0 | <0.03 | 50.0 | B | | — |

[a] Units in gal/ton.
[b] Units in % of initial bulk volume.
[c] A perpendicular and B parallel to the bedding plane.
[d] Units in millidarcy.

As noted, except for the two low-yield oil shales, naturally occurring porosities in the raw oil shales are almost negligible and do not afford access to gases. Porosity may exist to some degree in the oil shale formation, where fractures, faults, or other structural defects occur. It is also believed that a good portion of pores are either blind or very inaccessible. Crackling and fractures or other structural defects often create new pores and also break up some of the blind pores. It should be noted here that "closed" or "blind" pores are normally not accessible by mercury porosimetry even at high pressures.

**Permeability.** Permeability of raw oil shale is essentially zero because the pores are filled with a nondisplaceable organic material. Previous work by Tisot [2] showed that gas permeability, either perpendicular or parallel to the bedding plane, was not detected in most oil shale samples at a pressure differential across the cores of 3 atm of helium for 1 minute. In general, oil shale constitutes a highly impervious system. Thus, one of the major problems of any in situ retorting project is to create a suitable degree of permeability in the formation.

Of practical interest is the dependence of porosity or permeability on temperature and organic contents. Upon heating to 510°C, an obvious increase in oil shale porosity is noticed. These porosities, which vary from 3 to 61 vol. % [2] of the initial bulk oil shale volume, represented essentially the volumes occupied by the organic matter. Therefore, the oil shale porosity increases as the extent of pyrolysis reaction increases. In the low Fischer assay oil shales, structural breakdown of the cores is insignificant and the porosities are those of intact porous structures. However, in the high Fischer assay oil shales, this is not the case because structural breakdown and disaggregation become extensive, especially in the richer oil shale, and the mineral matrices no longer remain intact. Thermal decomposition of the mineral carbonates occurring around 815°C also results in an increase in porosity. The increase in porosity from low to high Fischer assay oil shales varies from 2.82 to 50% as shown in Table 6. These increased porosities constitute essentially the combined spaces represented by the loss of the organic matter and the decomposition of the mineral carbonates. Crackling of particles is also due to devolatilization or organic matter that increases the vapor pressure of large nonpermeable pores.

Gas permeability [2] is low in both planes of the mineral matrices from the three low Fischer assay oil shales heated to 815°C. As noted in Table 6, the mineral matrix from the 13.5 gal/ton oil shale has the highest permeability of 8.02 millidarcy. This value may be somewhat higher than the primary permeability of mineral matrix; that is, the permeabilities created by removing the organic matter and by thermally destroying the mineral carbonates are also included. Even though the oil shale cores used for these measurements have no visible fractures, minute fractures may have formed during heating to 815°C, which probably contributed to some secondary permeabilities. Permeabilities after heating to 815°C, for the oil shales that exceeded 13.5 gal/ton, are not given. In these oil shales, structural breakdown of the mineral matrices in a stress-free environment was so extensive as to preclude measurements of the permeabilities in high Fischer assay oil shales. [2]

In the case of eastern U.S. shales, especially Devonian oil shales, the decomposition of kerogen produces lighter hydrocarbons than those for other shales. This often results in a substantial increase in volatile pressure in the solid matrix, which leads to cracking of the solid structure. This is also why the oil yield from eastern oil shale pyrolysis is not an accurate measure of the organic matter content of the shale.

**Compressive strength.** The raw oil shales have high compressive strengths both perpendicular and parallel to the bedding plane [3]. After heating, the inorganic matrices of low Fischer assay oil shales retain high compressive strength in both perpendicular and parallel planes. This indicates that a high degree of inorganic cementation exists between the mineral particles constituting each lamina and between adjacent laminae. With increase in organic matter the compressive strength of the respective organic-free mineral matrices decreases, and it becomes very low in those rich oil shales. The results are shown in Figure 1 [4].

Also worthy of note is the existence of the structural transition point. Gradual expansion of oil shale in a stress-free environment was noted immediately upon application of heat. Around 380°C, the samples are seen to undergo drastic changes in compressive strength. The greater loss of compressive strength at the yield point and the low recovery on reheating for the richer oil shales are both attributed to extensive plastic deformation effects. The degree of plastic deformation thus seems to be directly proportional to the amount of organic matter in oil shale. The discontinuities in the pressure plot at temperatures below the yield point presumably arise from the evolution of pore water from the oil shale matrix. The well-defined transition point at 380°C, therefore, represents a pronounced change in the compressive strength of richer oil shale. It is also interesting to note that near this temperature most coals exhibit plastic properties.

**Thermal properties.** The term "thermal" is used here to represent parameters that are directly or indirectly related to the transport, absorption, or release of heat. Properties such as thermal conductivity, thermal diffusivity, enthalpy, density, and heat capacity fall into this definitive classification scheme. For materials that are thermally active, that is, those which undergo thermal decomposition or phase transformation (this is the case with oil shales in general), it is also necessary to characterize their thermal behavior by thermoanalytical techniques such as differential thermal analysis (DTA) and thermogravimetric analysis (TGA). Therefore, the DTA and TGA characteristics are also included in the group of thermal properties.

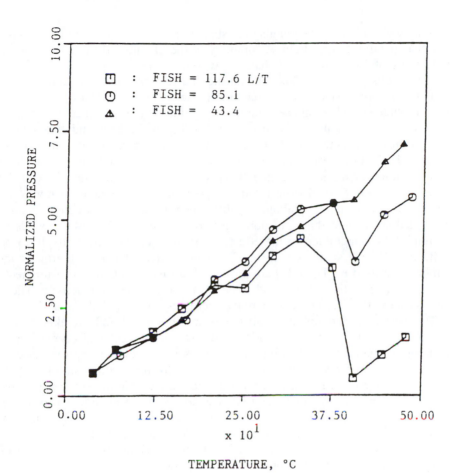

**Figure 1**  Variation of compressive strength with organic content. (Reprinted from Wang, Y., Master's thesis, University of Akron, OH, 1982, with permission.)

*Thermal conductivity.* Measurements of thermal conductivity of oil shale show that blocks of oil shale are anisotropic about the bedding plane. The measurements were made by techniques such as the transient-probe method [2], the thermal comparator technique [5], and the line source method [6]. The range of temperature and shale grades investigated in some instances is, however, quite limited [7, 8]. Some earlier studies, in addition, did not focus on the anisotropic nature of the heat conduction. Recent studies have shown that the thermal conductivity as a function of temperature, oil shale assay, and direction of heat flow, parallel to the bedding plane (parallel to the earth's surface for a flat oil shale bed) was slightly higher than the thermal conductivity perpendicular to the bedding plane. As layers of material were laid to form the oil shale bed, the resulting continuous strata had slightly higher resistance to heat flow perpendicular to the strata than parallel to the strata. A summary of the literature data on the thermal conductivity of Green River oil shale is given in Table 2.

The results shown in Table 2 indicate that the thermal conductivities of retorted and burned shales are lower than those of the raw shales from which they are obtained. This can be interpreted as due to the fact that the mineral matter is a better conductor of heat than the organic matter and the organic matter in turn is a better conductor than the voids created by its removal [6]. Although the first of these hypotheses is reasonable when one takes into account the contribution of the lattice conductivity to the overall value, the effect of the amorphous carbon formed from the decomposition of the organic matter could also be of importance in explaining the difference in thermal conductivity values for retorted shales and the corresponding burned samples. The role of voids in determining the magnitude of the effective thermal conductivity is likely to be significant only for samples with high organic content. Data for thermal conductivity measured by Tihen [6] are presented in Figure 2. Other measurements [11] were made of the effect of oil shale assay on thermal conductivity. These data also show that thermal conductivity decreases with oil shale assay [12].

The thermal conductivities of oil shales are, in general, only weakly dependent on temperature [12]; the majority of studies report a gradual decrease with increasing temperature [8, 13]. Extreme caution is, however, to be exercised in the interpretation of results at temperatures close to the decomposition temperature of the shale organic matter. This is due to the fact that the kerogen decomposition reaction (or pyrolysis reaction) is endothmeric in nature. For example, thermal conductivity values reported at temperatures around 400°C normally include the effects of decomposition of the shale organic matter, even though it has been claimed that this is not a critical parameter as long as the rate of temperature increase of the sample is relatively low.

**Table 2 Comparison of thermal conductivity values for Green River oil shales**

| Temperature range (°C) | Fischer assay | Plane | Thermal conductivity (J/m-sec-°C) | Reference |
|---|---|---|---|---|
| 38–593 | 7.2–47.9 | — | 0.69–1.56 (raw shales) | [6] |
| | | | 0.26–1.38 (retorted shales) | |
| | | | 0.16–1.21 (burnt shales) | |
| 25–420 | 7.7–57.5 | A | 0.92–1.92 | [9] |
| | | Mean value | 1.00–1.82 (burnt shales) | |
| 38–205 | 10.3–45.3 | A | 0.30–0.47 | [10] |
| | | B | 0.22–0.28 | |
| 20–380 | 5.5–62.3 | A | 1.00–1.42 (raw shales) | [5] |
| | | B | 0.25–1.75 (raw shales) | |

ᵃUnits of gal/ton.
ᵇA parallel and B perpendicular to the bedding plane.

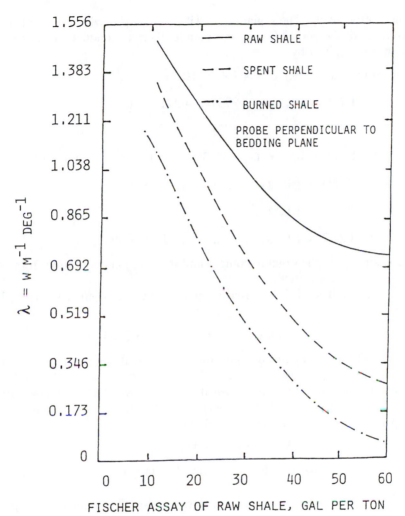

**Figure 2** Thermal conductivity of raw, spent, and burned oil shale (1 gal/ton = 4.18 cm³ kg). (Reprinted with permission from reference 13.)

The thermal conductivity values of oil shales show an inverse dependence on the organic matter content [6, 9]. Equations that have been proposed by various authors relating the thermal conductivity to the three parameters—temperature, organic content, and extent of kerogen conversion—are of the following form:

$$K = c_1 + c_2x_1 + c_3x_2 + c_4x_3 + c_5x_1x_2 + c_6x_2x_3 + c_7x_3x_1 + c_8x_1x_2x_3 \quad (1)$$

where $x_1$, $x_2$, and $x_3$ denote the organic content, kerogen conversion, and temperature, respectively. The equation of Tihen et al. [6], which was taken to be the average for the perpendicular and parallel thermal conductivities of raw and spent shales, is given by

$$K = (1 - x_2) \{1.9376 - 4.739 \times 10^{-2} x_1$$

$$+ 1.776 \times 10^{-3} (x_3 - 273) + 4.371 \times 10^{-4} x_1^2$$

$$- 4.885 \times 10^{-6} (x_3 - 273)^2$$

$$- 1.671 \times 10^{-5} x_1 (x_3 - 273)\} + x_2 \{1.680 - 5.204 \times 10^{-2} x_1 \quad (2)$$

$$- 1.003 \times 10^{-4} (x_3 - 273) + 4.951 \times 10^{-4} x_1^2$$

$$- 1.468 \times 10^{-9} (x_3 - 273)^2$$

$$+ 0.667 \times 10^{-5} x_1 (x_3 - 273)\}, \quad J/sec\text{-}m\text{-}°C$$

where $x_1$ and $x_2$ are the organic content and the kerogen conversion in fractions, and $x_3$ is in degrees Kelvin.

Prats and O'Brien [9] proposed a second-order polynomial in $(x_3 - 25)$ of the form

$$K = c_1 [1 - D_1 (x_3 - 25) + D_2 (x_3 - 25)^2] \exp(c_2 F) \quad (3)$$

where $F$ is Fischer assay in liters per ton, $K$ is thermal conductivity in W/m °C, and $x_3$ is shale temperature in degrees Celsius. In Eq. (3), $c_1$, $c_2$, $D_1$, and $D_2$ are again empirically obtained. As shown in Eq. (3), the thermal conductivity is a relatively simple function of the temperature and the organic content of shale. A simpler equation was proposed for Baltic shales [14].

$$K = 1.30/F + 0.06 + 0.003T \quad (4)$$

where $F$ is Fischer assay in liters per ton, $T$ is in degrees Celsius, and $K$ is in W/m°C.

When using simpler equations, one has to realize their limitations, especially in the case of extrapolation or interpolation of experimental data. Most of simple expressions normally have narrower ranges of validity.

The problem of thermal conduction through a bed of oil shale rubble is quite complex. This problem is similar to that of packed beds of randomly sized and oriented particles. Furthermore, it is difficult to generalize the size distribution of fractured underground beds of oil shale and accurately control the size and shape of particles when in situ, underground rubblization takes place. In this regard, various process ideas related to rubblization and heating the shales have been and will be generated.

Thermal diffusivity, $\alpha$, is defined as

$$\alpha = \frac{k}{\rho C_p} \quad (5)$$

where $k$, $\rho$, and $C_p$ denote the thermal conductivity, density, and heat capacity, respectively. Therefore, the thermal diffusivity has dimensions of $L^2t^{-1}$, just like mass and momentum diffusivities. For an oil shale particle to reach a predetermined temperature throughout the particle dimension, the required time may be estimated by

$$t = \frac{\rho C_p L_{ch}^2}{k} \tag{6}$$

where $L_{ch}$ is the characteristic length. Equation (6) is based on the isothermality criterion of $t_{ch}^* = 1.0$, in which the characteristic time becomes unity. A rough idea of the characteristic length may be explained by the "penetration depth" or a "major dimension." For a sphere, the characteristic length is calculated from

$$L_{ch} = \frac{\text{volume}}{\text{surface area}} = \frac{\frac{4}{3}\pi R^3}{4\pi R^2} = \frac{R}{3} = \frac{D}{6} \tag{7}$$

The same calculation can be carried out for a regular cylinder whose diameter is the same as its length:

$$L_{ch} = \frac{\text{volume}}{\text{surface area}} = \frac{2\pi R^3}{2\pi R(2R) + 2\pi R^2} = \frac{R}{3} = \frac{D}{6} \tag{8}$$

An analogous calculation can be made for determination of a characteristic dimension.

As can be readily seen from Eq. (6), the heat-up time required is proportional to the square of the characteristic dimension. In other words, successful operation of an in situ retort using the combustion retorting process depends quite strongly on how fine the rubblization of an oil shale bed can be made.

The dependence of the thermal diffusivity of Green River oil shale perpendicular to the bedding plane on the temperature and the Fischer assay was measured by Dubow et al. [11] and is shown in Figure 3.

The thermal diffusivity values show the same broad trends with variations in temperature and shale grade as do the thermal conductivities [12]. The thermal diffusivity decreases with increasing temperature and organic content in the oil shale. Thus, the retorted and burned shales show reduced thermal diffusivities compared with the raw shales [12], as shown in Table 3. Oil shale samples containing large amounts of pyrites ($FeS_2$) are likely to have high thermal diffiusivities, because that of pyrite itself is quite high.

*Heat capacity of oil shale.* Earlier work [15] on the specific heat of American oil shales is restricted to limited ranges of temperatures and shale grades. Later studies [16] reported heat capacity dependences on temperatures and shale grade, typified by equations of the type

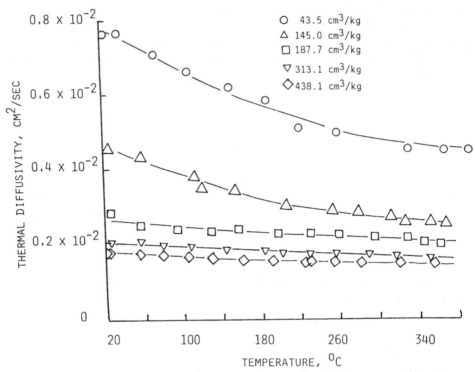

**Figure 3** Dependence of thermal diffusivity of oil shale perpendicular to the bedding plane on temperature and Fischer assay.

$$c = c_1 + c_2 x_1 + c_3 x_2 + c_4 x_1 x_2 \tag{9}$$

Again, it is difficult to generalize the heat capacity of oil shale in any sense, because of the vast heterogeneity of oil shales even within the same formation as well as among different formations.

Considerable increases in the values of heat capacity with increasing organic content have been observed [15, 16], although the relative contributions of the

## Table 3 Thermal diffusivity of Green River oil shale

| Temp. range (°C) | Fischer assay[a] | Thermal diffusivity ($10^2$ cm$^2$ sec$^{-1}$) | Measurement technique | Reference |
|---|---|---|---|---|
| 38–260 | 6.7–48.4 | 0.26–0.98 (raw shale) | Transient line probe | 6 |
| 38–482 | | 0.13–0.88 (retorted shale) | | |
| 38–593 | | 0.10–0.72 (burned shale) | | |
| 25–350 | 5.0–82.2 | 0.10–0.90 | Laser flash | 59 |

[a]Fischer assay is in gallons per short ton.

various oil shale constituents to the overall values are somewhat uncertain. Values of heat capacity for most oil shales are not available in the literature.

The heat capacity correlation given by Sohn and Shih [16] is

$$C_{ps} = \{(907.09 + 505.85x_1)(1 - x_2) + 827.06\, x_2\} \tag{10}$$
$$+ \{(0.6184 + 5.561x_1)(1 - x_2) + 0.92x_2\}(x_3 - 298) \quad J/kg\text{-}°C$$

Equation (10) gives at least some estimation of the heat capacity of oil shale.

Since the measurement of heat capacity of a solid sample is not an involved task, it may not be a bad idea to measure it when the information is needed.

***Enthalpy and heat of retorting.*** Wise et al. [17] measured the enthalpy of raw, spent, and burned oil shale from the Green River Formation. The spent shale had oil produced by retorting but retained a char residue. However, burned shale had virtually all organic matter removed. Some of their results are shown in Figure 4.

The enthalpy data can be represented as a function of temperature and Fischer assay of oil shale grade as

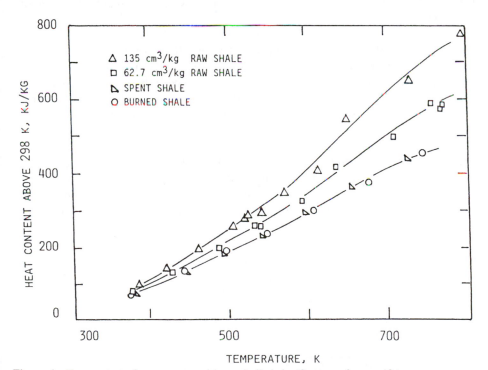

**Figure 4** Heat content of raw, spent, and burned oil shale. (Source: reference 13.)

$$\Delta H = a + bT + cT^2 + dT^3 + eT^4 + fT^2F \qquad (11)$$

where $T$ is temperature and $F$ is oil shale Fischer assay. All the coefficients are also given in [16]. The specific heat can be determined by differentiation,

$$C_{p,s} = \left[ \frac{\partial H_s}{\partial T} \right]_P \qquad (12)$$

These data are in good agreement with earlier data of Shaw [18] and Sohn et al. [19].

The available data for heat of retorting of Green River oil shales are listed in Table 4. The reported values for the heat of retorting show the expected increase with increasing shale assay and temperature; the disparity in the range of values observed by different investigators possibly reflects differences in the composition of the shale samples. It should be borne in mind that the presence of minerals that decompose at temperatures below the range at which the organic matter is thermally extracted would increase the heat requirements for processing shales containing these minerals. Thus, it has been estimated that shales containing nahcolite and dawsonite would require an additional 117 cal/g (or, 490 J/g) and 215 cal/g (or 890 J/g), respectively [20]. Heat requirements for retorting oil shales containing 17% analcite would be increased by about 6% [21].

***Density or specific gravity.*** The density of Green River oil shale was measured by Tisot [2] and found to be in the range 1.8 to 2.0 g/cm³. Later, some efforts were made to correlate the oil yield from oil shale with the specific gravity [13]. This idea may have some practical significance, because oil yield is usually a constant fraction of organic content and oil shale density is dependent on organic content.

***Self-ignition temperature.*** The self-ignition temperature is the temperature at which oil shale samples ignite spontaneously. There is no standardized procedure for making this measurement. However, if a consistent measurement of this temperature is made for oil shale, it can provide valuable information regarding the fuel characteristics of the shale.

**Table 4  Values reported for heat of retorting of Green River oil shales**

| Heat of retorting (kJ/kg) | Fischer assay (gal/ton) | Reference |
|---|---|---|
| 238–878 | 23.5–46.7 | [19] |
| 581–699 | 8.0–32.8 | [20] |
| 335 | 25.6 | [22] |

Reproduced with permission from Wang, Y., M. S. thesis, The University of Akron, Akron, OH, 1982.

The spontaneous ignition temperature of oil shale has been measured and characterized under a variety of conditions by Allred [22]. This information regarding the self-ignition temperature is important, because it governs not only the initiation of the combustion retorting process but also the dynamics of oil shale retorting by the advancing oxidation zone. Branch [23] explains its importance in his article. In the countercurrent combustion retorting process, the combustion front moves toward the injected oxidizer, whereas in the cocurrent process the front moves in the same direction as the oxidizer. Therefore, the spontaneous ignition temperature of the raw oil shale should be below the oil shale retorting temperature in the countercurrent process. In the cocurrent combustion process, char remaining after retorting is burned to sustain the retort by providing the necessary energy.

The self-ignition temperatures of Colorado oil shale have been measured over a wide range of total pressure and oxygen partial pressure [22]. With nitrogen as a diluent, the ignition temperature was found to depend on oxygen partial pressure but not significantly on total pressure. The temperature at which ignition could occur was also found to correlate with the temperature at which methane and other light hydrocarbons are devolatilized from the oil shale. Higher ignition temperatures were required for lower oxygen partial pressures, and the lowest ignition temperature of 450 K was considerably lower than the oil production temperature of about 640 K. As shown in Figure 5, the ignition of Colorado oil shale in air at pressures between $0.3 \times 10^5$ and $14 \times 10^5$ Pa shows behavior similar to that of cool flame oxidation of light hydrocarbons. This experimental evidence strongly suggests that ignition of oil shale may be associated with oxidation of gaseous hydrocarbons evolved from the oil shale.

Joshi [24] did an interesting study of the self-ignition temperature (SIT). He defined the self-ignition temperature as the temperature at which the shale bursts into flames, within 360 seconds after being introduced into a preheated, isothermal retorter. His SIT measurements were made at oxygen partial pressure of 0.21 atm, that is, at atmospheric total pressure. Joshi's approach was different from that of Allred, in that different oil shales can be characterized by the self-ignition temperature. In other words, the self-ignition temperature defined by Joshi can be used more like a physical property. It was found that the SIT depends on the Fischer assay of oil shale. The higher the Fischer assay, the lower the SIT in general. Figure 6 depicts graphically the relation between SIT and Fischer assay.

Joshi's results also suggest that the ignition of oil shale is associated with the oxidation of gaseous hydrocarbons evolved from oil shale. The ignition of the shale samples was always preceded by a slight exploding sound, which strongly suggests that the ignition occurred only when the vapor pressure of the gaseous hydrocarbons evolved reached a certain level. Further tests by Joshi [24] revealed that the SIT is also a function of particle size. Tests were carried out on particles of size −4+8 mesh to −60 mesh. It was also found that Colorado

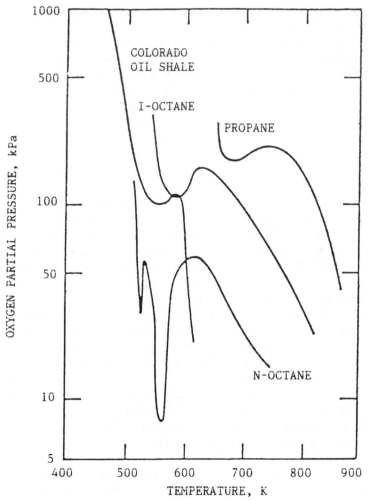

**Figure 5** Self-ignition temperature of raw oil shale, *i*-octane, *n*-octane, and propane. (Source: Reference 13.)

shale particles of size −40+60 mesh and smaller did not burst into flames under the defined conditions. The same phenomenon was observed for Cleveland shale No. 2, with the limiting particle size being −20+40 mesh (which did not burst into flames). It has been stated that as the particle size becomes smaller the diffusional limitations of the product vapor decrease substantially in magnitude. As a result, there is a continuous diffusion of product vapor (gaseous hydrocarbons) from the rock matrix to the surface and from the surface to the boundary layer.

As a result, heating induces a buildup of gaseous hydrocarbons inside the rock matrix, which ultimately causes cracking of the shale particles. The con-

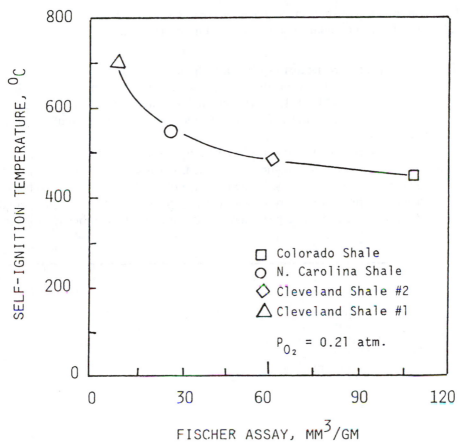

**Figure 6**   Self-ignition temperature profile as a function of Fischer assay. (Reprinted from Joshi, R., Master's thesis, University of Akron, Akron, OH, with permission.)

centration of the gaseous hydrocarbons on the surface immediately after cracking is high enough to stimulate ignition. As mentioned earlier, the SIT data are very useful, because the ignition temperature governs the initiation of the combustion process and also the dynamics of in situ oil shale retorting. Furthermore, the data are valuable because they give an indication of the explosibility of oil shale dust in oil shale mining operations. However, such a hazard is considerably less significant and less likely than that encountered in coal mining [2, 9].

## 7.1.2 Thermal Characteristics of Oil Shale Minerals

Thermal or thermoanalytical methods such as thermogravimetric analysis and differential thermal analysis are particularly useful for characterization of the thermal behavior of oil shales and oil shale minerals. The use of TGA and DTA has been well established in the areas of coal and polymer research, because a relatively simple procedure generates valuable information.

**Thermoanalytical properties of oil shale.** Figures 7 and 8 show the effect of the surrounding atmosphere on the thermal behavior of Green River oil shale [12]. Figure 7 shows the DTA curve in an inert atmosphere of flowing $N_2$, whereas Figure 8 shows the DTA curve in the presence of air, that is, in an oxidative atmosphere.

The peak corresponding to kerogen decomposition is seen to be endothermic in nature. This endothermic nature is expected, because the pyrolysis reaction requires input of thermal energy. In the presence of air, however, two exotherms are apparent as shown in Figure 8, with the first peak at 439°C and the second at 500°C. The first exothermic peak may be attributed to the combustion of light

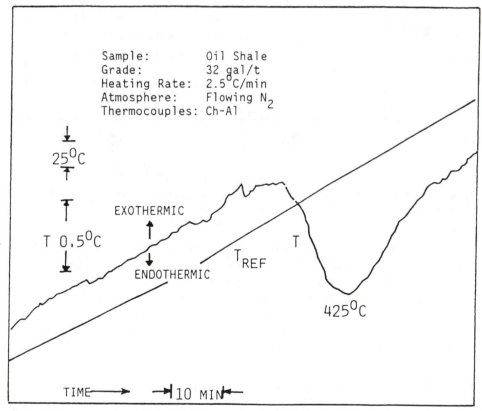

**Figure 7** Effect of surrounding atmosphere on the thermal behavior of Green River oil shale: DTA in an inert atmosphere of flowing $N_2$. (Source: Reference 25.)

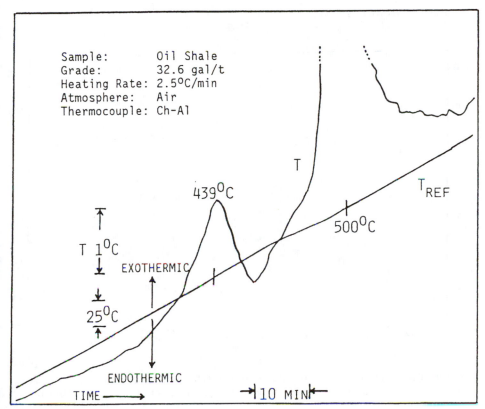

**Figure 8** Effect of surrounding atmosphere on the thermal behavior of Green River oil shale: DTA in the presence of air. (Source: Reference 25.)

hydrocarbon fractions from the shale organic matter; the second exotherm appears to be from the burnoff of carbonaceous char [12].

Figure 9 shows the TGA curve of Stuart shale of Queensland, Australia. The sample used in this experiment was from the Kerosene Creek member and supplied by Southern Pacific Petroleum NL. The total mass loss by the TGA from 25 to 600°C was 12.2%, which includes (1) evaporative loss of moisture, (2) kerogen decomposition and devolatilization, and (3) thermal decomposition of carbonates and $CO_2$ vaporization.

**Thermal characteristics of oil shale minerals.** Various carbonates existing in the Green River oil shales, such as ferroan, ankerite, and dawsonite, were identified and quantified by use of thermal analysis techniques [26]. A quantitative determination method for nahcolite and trona in Colorado oil shales was proposed by Dyni et al. [27].

Sample:   STUART  S

TGA

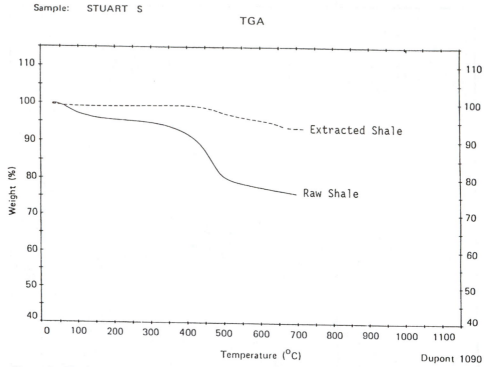

**Figure 9**  Thermogravimetric analysis of the Stuart shale before and after extraction. (From Kesavan, S., *Fuel Sci. Technol. Int.* 6(5): 505, 1988. Reprinted by permission of Marcel Dekker, New York.)

Rajeshwar et al. [23] summarized thermal characteristics of minerals commonly found in oil shale deposits, and their results are shown in Table 5.

Regarding the decomposition of dawsonite, there are some conflicting beliefs [12]. One belief is that dawsonite decomposes at 370°C according to the reaction:

$$NaAl(OH)_2CO_3 \longrightarrow H_2O + CO_2 + NAlO_2 \qquad (13)$$

Another is based on the investigation of thermal behavior of dawsonite at temperatures between 290 and 330°C, and the reaction taking place is believed to be [28]

$$2NaAl(OH)_2CO_3 \longrightarrow Na_2CO_3 + Al_2O_3 + H_2O + CO_2 \qquad (14)$$

Yet another is the belief that dawsonite decomposes in two steps. In the first step, it has been found that between 300 and 375°C, crystalline dawsonite decomposes with the evolution of all the hydroxyl groups and two-thirds of the carbon dioxide, leaving an amorphous residue. In the second step, the balance of $CO_2$ is released over the range 360 to 650°C producing crystalline $NaAlO_2$.

**Table 5 Thermal characteristics of minerals commonly found in oil shale deposits**

| Minerals | Chemical formula | Type of thermal reaction | DTA peak temperature (°C) |
|---|---|---|---|
| Calcite | $CaCO_3$ | Dissociation | 860–1010 |
| Dolomite | $CaMg (CO_3)_2$ | Dissociation | 790, 940 |
| Analcite | $NaAlSi_2O_6 \cdot H_2O$ | Dehydration, dissociation | 150–400 |
| Shortite | $Na_2Ca_2(CO_3)_3$ | Dissociation | 470 |
| Trona | $2Na_2CO_3 \cdot NaHCO_3 \cdot 2H_2O$ | Dissociation, dehydration | 170 |
| Pyrite | $FeS_2$ | Oxidation, dissociation | 550 |
| Potassium feldspar | $KAlSi_3O_8$ | Dissociation | — |
| Gaylussite | $CaNa_2(CO_3)_2 5H_2O$ | Dehydration, crystallographic transformation, melting | 145, 175, 325, 445, 720–982 |
| Illite | $KAl_4Si_7AlO_{20}(OH)_4$ | Dehydroxylation | 100–150, 550, 900 |
| Plagioclase | $NaAlSi_3O_8–CaAl_2Si_2O_8$ | Dissociation | — |
| Nahcolite | $NaHCO_3$ | Dissociation | 170 |
| Dawsonite | $NaAl(OH)_2CO_3$ | Dehydroxylation, dissociation | 300, 440 |
| Gibbsite | $\gamma\text{-}Al(OH)_3$ | Deydroxylation | 310, 550 |
| Ankerite | $Ca(Mg,Mn,Fe)(CO_3)_2$ | Dissociation | 700, 820, 900 |
| Siderite | $FeCO_3$ | Oxidation, dissociation | 500–600, 830 |
| Albite | $NaAlSi_3O_8$ | Dissociation | — |
| Quartz | $SiO_2$ | Crystallographic transformation | ~ 575 |

The dominant mineral constituent in oil shale is dolomite. Dolomite is approximately a one-to-one mixture of magnesite ($MgCO_3$) and calcite($CaCO_3$). Therefore, the chemical formula of dolomite is expressed as either $CaCO_3 \cdot MgCO_3$ or $CaMg(CO_3)_2$. Upon heating, dolomite undergoes a two-stage thermal decomposition reaction known as calcination.

$$CaCO_3 \cdot MgCO_3 \xrightarrow{\Delta} CaCO_3 \cdot MgO + CO_2 \qquad (15)$$

$$CaCO_3 \cdot MgO \xrightarrow{\Delta} CaO \cdot MgO + CO_2 \qquad (16)$$

The first reaction is called half-calcination of dolomite, whereas the second is called full calcination of dolomite. The equilibrium decomposition temperature, where $K_a = 1$, is 380°C for magnesite decomposition and 890°C for calcite decomposition.

## 7.1.3 Electric Properties of Oil Shale

Electric properties also change as functions of temperature and other variables. Both AC and DC methods can be employed in measuring electric properties of oil shale. In general, AC techniques are preferable in view of their capability to detect and resolve various polarization mechanisms in the material.

**Electric resistivity.** Measurements on various types of oil shales in DC electric fields have shown an exponential decrease in resistivity values as a function of temperature [4, 12, 29]. These trends are typically characteristic of ionic solids that conduct by thermally activated transport mechanisms. The presence of various minerals in the oil shale rock matrix makes it difficult to identify conclusively the current-carrying ions in the material, although the close correspondence of activation energies at high temperatures (>380°C) with those typically observed for carbonate minerals seems to indicate that carbonate ions could be a major current-carrying species [4, 12].

However, estimates made from such data are at best speculative and must be considered with due caution [12]. The chemical change in the oil shale material due to heating could also influence its conduction behavior. Thus, changes in the resistivity (from $10^{10}$ ohm-cm at room temperature to 10 ohm-cm at 900°C) of Russian shales were attributed to the thermal decomposition of oil shale kerogen [29]. Figure 10 shows the frequency-dependent behavior of electric resistivity as a function of reciprocal temperature for a 117 L/ton Fischer assay sample of raw Green River oil shales. The minima in the resistivity curves are observed at temperatures in the range 40 to 210°C and are due to gradual loss of free moisture and bonded water molecules to the clay particles in the shale matrix. Figure 11 shows the same behavior for the reheated materials; however, the trends of the curves are quite different from those for raw shales.

For these experiments, the shales were cooled back to room temperature and reheated back to approximately 500°C. The curves of Figure 11 show the usual Arrhenius behavior typical of ionic solids, as mentioned before. It can be seen that there is no minimum or peak in the resistivity data and that the results are attributable to thermally activated conduction.

**Dielectric constants.** An extensive review of the dielectric constant of oil shale is given in [12]. The dielectric constant of oil shales also exhibits a functional dependence on temperature and frequency. Anomalously high dielectric constants are observed for oil shales at low temperatures, and these high values are attributed to electrode polarization effects by some investigators [30]. A more likely explanation is the occurrence of interfacial polarization (e.g., Maxwell-Wagner type) in these materials arising from the presence of moisture and as a result of accumulation of charges at the sedimentary varves in the shale.

Figure 12 shows the variation of dielectric constant with the number of heating cycles for several grades of Green River oil shales [31]. Each heating cycle consisted of heating the sample at 110°C for 24 hours and cooling back to room temperature prior to testing. A noticeable decrease in dielectric constant

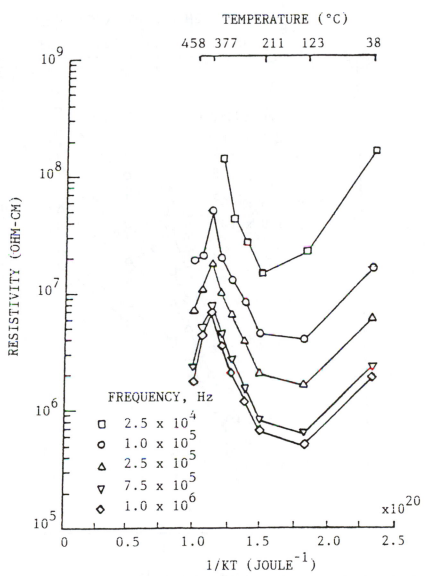

**Figure 10** Frequency-dependent behavior of electrical resisivity (ρ) as a function of reciprocal temperature for a 117 L/t shale sample. (Source: Reference 25.)

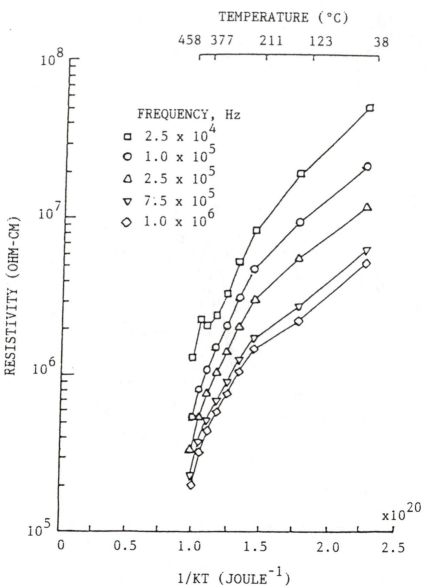

**Figure 11** Same plot as in Figure 10 for the sample reheated in a second cycle. (Source: Reference 25.)

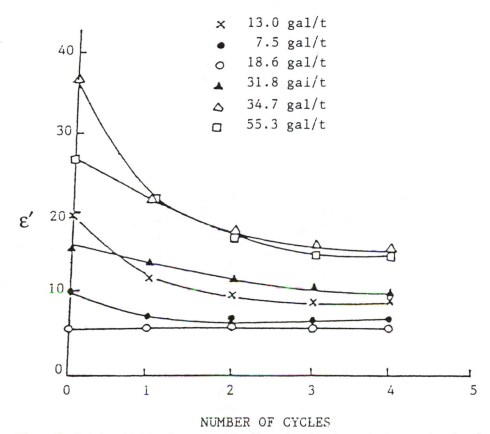

NUMBER OF CYCLES

**Figure 12** Variations of dielectric constant ($\epsilon'$) with number of heating cycles for several grades of Green River oil shales. (Source: Reference 25.)

observed with each subsequent drying cycle is evident. Figure 13 shows the variation of dielectric constant with frequency and thermal treatment for Green River oil shales [31]. The degree of frequency dispersion at each heating cycle attests to the appreciable effect of moisture on interfacial polarization mechanisms in oil shale.

Figures 14 and 15 show the variation of the dielectric constant of Green River oil shales as a function of frequency and temperature at low temperature (<250°C) and at high temperature (>250°C), respectively [32]. A general explanation may be that the dielectric constant decreases with increasing temperature up to 250°C and thereafter increases again, attaining values comparable to those observed initially for the raw shales. The initial decrease may be due to the gradual release of absorbed moisture and chemically bonded water from the shale matrix.

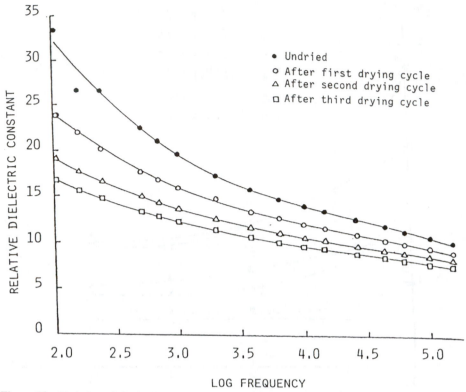

**Figure 13** Variation of $\epsilon'$ with frequency and thermal treatment for Green River oil shales. (Source: Reference 25.)

However, the subsequent increase may be due to more complex factors, including (1) increased orientational freedom of the kerogen molecules, (2) buildup of carbon in the shale, and (3) presence of a space-charge layer in the material at high temperatures. Further details can be found in [12].

## 7.1.4 Molecular Consideration of Kerogen

In applying chemical and engineering principles to the decomposition of oil shale, difficulties are encountered because of the lack of structural and molecular understanding of kerogen. A model of kerogen structure has been proposed by Yen [33], and the formula is approximately $C_{220}H_{330}O_{18}N_2S_4$. If we use this formula as a representative structure for the kerogen molecule, its molecular weight becomes 3414. In other words, kerogen has a macromolecular structure and a high molecular weight.

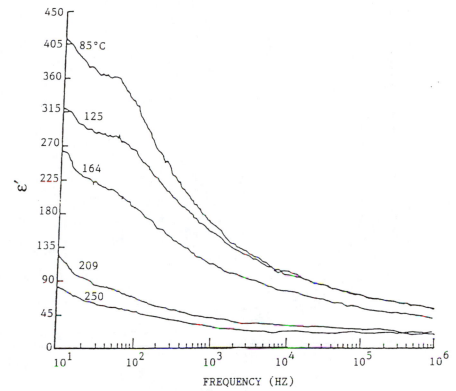

**Figure 14** Frequency and temperature dependence of $\epsilon'$ at low temperatures ($<250°C$). (Source: Reference 25.)

**Derivation of stoichiometric coefficient.** Because the structure of kerogen cannot be represented as a uniquely defined one, a stoichiometric equation for kerogen decomposition is not a routine. If we write

$$\text{Kerogen} \longrightarrow \text{oil vapors} + \text{gaseous products} \qquad (17)$$

the equation cannot be taken as a stoichiometric equation, since the atomic balance on constituent atoms is not satisfied. The following discussion is intended to provide a theoretical bridge between a qualitative equation and a stoichiometric equation.

The initial mass of organic carbon per cubic meter of particle volume is

$$m_1 = \rho_s \omega, \quad \text{kg/m}^3$$

$$= \rho_s \omega 1000/M_k, \quad \text{mol/m}^3 \qquad (18)$$

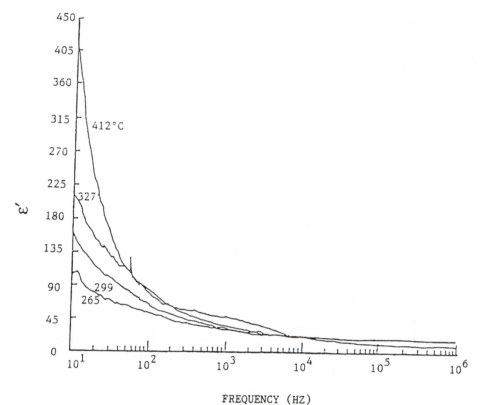

FREQUENCY (HZ)

**Figure 15** Frequency and temperature dependence of $\epsilon'$ at high temperatures (>250°C). (Source: Reference 25.)

where $\rho_s$ is the density of oil shale particle in kg/m$^3$, $\omega$ is the mass fraction of kerogen in the particle, and $M_k$ is the molecular weight of kerogen in g/mol. The total mass of oil recoverable (conventionally) from oil shale per cubic meter of particle volume is

$$m_2 = F \left[\frac{1}{\text{M/T}}\right] \rho_s \left[\frac{\text{kg}}{\text{m}^3}\right] \rho_o \left[\frac{\text{g}}{\text{cm}^3}\right] = F\rho_s\rho_o \left[\frac{\text{g}}{\text{m}^3}\right]$$

$$= F\rho_s\rho_o/M_p \quad \text{gmol/m}^3$$

(19)

where $F$ is the Fischer assay of oil shale in liters per metric ton, $\rho_o$ is the density of shale oil in g/cm$^3$, and $M_p$ is the average molecular weight of condensable product in g/mol.

Assuming the kerogen decomposes into oil vapors and gaseous products completely at moderate decomposition temperatures, the stoichiometric coefficient is

$$\alpha = \frac{F\rho_s\rho_o/M_p}{\rho_s\omega 1000/M_k}$$ (20)

$$= \frac{F\rho_o M_k}{\omega M_p 1000}$$

where $\alpha$ is a theoretically obtained stoichiometric coefficient for reaction (17).

**Relation between Fischer assay and mass fraction of kerogen.** The empirical correlation of Fischer assay of oil shale is given by Cook [34].

$$F = 2.216wp - 0.7714 \quad \text{gal/ton}$$ (21)

where $F$ is the Fischer assay estimated in gallons of oil recoverable per ton of shale and $wp$ is the weight percent of kerogen in the shale. In using this equation, care must be exercised, as the mass fraction of kerogen in the shale is dependent on the measurement technique. For some shales, the maximum amount of oil recoverable via supercritical extraction or $CO_2$ retorting is significantly higher than the Fischer assay value.

It is also conceivable that such a correlation can be sensitive to the type of oil shale retorted as well as dependent on the extraction process chosen. Anyway, there is still little doubt that the Fischer assay of any oil shale is strongly correlated with the kerogen content of the shale.

### 7.1.5 Boiling Range Distributions of Various Shale Oils

Shale oils or oil shale crudes are obtained by extraction of oil shale via various physicochemical processes. The characteristics of shale oils are important in devising a process for upgrading shale oils as well as in finding better uses for them. In particular, distillation properties of shale oils are crucially important not only for their refining but also for their upgrading. In this section, some of the work pertinent to the characterization of shale oils is reviewed and discussed.

**Analytical methods.** In characterizing hydrocarbon mixtures for specification or other purpose, a precise analytical distillation—for example, 20 to 50 plates—may be needed. However, it is extremely costly and tedious to carry out such a distillation on a relatively large scale. Gas chromatography can be employed to obtain essentially a boiling point analysis with a separating column at a constant temperature. However, this technique is restricted to a rather narrow boiling range, because lighter components emerge too soon and tend to overlap and heavy components emerge very late, producing relatively wide bands or remaining in the column.

The new technique of temperature programming of the separating column makes a wide-range, single-stage analysis possible [25]. By using a column packing that separates according to boiling point and by precise programming of the column temperature, the boiling range for various peaks can be determined from times or temperatures of emergence.

A standard method for boiling range distribution of petroleum fractions by gas chromatography has been developed by the American Society for Testing and Materials and is given in ASTM D2887. Following are the brief descriptions of this procedure.

### ASTM D2887 procedure [35]

**Scope.** This standard method covers determination of the boiling range distribution of petroleum products. The method is generally applicable to petroleum fractions with a final boiling point of 1000°F (538°C) or lower at atmospheric pressure as measured by this method.

**Summary of method.** The sample is introduced into a gas chromatographic column that separates hydrocarbons in order of boiling points. The column temperature is raised at a reproducible rate and the area under the chromatogram is recorded throughout the run. Boiling temperatures are assigned to the time axis from a calibration curve, obtained under the same conditions by running a known mixture of hydrocarbons covering the boiling range expected in the sample. From these data the boiling range distribution may be obtained.

**Initial and final boiling points.** The initial boiling point (IBP) is the point at which a cumulative area count equals 0.5% of the total area under the chromatogram. The final boiling point (FBP) is the point at which a cumulative area count equals 99.5% of the total area under the chromatogram. The normal boiling point (NBP) is the point at which the vapor pressure reaches 760 mm Hg, or 1 atm.

**Apparatus for boiling range distribution.** A gas chromatograph equipped with a thermal conductivity detector is typically used for the experiment on gasoline fractions and is also recommended by the ASTM Standard procedure. For all other types of samples, either a thermal conductivity detector (TCD) or a flame ionization detector (FID) may be used. The detector must have sufficient sensitivity to detect 1.0% dodecane with a peak height of at least 10% of full scale on the recorder under the conditions prescribed in this method, without loss of resolution. Practically any column can be used, provided, under the conditions of the test, separations are in order of boiling points and the column resolution ($R$) is at least 3 and not more than 8. Because a stable baseline is essential for this method, matching dual columns are required to compensate for column bleed, which cannot be eliminated completely by conditioning alone.

The temperature programming must be over a range that is sufficient to establish a retention time of at least 1 minute for the IBP and to elute the entire sample. A microsyringe is needed for sample injection and a flow controller is required to hold carrier gas flow constant to $\pm 1\%$ over the full operating temperature range. The carrier gas is either helium or hydrogen for a TCD; nitrogen, helium, or argon may be used with a FID. The calibration mixture to be used is a mixture of hydrocarbons of known boiling points covering the boiling range of the sample. At least one compound in the mixture must have a boiling point lower than the initial boiling point of the sample in order to obtain an accurate distribution of boiling range.

In order to test column resolution, prepare a mixture of 1% each of $C_{16}$ and $C_{18}$ *n*-paraffin in a suitable solvent such as octane. Inject the same volume of this mixture as used in analyses of samples and obtain the chromatogram. As shown in Figure 16, calculate resolution ($R$) from the distance between the $C_{16}$ and $C_{18}$ *n*-paraffin peaks at the peak maxima, $d$, and the width of the peaks at the baseline, $Y_1$ and $Y_2$.

$$R = [2(d_1 - d_2)]/(Y_1 + Y_2) \tag{22}$$

Resolution $R$, based on Eq. (22), must be at least 3 and not more than 8.

Tables 6 and 7 show the typical operating conditions and the boiling points of *n*-paraffins, respectively. These tables provide valuable information regarding the boiling range properties.

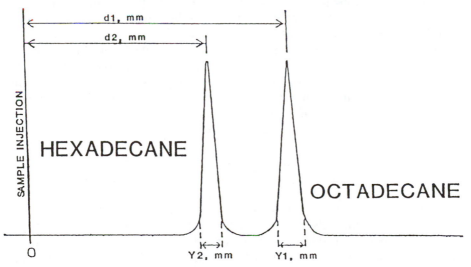

**Figure 16**   Column resolution ($R$). (Source: Reference 34.)

## Table 6 Typical operating conditions

|  | 1 | 2 | 3 | 4 |
|---|---|---|---|---|
| Column length, ft | 4 | 5 | 2 | 2 |
| Column ID, in. | 0.188 | 0.09 | 0.188 | 0.188 |
| Liquid phase | OV-1 | SE-30 | UC-W98 | SE-30 |
| Percent liquid phase | 3 | 5 | 10 | 10 |
| Support material | Sa | Gb | Gb | Pd |
| Support mesh size | 60/80 | 60/80 | 60/80 | 60/80 |
| Initial column temperature, deg C | −20 | −40 | 50 | 50 |
| Final column temperature, deg C | 360 | 350 | 350 | 390 |
| Programming rate, deg C/min | 10 | 6.5 | 8 | 7.5 |
| Carrier gas | He | He | $N_2$ | He |
| Carrier gas glow rate, ml/min | 40 | 30 | 60 | 60 |
| Detector | TCD | FID | FID | TCD |
| Detector temperature, deg C | 360 | 360 | 350 | 390 |
| Injection port temperature, deg C | 360 | 370 | 200 | 390 |
| Sample size, μl | 4 | 0.3 | 1 | 5 |
| Column resolution (R) | 5.3 | 6.4 | 6.5 | 3 |

[a]Diatoport S—silane treated.
[b]Chromosorb G (AW-DMS).
[c]On column injection.
[d]Chromosorb P. acid washed.
Reprinted with Permission from ASTM Standards D2887, American Society of Testing Materials.

## 7.2 OIL SHALE AS SYNTHETIC FUEL SOURCE

Interest in retorting oil from oil shale in order to produce a competitively priced synthetic fuel has intensified after the oil embargo of the 1970s. With more stable petroleum crude prices as well as governmental policy changes in 1990s, the commercial interest, once very high, has substantially cooled. However, it should be noted that oil shales have occasionally been used as solid fuels in certain areas for a long time and the research has quite a long history. Mixed with sediment, shale forms a tough, dense rock ranging in color from light tan to black. The Ute Indians, on observing outcroppings burst into flames after being hit by lightening, referred to it as "the rock that burns." Oil shale has been given various names in different regions.

Oil shales are widely distributed throughout the world with known deposits in every continent. Shales have been used in the past as a source of liquid fuel all over the world including Scotland, Sweden, France, South Africa, Australia, USSR, China, Brazil, and the United States. However, the oil shale industry has experienced several up-and-down trends, for various reasons.

Even though there is no written history, oil shales are believed to have been used directly as solid fuels in various areas of the world. As early as the 1850s, shale oil was being touted as a replacement for the wood that America depended on for its energy. The oil shale industry in the United States was an important part of the U.S. economy prior to the discovery of crude oil in 1859. As Colonel

## Table 7  Boiling points of *n*-paraffins

| Carbon no. | deg F | deg C | Carbon no. | deg F | deg C |
|---|---|---|---|---|---|
| 2 | −127 | −89 | 24 | 736 | 391 |
| 3 | −44 | −42 | 25 | 755 | 402 |
| 4 | 31 | 0 | 26 | 774 | 412 |
| 5 | 97 | 36 | 27 | 792 | 422 |
| 6 | 156 | 69 | 28 | 809 | 432 |
| 7 | 209 | 98 | 29 | 825 | 441 |
| 8 | 258 | 126 | 30 | 841 | 450 |
| 9 | 303 | 151 | 31 | 858 | 459 |
| 10 | 345 | 174 | 32 | 874 | 468 |
| 11 | 385 | 196 | 33 | 889 | 476 |
| 12 | 421 | 216 | 34 | 901 | 483 |
| 13 | 456 | 235 | 35 | 916 | 491 |
| 14 | 488 | 253 | 36 | 928 | 498 |
| 15 | 519 | 271 | 37 | 941 | 505 |
| 16 | 548 | 287 | 38 | 958 | 512 |
| 17 | 576 | 302 | 39 | 964 | 518 |
| 18 | 602 | 317 | 40 | 977 | 525 |
| 19 | 627 | 331 | 41 | 988 | 531 |
| 20 | 651 | 344 | 42 | 999 | 537 |
| 21 | 674 | 356 | 43 | 1009 | 543 |
| 22 | 696 | 369 | 44 | 1018 | 548 |
| 23 | 716 | 380 | | | |

"$C_1$–$C_{30}$ values from Selected Values of Hydrocarbons and Related Compounds, API Project 44, Loose-Leaf Data Sheet: Table 20a–e (Part 1), April 30, 1956. $C_{21}$ to $C_{42}$ deg F values calculated from deg C values reported in Vapor Pressures and Boiling Points of High Molecular Weight Hydrocarbons, $C_{21}$ to $C_{100}$, Report of Investigation of API Project 44, August 15, 1965.
Reprinted from ASTM Standards D2887-73 with Permission, American Society of Testing Materials.

Drake brought in his first oil well in Titusville, Pennsylvania, shale oil was gradually forgotten about and virtually disappeared from consideration with the availability of vast supplies of inexpensive fuel. Similarly, Scotland had an operating shale industry from 1850 to 1864, when the low price of foreign crude forced it to cease its operation. Interestingly, British Petroleum (BP) was formed as a shale oil company. In Russia, oil shale from Estonia once supplied fuel gas for Leningrad.

Interest in oil shale revived in the 1920s as domestic reserves of crude oil declined, but discoveries of large quantities of oil in Texas again set aside the hopes of an embryonic oil shale industry. Serious interest in oil shale revived again in the 1970s and 1980s, when the Arab oil embargo gave rise to concerns about the world energy supply. In the 1980s the stable crude price again became a reason for somewhat diminishing interest in the field. It remains to be seen whether the desire for energy self-sufficiency will again swing the balance in favor of the development of the western U.S. oil shales. At present, oil shale is exploited in several countries—Israel, Brazil, Australia, USSR, China, and the United States. In the USSR, oil shale has been fed directly into a power station for electricity generation; in China, oil production has been estimated to be about 40,000–50,000 bpd.

In the United States, President Carter signed into law the Energy Security Act, S.932, on June 30, 1980. This legislation was intended to help create 70,000 jobs a year to design, build, operate, and supply resources for synthetic fuel plants and for production of biomass fuels. The Act established the Synthetic Fuels Corporation. Different directions under President Reagan and relatively stable oil prices made the synthetic fuel industry less attractive. Under President Bush's administration, production and development of synthetic fuels became strategically less important than clean coal technology (CCT) and acid rain control. Under President Clinton's administration, this deemphasizing trend has continued in favor of budget deficit reduction.

Oil and gas (i.e., fossil fuel fluids) accounted for one-third of the total energy consumed in the United States by the late 1920s. By the mid-1940s, oil and gas began to provide half of the U.S. energy needs. They account for three-fourths of U.S. energy needs today. The demand is still increasing. Consequently, modern society's unprecedented appetite for energy, without new discoveries of petroleum reservoirs, will make it necessary to supplement supplies of domestic energy from crude oil and natural gas with synthetic fuels such as those from oil shale or coal.

Market forces will greatly affect the development of oil shale. Besides competing with conventional crude oil and natural gas, shale oil will have to compete favorably with coal-derived fuels for similar markets [36]. The liquid fuels derived from coal will be methanol, other products of indirect liquefaction, Fischer-Tropsch hydrocarbons, or oxygenates. Table 8 shows synfuel products and markets. Table 9 shows the areas involved in major synfuels development by various countries.

Depending on the relative level of success in synfuel or alternative energy development, the energy consumption patterns in the twenty-first century may be affected. More emphasis undoubtedly will be placed on clean energy development as well as environmentally clean utilization of conventional fuel.

# 7.3 CONSTRAINTS IN SHALE OIL PRODUCTION

In commercial exploitation of shale oils, one can be faced with various constraints that represent possible detering factors. They are technological, economic (or financial), institutional, environmental, socioeconomic, and water availability constraints. The Office of Technology Assessment (OTA) analyzed requirements for achieving each of the production goals by 1990, given the present state of knowledge and the current regulatory structure [3]. Table 10 shows the factors identified as potentially hindering reaching the goals, as assessed by OTA. In this table the original target year of 1990 is used without

**Table 8 Synfuel products and markets [36]**

| Product | Technology status | Market | Commercial plant start-up (probable) |
|---------|-------------------|--------|--------------------------------------|
| Shale oil | Pilot plants up to 2000 ton/d | Middle distillates (jet, diesel fuel) | |
| Coal liquids | Direct liquefaction in pilot plants (250 ton/d) | Light and middle distillates, petrochemical feedstocks | Mid-1990s |
| | Indirect liquefaction proven in SASOL | | In commercial production |
| Methanol | Coal gasification proven in large-scale plants; low-pressure methanol synthesis process actively being used based on syngas from natural gas | Chemical feedstock, gas, turbine, fuel, gasoline replacement, off-peak energy generation | Very active; large capacities; since early 1920s |

alteration. The constraints judged to be "moderate" will hamper but not necessarily preclude development; those judged to be "critical" could become more severe barriers. When it was inconclusive whether or to what extent certain factors would impede development, they were called "possible" constraints.

## 7.3.1 Technological Constraints

Oil shale can be retorted by either aboveground (ex situ) or in situ processing. In aboveground processing, the shale is mined and then heated in retorting vessels. In a true in situ (TIS) process, an oil shale deposit is first fractured by

**Table 9 Synfuels development in various countries**[a]

| Country | Coal | Shale | Tar sands | Biomass |
|---------|------|-------|-----------|---------|
| United States | X | X | — | X |
| USSR | X | X | — | — |
| Canada | X | — | X | — |
| Australia | — | X | — | — |
| South Africa | X | — | — | — |
| Europe | X | — | — | — |
| Brazil | X | X | — | X |
| China | X | X | — | — |
| Japan | X | — | — | X |
| Korea | X | — | — | — |
| Israel | — | X | — | — |

[a]Note: prediction of future production has not been made, because of its high uncertainty.

**Table 10 Constraints to implementing shale oil production targets**

| Possible deterring factors | Severity of impediment to 1990 production target (bbl/d) | | | |
|---|---|---|---|---|
| | 100,000 | 200,000 | 400,000 | 1 million |
| **Technological** | | | | |
| Technological readiness | None | None | None | Critical |
| **Economic and financial** | | | | |
| Availability of private capital | None | None | None | Moderate |
| Marketability of the shale oil | Possible | Possible | Possible | Possible |
| Investor participation | None | Possible | Possible | Possible |
| **Institutional** | | | | |
| Availability of land | None | None | Possible | Critical |
| Permitting procedures | None | None | Possible | Critical |
| Major pipeline capacity | None | None | None | Critical |
| Design and construction services | None | None | Moderate | Critical |
| Equipment availability | None | None | Moderate | Critical |
| **Environmental** | | | | |
| Compliance with environmental regulations | None | None | Possible | Critical |
| **Water availability** | | | | |
| Availability of surplus surface water | None | None | None | Possible |
| Adequacy of existing supply systems | None | None | Critical | Critical |
| **Socioeconomic** | | | | |
| Adequacy of community facilities and services | None | Moderate | Moderate | Critical |

Source: Office of Technology Assessment.

explosives and then retorted underground. Modified in situ (MIS) processing is a more advanced in situ technology in which a portion of the deposit is mined and the rest is rubblized by explosives and retorted underground. The crude shale oil can be burned as a boiler fuel, or it can be further converted into syncrude by adding hydrogen. Issues to be answered include:

1. What are the advantages and disadvantages of different mining and processing methods?
2. Are the technologies ready for large-scale applications?
3. What are major areas of uncertainty in the technology?
4. Are the technologies for process optimization available?
5. Are there sufficient scale-up data from pilot and demonstration scale plant operation?
6. What is the possibility of observing some technological breakthrough in the process?
7. Are there sufficient data regarding the physical, chemical, and geological properties of oil shale?

## 7.3.2 Economic and Financial Constraints

Undoubtedly, an oil shale plant will be quite costly, and the product oil will also have to be expensive based on the current energy price structure. Depending on the world petroleum prices, shale oil may be competitive in the foreseeable future. However, the long-term profitability of the industry could be impeded by future pricing strategies of competing fuels, which are normally uncontrollable from the oil shale industry's standpoint. This concern makes the risk level of an oil shale industry higher. Even though the cost breakdown is difficult to make, a typical cost distribution for an oil shale project is shown in Table 11.

It can be seen from the table that the mining cost is a good share of the total operating cost. Considering the mining and retorting cost, the cost to obtain the shale oil crude is approximately 70% of the total operating cost, making the

**Table 11  Typical cost distribution for an oil shale project**

|  | Construction (%) | Operation (%) |
|---|---|---|
| Mining (crushing and spent shale disposal) | 16 | 43 |
| Retorting | 37 | 28 |
| Upgrading | 22 | 29 |
| Utilities and off sites | 25 | — |
| Total | 100 | 100 |

Reproduced from Taylor, R. B. *Chem. Eng.*, September 7:63–71, 1981, with permission.

crude less competitive economically. Most oil shale processes have energy efficiencies ranging from 58 to 63%. Table 12 shows some comparative information regarding the energy efficiencies of various synfuel processes.

Improving the energy efficiency without increasing the capital and operational cost is another task that has to be attacked by the process development team. Issues related to economics of oil shale processing are:

1. What are the economic and energy-supply benefits of oil shale development?
2. What are the environmental and ecological impacts of oil shale development?
3. What are the negative economic effects of establishing the industry?
4. How much will oil shale facilities cost?
5. What is the return on investment, especially in the long term?
6. At what level of petroleum process is shale oil competitive?
7. Overall, is oil shale competitive?

The OTA's assessments were more conservative than those of many developers. Cost analysis based on the current statistics will give us a better answer, because the process economic situations always change.

### 7.3.3 Environmental Constraints

The oil shale deposits found in the Green River Formation in the states of Colorado, Wyoming, and Utah are the largest in the United States. These deposits are estimated to contain about 1800 billion barrels of recoverable oil. Due to the vast resources and the high oil-yielding quality of shales, this region is the most attractive to oil shale industries. However, the technology used in mining and processing of oil shale must answer to questions of environmental effects. The Devonian-Mississippian Eastern black shale deposits are widely distributed between the Appalachians and Rocky Mountains. Even though these oil shales represent a large resource, they are generally much lower in grade than Green River Formation oil shales [7].

Several factors affecting environmental constraints in exploiting the Green River Formation are discussed as an example case.

**Table 12 Energy Efficiencies of Synfuel Processes**

| Process | Efficiency (%) |
| --- | --- |
| Lurgi pressure gasification | 70 |
| Lurgi pressure gasification followed by shift methanation | 63 |
| Fischer-Tropsch synthesis | 40 |
| Low-pressure methanol synthesis | 49 |
| Methanol synthesis followed by MTG | 45 |
| Shale oil processes | 58–63 |

1. Nature of region. The Upper Colorado Region, which is the upper half of the Colorado River Basin, is traditionally "western rural" and consists of sparsely vegetated plains. The population density is also low, approximately three persons per square mile.
2. Water availability. A rate-limiting factor in further development of the area is the availability of water. Water of the Colorado River could be made available for depletion by oil shale. An important factor that must be taken into consideration in any water-use plan, is the potential salt loading of the Colorado River. The average annual salinity is anticipated to increase. The economic damages associated with these higher salinity levels could be significant and have been the subject of extensive economic studies.
3. Other energy and mineral resources. The tristate oil shale area has extensive energy resources other than oil shale. Natural gas recoverable in this area is estimated at 85 trillion ft$^3$, whereas crude oil reserves are estimated at 600 million barrels and coal deposits at 6–8 billion tons. Furthermore, 27 billion tons of alumina and 30 billion tons of nahcolite are present in the central Piceance Creek Basin. These minerals may be mined in conjunction with oil shale.
4. Ecology. Ecologically, the tristate region is very valuable. Due to the sparse population density, the region has retained its natural character. Animals include antelope, bighorn sheep, mule deer, elk, black bear, moose, and mountain lion. However, there is little fishery habitat in the oil shale areas, even though the Upper Colorado Region includes 36,000 acres in natural lakes.
5. Sources of fugitive dust. Such operations as crushing, sizing, transfer conveying, vehicular traffic, and wind erosion are sources of fugitive dust.
6. Gaseous emissions. Gaseous emissions such as $H_2S$, $NH_3$, CO, $SO_2$, and NO and trace metals are sources of pollution. Such emissions are at least conceivable in oil shale processing operations.
7. Recreation. Outdoor recreation in the tristate oil shale region is considered to be of high quality, due to the vastness of the essentially pristine natural environments and to the scenic and ecological richness of the area. Keeping the area scenically beautiful and preserving the high-quality natural resources must be taken into consideration when oil shale in the region is exploited. Such consideration is generally true with other oil shale regions in the world.

## 7.4 RESEARCH NEEDS IN OIL SHALE

The synthetic crude reserves in oil shale are sufficient to meet U.S. consumption for several centuries at current rates of liquid fuel utilization. Raw shale oil is the product of the retorting process and, like petroleum crude, is highly paraffinic. However, it contains fairly high levels of sulfur, nitrogen, and oxygen as well as olefins, and it requires substantial upgrading before it can be substituted for refinery feed. Sulfur removal down to a few parts per million (ppm) is necessary in order to protect multimetallic reforming catalysts. Removing nitrogen, which also poisons cracking catalysts, from condensed heterocyclics requires a good technology that uses less hydrogen.

This problem stresses an understanding of oil shale on a molecular level. Research needs in this section address five general classes of problems: chemical characterization of the organic and inorganic constituents, correlations of physical properties, recovery processes of synthetic fuels, refining of crude oil, and environmental and toxicological problems.

## 7.4.1 Chemical Characterization

Better analytical techniques will have to be developed in order to obtain the information needed to better understand the oil shale technology as well as to develop new technology for utilizing shale. Most of the analytical methods have a long history in petroleum analytical chemistry, but their validity for the shale application often becomes questionable. Basic questions to be answered include:

1. Where are the organic heteroatoms?
2. How are they bonded into the basic carbon structure?
3. What are the model compounds for sulfur and nitrogen sources?
4. What is the aromaticity level?
5. What is the ratio of alkanes to alkenes?
6. What are the inorganic ingredients?

One way of characterizing oil shale is via separation of organics by extraction. Most commonly used techniques include gel permeation chromatography for molecular weight distribution, gas chromatography (GC) using both packed columns and glass capillaries for product oil distribution, simulated GC distillation, and liquid chromatography.

Mass spectrometry (MS) can be extremely useful. In particular, GC/MS is good for product identification and model compound studies. Pyrolysis GC/MS can be used for examining shale decomposition reactions.

Elemental analysis of oil shale is like that of petroleum and coal. Tighter and more specific requirements for C, H, O, N, and S analysis are needed for oil shale, as organic carbon in the shale should be distinguished from inorganic carbon in carbonate materials. Analyses for C, H, N, and S, which frequently used for coal analysis, can also be used for oil shale.

## 7.4.2 Correlation of Physical Properties

Physical properties of oil shale should be characterized via electrical and conductive measurements, scanning microscopy, spectroscopic probes, and all other conventional methods. Especially the correlation between the physical properties and the conversion is useful for designing a pilot-scale or commercial retort. The $^{13}$C-NMR work should be expanded to provide a detailed picture of the various chemical forms encountered. Electron spin resonance (ESR) studies of carbon radicals in oil shale could also provide invaluable clues about the conversion process.

Correlations of physical properties that are applicable for diverse oil shales are especially useful. Predictive forms of correlations are powerful in engineering design calculations as well as in good database establishment.

## 7.4.3 Mechanisms of Retorting Reactions

Kinetics of oil shale retorting have been studied by various investigators. However, details of reaction mechanism have not been generally agreed upon. This is why all the problems associated with in situ pyrolysis require primarily field experiments rather than small laboratory-scale experiments. The retorting process itself can be improved when its chemical reaction mechanisms are fully known.

## 7.4.4 Heat and Mass Transfer Problems

Oil shale rocks are normally low in both porosity and permeability. Therefore, it is important to know the combined heat and mass transfer processes of a retort system. The processes of heat and mass transfer in retort operation affect the process operating cost significantly, because the thermal efficiency is directly related to the total energy requirement, and the mass transfer conditions directly affect the recovery of oil and gas from a retort. Various process technologies differ from each other in the method of heat and mass transfer and there is always room for further improvement. Therefore, understanding of this transport process is essential in mathematical modeling of a retort system.

## 7.4.5 Upgrading Shale Oil Crudes

As mentioned earlier, prerefining of shale oil crude is necessary to reduce sulfur and nitrogen levels and contamination by mineral particulates. Research opportunities with great significance exist in heteroaromatics removal, final product quality control, and molecular weight reduction.

Raw shale oil has relatively high pour point of 75–80°F, compared with −30°F for Arabian Light. Olefins and diolefins may amount to as much as one-half of the low-boiling fraction of 600°F or lower, leading to the formation of gums.

Raw shale oil typically contains 0.5 to 1.0% oxygen, 1.5 to 2.0% nitrogen, and 0.15 to 1.0% sulfur. Sulfur and nitrogen removal must be very complete, as these compounds poison most of the catalysts used in refining and their oxides ($SO_x$ and $NO_x$) are highly publicized pollutants.

Because raw shale oil is a condensed overhead product of pyrolysis, it does not contain the same kinds of macromolecules found in petroleum and coal residuum. Conventional catalytic cracking, however, is an efficient technique for molecular weight reduction. It would be a great idea to develop a new cracking catalyst that is more resistant to basic poisons (N, S). In doing so, the research should focus on reduction of molecular weight of shale oil crude with low consumption of hydrogen.

### 7.4.6 By-Product Minerals from U.S. Oil Shale

Diverse kinds of carbonate and silicate minerals occur in oil shale formations. Trona beds [$Na_5(CO_3)(HCO_3)_3$] in Wyoming are a major source of soda ash, and nahcolite [$NaHCO_3$] is a potential by-product of oil shale mining in Utah and Colorado.

Precious metals and uranium are contained in good amounts in eastern U.S. shales. These mineral resources may not be recovered in the near future, because a commercially favorable recovery process has not yet been developed. However, it should be mentioned that there are many patents on recovery of alumina from dawsonite-bearing beds by leaching, precipitation, calcination, and so forth.

### 7.4.7 Characterization of Inorganic Matters in Oil Shale

The analytical techniques applied to the characterization of the inorganic constituents in oil shale include X-ray diffraction (XRD), thermogravimetric analysis (TGA), scanning electron microscopy (SEM), and transmission electron microscopy (TEM). Electron microprobe analysis (EMPA) may provide useful information on the phases encountered.

## 7.5 OIL SHALE RETORTING AND EXTRACTION PROCESSES

The organic matter in oil shale contains both bitumen and kerogen. The bitumen fraction is soluble in most organic solvents, and it is not difficult to extract from oil shale. Unfortunately, the readily soluble bitumen content in oil shale is only a minor portion, whereas insoluble kerogen accounts for a major portion of oil shale organic matter. Furthermore, the oil shale kerogen is nearly inert to most chemicals, making some reactive processes less efficient and more difficult.

In terms of the extraction philosophy, several approaches can be considered for the extraction of oil (organic matter) from the mineral matrix (inorganic rock matrix): (1) drastically break the chemical bonds of the organics, (2) mildly degrade or depolymerize the organics, and (3) use solvents that have extraordinarily strong solvating power, such as supercritical fluids. The first approach is widely used in industrial applications, because high-temperature pyrolysis drastically breaks the bonds of the organics. Retorting processes belong to the category, and such processes have a long history. The second approach may be achieved by a biochemical process. The third approach can be achieved by supercritical fluid extraction, which is based on the strong solvent power of a fluid in its supercritical region.

In the extraction of shale oil from oil shale, not only chemical properties but also physical properties of oil shale play important roles. The low porosity, low permeability, and tough mechanical strength of oil shale rock matrix make the extraction process less efficient by making the transport of products much harder. Both heat and mass transfer conditions also crucially affect the process economics as well as the process efficiency.

The oil shale retorting processes can be classified as ex situ (or surface, off the sites) and in situ (or subsurface, within the existing formation). As the names imply, ex situ processes are carried out above the ground after the shale is mined and crushed, whereas in situ processes are carried out under the ground, thus not requiring mining of all the shale.

In this section, various processes suggested and demonstrated for oil shale extraction are reviewed. Engineering and technological aspects of the processes are discussed.

## 7.5.1 Ex Situ Retorting Processes

In an ex situ process, shale rock is mined, either on the surface or underground, crushed, and then conveyed to a retorter, which is subjected to temperatures ranging from 500 to 550°C at which the chemical bonds linking the organic compounds to the rock matrix are broken and the kerogen molecules are pyrolyzed, yielding simpler and lighter molecules.

The advantages of ex situ processes include the following:

1. The efficiency of organic matter recovery has been demonstrated to be high, up to 70–90% of the organic content of the retorted shale. Therefore, the amount of wasted organic matter (i.e., the unextracted portion) will be minimized.
2. Control of process variables is possible and relatively easy. The effects of undesirable process conditions can be minimized.

3. Product recovery becomes easy, once oil is formed.
4. The process units are general enough to be used repeatedly for retorting.

The disadvantages of ex situ processes are:

1. The operation cost is normally high, because oil shale has to be mined, crushed, transported, and heated.
2. Spent shale disposal, underground water contamination, and revegetation problems are yet to be solved convincingly.
3. The processes are somewhat limited to rich shale resources accessible for mining.
4. The capital investment for large-scale units is very high.

The liberated compounds from oil shale retorting include gas and oil, which are collected, condensed, and upgraded to a liquid product that is roughly equivalent to crude oil. This oil can be transported by a pipeline to a refinery, where it is refined into the final product. In this section, various ex situ processes are reviewed.

**U.S. Bureau of Mines gas combustion retort.** The first major experimental retort was built and operated by the U.S. Bureau of Mines. From 1944 to 1956, pilot plant investigations of oil shale retorting were carried out by the U.S. Bureau of Mines at facilities several miles west of Rifle, Colorado. Among the numerous types of retorts, gas combustion retorting gave the most promising results and was studied extensively [37]. This retort is a vertical, refractory-lined vessel through which crushed shale moves downward by gravity, countercurrent to the retorting gases. Recycled gases enter the bottom of the retort and are heated by the hot retorted shale as they pass upward through the vessel. Air and additional recycle gas (labeled as dilution gas) are injected into the retort through a distributor system at a location approximately one-third of the way up from the bottom and are mixed with rising hot recycled gas. Figure 17 shows a schematic of a gas combustion retort. Combustion of the gases and of some residual carbon heats the shale immediately above the combustion zone to retorting temperature. Oil vapors and gases are cooled by the incoming shale, and the oil leaves the top of the retort as a mist. The oil vapors and mists are subsequently chilled to produce liquid oil products that have to be upgraded.

The retorter is similar to a moving bed reactor in its operating concepts.

**The TOSCO-II oil shale process.** The TOSCO-II oil shale retorting process was developed by the Oil Shale Corporation. An article by Whitcombe and Vawter [38] describes the process in detail and also presents economic projections for production of crude shale oil and hydrotreated shale oil. As an oil shale

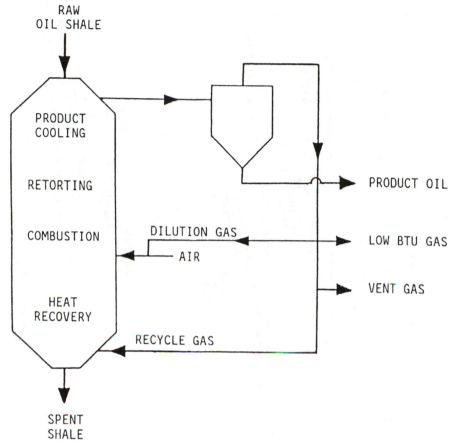

**Figure 17**   Gas combustion retort.

process, the TOSCO process is one of the few complete processes for production of shale oil.

***Process description.***   Oil shale is crushed and heated to approximately 480°C by direct contact with heated ceramic balls. At this temperature, the organic material in oil shale rapidly decomposes to produce hydrocarbon vapor. Subsequent cooling of this vapor produces crude shale oil and light hydrocarbon gases. Figure 18 shows a schematic diagram of the process.

The pyrolysis reaction takes place in a retorting kiln (referred to as the pyrolysis reactor or retorter) shown in the central portion of the schematic. The feed streams to the retort are $\frac{1}{2}$-in.-diameter ceramic balls heated to about 600°C, and preheated shales are crushed to a size of $\frac{1}{2}$-in. or smaller. The rotation of the retort mixes the materials and causes a high rate of heat transfer from the

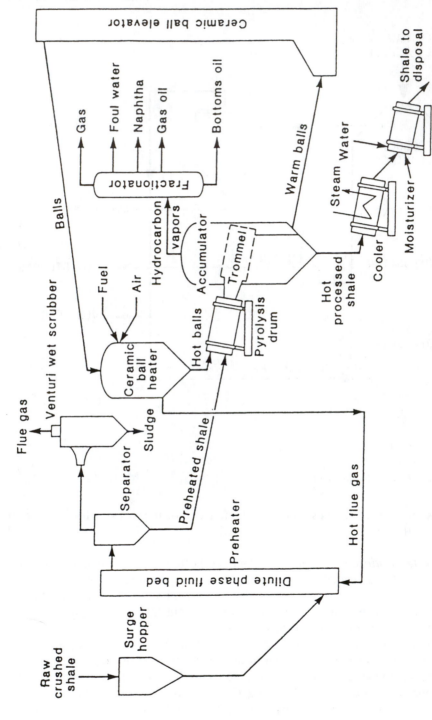

**Figure 18** TOSCO II process (The Oil Shale Corporation, Los Angeles, CA). (Reproduced with permission from Probstein, R. F., and R. E. Hicks. *Synthetic Fuels*. New York: McGraw-Hill, 1982.)

ceramic balls to the shale. At the discharge end of the retort, the ceramic balls and shale are at the same temperature and the shale is fully retorted.

The hydrocarbon vapor formed by the pyrolysis reaction flows through a cyclone separator to remove entrained solids and then into a fractionation system that is similar to the primary fractionator of a catalytic cracking unit. From this stage, oil vapor produces heavy oil, distillate oils, naphtha, and light hydrocarbon gases.

The ceramic balls and spent shale flow from the retort into a cylindrical trommel screen. Spent shale passes through the screen openings into a surge hopper. The ceramic balls flow across the screen and into a bucket elevator for transport to the ball heater, where they are heated by direct contact with flue gas. The ceramic balls are then recycled to the retort.

Spent shale, discharged from the retort at 480°C, is first cooled in a rotating vessel containing tubes in which water is vaporized to produce high-pressure steam. It flows into another rotating vessel in which it is further cooled by direct contact with water. The water flow is controlled so that the spent shale from the vessel contains 12% moisture by weight. The moisture is added to control dust emissions and to make the spent shale suitable for compaction before disposal.

Preheating of oil shale is achieved by direct contact of the crushed shale with the flue gas effluent from the ball heater. The principal gaseous effluent from the process is the flue gas used to heat the ceramic balls and to preheat the shale. The process includes a wet scrubber system to control the particulate content of the gas and an incinerator to control its hydrocarbon content. Emissions of $SO_x$ and $NO_x$ are controlled by the choice of fuels used in the process as well as the firing temperatures of process heaters. The process effectively uses the concept of energy integration for minimization of energy cost.

*Process yield.* Tests in a pilot plant and semiworks have shown that the TOSCO-II process recovers nearly 100% of the recoverable hydrocarbon in oil shale as determined by the Fischer assay procedure. It should be noted that the Fischer assay does not represent 100% of total organic matter recoverable from shale and that the TOSCO process is an effective scale-up of the Fischer assay procedure. Table 13 shows results from a 7-day, continuous operation of the semiworks plant, as presented in reference 38. The average plant yield during this period was 322.1 lb of hydrocarbons per ton of oil shale processed, approximately 1.7% greater than by Fischer assay of the average shale sample used for the period [38].

*Gas and crude shale oil product.* Table 14 shows a typical analysis of the $C_4$ and higher hydrocarbons produced in the TOSCO-II retort. The effluent gas is practically free of nitrogen and contains a good amount of carbon dioxide produced by pyrolysis.

Table 15 shows the properties of shale oil ($C_5$ and heavier fractions) produced by the TOSCO-II retorting process. The average sulfur level is 0.7%,

## Table 13 The TOSCO-II semiworks plant yield data

| Hydrocarbons | Plant yield (lb/ton) | Fischer assay yield (lb/ton) |
|---|---|---|
| $C_1$–$C_4$ | 49.6 | 24.3 |
| $C_5$ and heavier | 272.5 | 292.4 |
| Total hydrocarbons | 322.1 | 316.7 |
| Other gaseous products | | |
| $H_2$ + CO | 4.5 | 3.7 |
| $CO_2$ + $H_2O$ | 32.7 | 31.3 |

(Reproduced with permission from Whitcombe and Vawter, Table 4.1, p. 51, in *Science and Technology of Oil Shale*, edited by Y. F. Yen, Ann Arbor Science Publishers, Ann Arbor, MI, 1976.)

whereas the average nitrogen content is 1.9%, which is quite high compared with that of conventional crude oils. In conventional crude oils, even one of the most nitrogen-rich crudes does not exceed 1.0 wt.% nitrogen. The principal objective of the hydrotreating facilities is removal of nitrogen compounds that are poisonous to catalysts of many upgrading processes, including reforming, cracking, and hydrocracking.

*Process units.* Figure 19 is a block flow diagram of the TOSCO-II process for commercial operation [38]. The commercial plant includes two hydrotreating units. The first is the distillate hydrotreater that processes the 200–500°C oil formed in the retorter plus components with similar boiling ranges formed in the coker. The second hydrotreater processes $C_5$ to 200°C naphtha formed in the retort, the coker, and the distillate hydrotreater. The distillate hydrotreater is

## Table 14 Typical analysis of $C_4$ and lighter gases from the semiworks plant

| Component | Wt. % |
|---|---|
| $H_2$ | 1.50 |
| CO | 3.51 |
| $CO_2$ | 33.08 |
| $H_2S$ | 5.16 |
| $CH_4$ | 11.93 |
| $C_2H_4$ | 8.67 |
| $C_2H_6$ | 8.43 |
| $C_3H_6$ | 11.08 |
| $C_3H_8$ | 5.45 |
| $C_4$'s | 11.19 |
| Total | 100.00 |

Source: Reference 38.

**Table 15 Properties of crude shale oil from TOSCO-II retorting process**

| Boiling ranges*a* and components | Vol. % | °API | Wt. % S | N |
|---|---|---|---|---|
| C$_5$–204 | 17 | 51 | 0.7 | 0.4 |
| 204–510 | 60 | 20 | 0.8 | 2.0 |
| 510+ | 23 | 6.5 | 0.7 | 2.9 |
| Total | 100 | 21 | 0.7 | 1.9 |

*a*Boiling ranges are given in degress celsius.
Source: Reference 38.

designed to reduce the nitrogen content of the 200°C plus product from the unit to less than 1000 ppm. The naphtha hydrotreater is designed for a product nitrogen content of about 1 ppm. Sulfur removal is nearly complete in each of the hydrotreating units [38]. The product compositions shown in Table 14 can be altered by changing the fuels selected for burning in the process facilities. The production of C$_5$-plus fractions can be increased by burning the C$_3$ products in place of hydrotreated oil. Table 16 shows properties of the C$_5$-510°C fraction of hydrotreated shale oil, which is a blend of sulfur-free distillate products.

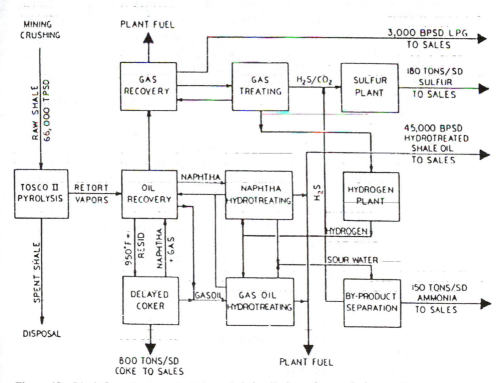

**Figure 19** Block flow diagram—hydrotreated shale oil plant. (Source: Reference 3.)

**Table 16 Properties of typical hydrotreated shale oil**

| Boiling ranges[a] and components | Vol. % | °API | Nitrogen (ppm) |
|---|---|---|---|
| C$_5$–204 | 43 | 50 | 1 |
| 204–361 | 34 | 35 | 800 |
| 361–EP | 23 | 30 | 1200 |
| Total | 100 | 40 | |

[a]Boiling ranges are given in degrees Celsius.
Source: Reference 38.

Refining such hydrotreated oils is relatively simple and requires an atmospheric distillation unit and a reformer. It would produce gasoline, sulfur-free light distillate fuels (No. 1 and No. 2 heating oils and diesel fuel), and a sulfur-free heavier distillate fuel oil suitable for use as industrial fuel oil [38].

*Spent shale disposal.* The spent shale produced by the TOSCO-II process is a fine-grained, dark material constituting about 80 wt. % of the raw shale feed. It contains on the average 4.5% of organic carbon via hydrogen-deficient char formation. The mineral constituents of spent shale, consisting of principally dolomite, calcite, silica, and silicates, are largely unchanged by the retorting process, except that carbonates have decomposed to oxides. During the retorting process, significant size reductions take place yielding most spent shale finer than 8 mesh and 60 wt. % of it finer than 200 mesh.

The spent shale disposal technology was seriously tested in 1965 after completion of the 1000 ton/day semiworks plant at Parachute Creek. Small revegetation test plots were constructed in 1966 to evaluate plant growth factors and plant species, and in 1967 the first field demonstration revegetation plot was constructed and seeded. Extensive off-site investigations have been carried out, including the spent shale permeability and a quality test of water runoff from spent shale embankments.

**The Union Oil retorting process.** The Union Oil process is one in which the heat needed for retorting is provided by combustion of coke inside the retort. The shale, which is fed from the bottom, is pushed up by means of a specially developed rock pump. The resulting oil is siphoned out from the bottom of the retort. According to a process described by Deering [39], spent, coke-containing shale derived from a gas-heated eduction zone is passed through a combustion-gasification zone countercurrently to an upflowing mixture of steam and oxygen-containing gas to effect partial combustion of the coke on the spent shale. The resulting heat of combustion is used to effect concurrent endothermic gasification reactions of steam with unburned coke, that is, $C + H_2O = CO + H_2$. Figure 20 shows a schematic of this process. The shale feed rate varies considerably, depending on the size of retort and the desired holding time.

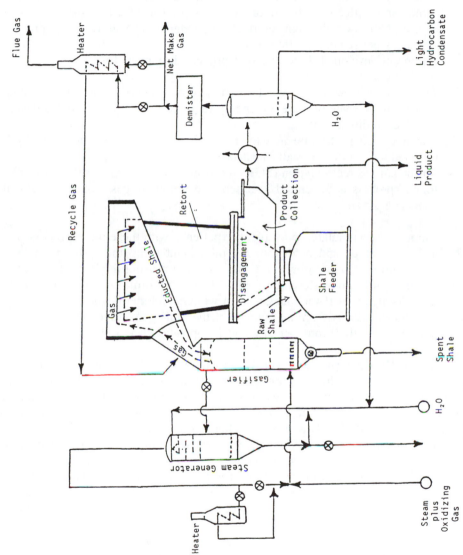

**Figure 20** Schematic of Union Oil's retorting-gasification process. (Based on U.S. Patent 4,010,092.)

337

A significant aspect of the process is that even if the recycled water-gas contains hydrogen and must pass through the combustion-gasification zone in which hydrogen-burning temperatures prevail, the overall yield of hydrogen is not significantly affected. This retort uses lump shale of about the same size range as in the gas combustion retort. Another important feature of the process is that it does not require cooling water.

Union Oil developed three different process designs as follows:

1. The A retort is the one in which internal combustion of the gas and residual char from the shale provide the energy required for the process. This design is based on direct heating.
2. The B design is the one in which the oil shale is heated indirectly by a recycled stream of externally heated gas.
3. The steam-gas recirculation (SGR) retort is the one in which the heat carrier for the process is generated in a separate vessel by gasifying the residual char with air and steam.

Union had accumulated pilot plant experience with these three processes: a 2 ton/day prototype and a 50 ton/day pilot plant using the A retort at Wilmington, California; a nominal 6 ton/day pilot retort using the B mode; and the SGR mode at the Union Research Center, Brea, California. A larger scale pilot plant was located near Parachute Creek in western Colorado, where the oil shale could be mined readily from an outcrop of the Mahogany zone. The pilot plant had a capacity of 1000 tons per day and its operation was fully successful.

**The Lurgi-Ruhrgas process.** The Lurgi-Ruhrgas (L-R) process "distills" hydrocarbons from oil shale by bringing raw shale in contact with a hot fine-grained solid heat carrier. The ideal heat carrier is the spent shale; but rich shales, which deteriorate into a fine powder, must be supplemented by more durable materials like sand. The L-R process schematic is given in Figure 21.

The pulverized oil shale and heat carriers are brought into contact in a mechanical mixer such as a screw conveyor. In pilot plant tests, the shale was first crushed to a maximum size of one-fourth to one-third of an inch, but larger commercial units might process particles as large as half an inch [39]. The oil vapor and gas are cleaned of dust in a hot cyclone, and the liquid oil is separated by condensation.

Retorted shale from the mixer passes through a hopper to the bottom of a lift pipe, with the dust from the cyclone. Preheated air introduced at the bottom of the pipe carries the solids up to the surge bin. Solids are heated by combustion of the residual char in the shale to approximately 550°C. When the residual char is not sufficient for this, fuel gas is added. In the surge bin, the hot solids separate from the combustion gases and return to the mixer, where they are brought in contact with fresh oil shale, completing the cycle.

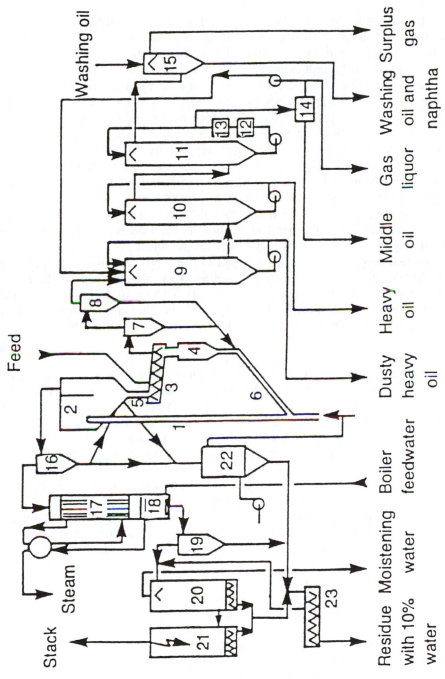

**Figure 21** Lurgi-Ruhrgas retort system. (Reproduced with permission from Matar, S., *Synfuels*, Pennwall Publishing Company, Tulsa, OK, 1982.)

Pilot plant tests have produced high yields, exceeding 100% Fischer assay from Colorado shale, at approximately 30 gallons per ton of shale. Because no combustion occurs in the mixer retort in this process, the product gas from the mixer has a relatively high calorific value (CV).

The L-R process can operate with very fine to medium particle sizes and, therefore, can be modified for a variety of shale feedstocks. The system is mechanically simple except for the mixer, which may be of some concern in designing because it must operate reliably in a harsh environment. However, the movement of dust through the system causes two major problems. One is the accumulation of combustible dust in the transfer lines, increasing the likelihood of fires and plugging. The other is entrainment of dust in the oil produced. (Even though most of the dust is removed in the hot cyclone, some is inevitably carried over in the retort product.) When the crude oil from the process is fractionated, the dust concentrates in the heaviest fraction and can be dealt with by diluting and filtering this fraction or by recycling it to the mixer.

The L-R process, originally developed in the 1950s for the low-temperature flash carbonization of coal, was tested on European and Colorado oil shale in a 20 ton/day pilot plant at Herten, Germany. Two 850 ton/day pilot plants for carbonizing brown coal were built in Yugoslavia in 1963, and a large plant that uses the L-R process to produce olefins by cracking light oils was built in Japan [39, 40].

**Superior's multi-mineral process.** The multimineral process was developed by Superior Oil and, in addition to synthetic gas and oil, it produces nahcolite ($NaHCO_3$), alumina ($Al_2O_3$), and soda ash [41]. A schematic of Superior Oil's multimineral retort system is shown in Figure 22. The process is a four-step operation for oil shale that contains recoverable concentrations of oil, nahcolite, and dawsonite [a sodium-aluminum salt, $Na_3Al(CO_3)_3 \cdot Al(OH)_3$]. Superior Oil operated a pilot plant of this type in Cleveland, Ohio.

The nahcolite is in the form of discrete nodules that are more brittle than the shale. It is recovered by secondary crushing and screening, followed by a specialized process called photosorting that recovers nahcolite product of greater than 80% purity. After the removal of nahcolite, the shale is retorted using the McDowell-Wellman process [39]. The unique, continuous-feed, circular-moving-grate retort used in this process is a proven, reliable piece of hardware that provides accurate temperature control, separate process zones, and a water seal that eliminates environmental contamination. This nahcolite has been tested as a dry scrubbing agent to absorb sulfurous and nitrous oxides.

The dawsonite in the shale is decomposed in the retort to alumina and soda ash. After the shale is leached with recycled liquor and makeup water from the saline subsurface aquifer, the liquid is seeded and the pH lowered to recover the alumina. This alumina can be extracted at a competitive price with alumina from bauxite. The soda ash is recovered by evaporation. The leached spent shale is

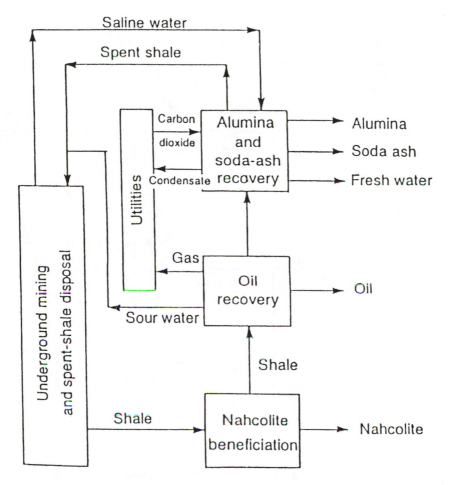

**Figure 22** Superior Oil's multimineral retort system. (Reproduced with permission from *Chemical and Engineering News*, January 10, 1977, American Chemical Society, Washington, DC.)

then returned to the process. The by-products may make the process even more economically attractive.

**The Paraho gas combustion process.** The Paraho retort is a stationary, vertical, cylindrical, refractory-lined kiln of mild steel by Paraho Development Corp. The raw shale enters at the top and is brought to temperature by a countercurrent flow of combustion gases. A schematic of the gas combustion retort system is shown in Figure 23. The shale is fed in at the top along a rotating "pantsleg" distributor and moves down through the retort. The rising stream of hot gas breaks down the kerogen to oil, gas, and residual char. The oil and gas are drawn

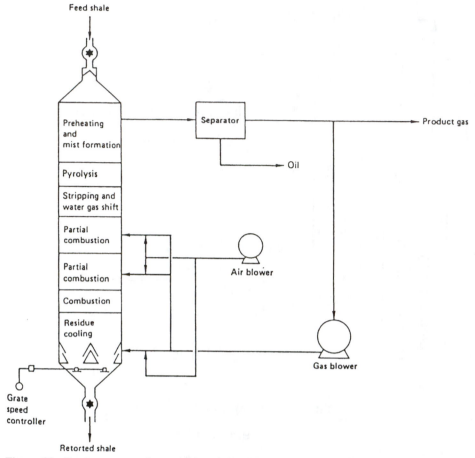

**Figure 23** Paraho processw for extracting shale oil. (Reproduced with permission from *Synthetic Fuels Data Handbook*, Cameron Engineers, Denver, CO, 1975.)

off, and the residual char burns in the mixture of air and recycled gas. By injecting part of the gas-air mixture through the bottom of the kiln, much of the sensible heat in the spent shale is recovered. The retort temperature is controlled by adjusting the compositions of the gas-air mixtures to the preheat and combustion zones [39]. Shale oil vapors flow upward and pass at a moderate temperature to an oil recovery unit. The end products are shale oil and low-Btu gas, and a typical analysis of raw shale oil is shown in Table 17. The produced shale oil can be upgraded to a crude feedstock.

The heavy naphtha cut (88–178°C) from the treated oil has a higher octane rating and lower sulfur than a comparable Arabian crude fraction. The diesel fraction (178–341°C) is identical to comparable fractions from other sources, so the heavy cut can be used as a feedstock for cracking units [41].

**Table 17 Typical analysis of Paraho retort shale oil**

| | |
|---|---|
| API gravity, °API | 19.70 |
| Nitrogen, wt. % | 2.18 |
| Conradson carbon, wt. % | 4.50 |
| Ash, wt. % | 0.06 |
| Sulfur, wt. % | 0.74 |
| Pour Point, °C | 26.0 |
| Viscosity, cp, 38°C | 256.0 |

Research and development on the Paraho retort, initiated by Paraho Development Corporation (the parent company of Development Engineering, Inc.) in August 1973, continued until April 1976 under the sponsorship of 17 energy and engineering companies. In 1978 Paraho delivered 100,000 barrels of raw shale oil to the U.S. Navy for defense testing purposes [39].

The Paraho process can handle shale particle sizes of at least 3 in., keeping crushing and screening costs at a minimum, yet achieving high conversion of better than 90% of Fischer assay. By burning the residual char and also recovering sensible heat from the spent shale, high thermal efficiency can be attained. The process is mechanically simple, requiring little auxiliary equipment. No water is required for product cooling.

**Petrosix retorting process.** The Petrosix retorting process was developed in Brazil by Petrobras to use oil shale deposits in the Irati belt, which extends 1200 km [41]. The estimated deposit for this area is 630 billion barrels of oil, 10 MM tons of sulfur, 45 MM tons of liquefied gas, and 22 billion $m^3$ of fuel gas.

This process has an external heater that raises the recycle gas temperature to approximately 700°C. The fuel for the heater can be gas, liquid, or solid. Shale is crushed in a two-stage stem that incorporates a fine rejection system. This retort also has three zones, high, middle, and low.

Crushed shale is fed by desegregation feeders to the retort top and then is forced downward by gravity countercurrent to the hot gases. In the middle zone, hot recycle gas is fed at 700°C. Shale oil in the mist form is discharged from the upper zone and is passed to a battery of cyclones and on to an electric precipitator to coalesce. The shale is then recovered and some gas is recycled to the lower zone, to adjust the retort temperature [41]. The remaining gas is treated in a light ends recovery section; then it is sweetened and discharged as liquefied petroleum gas (LPG). Figure 24 shows a schematic of this process, and Table 18 shows some properties of shale oil produced by this process.

**Chevron retort system.** A small pilot unit with a shale feed capacity of 1 ton/day was developed by Chevron Research. This process uses a catalyst and fractionation. The pilot operates on a staged, turbulent-flow bed process that re-

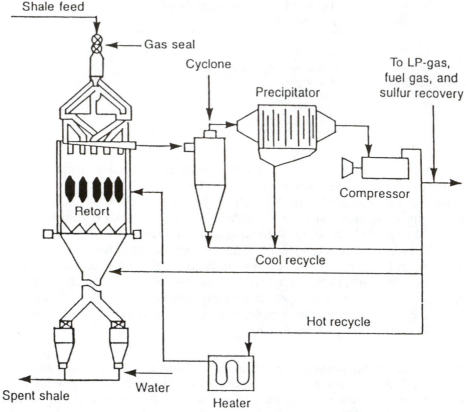

**Figure 24** Petrosix retort system. (Reproduced with permission from Matar, S., *Synfuels*, Penwell Publishing Co., Tulsa, OK, 1982.)

portedly uses the shale completely. Figure 25 shows a schematic of this process, which is also called the shale oil hydrofining process.

**Moving bed retorting process.** A U.S. patent of Barcellos[42] describes a moving bed retorting process for obtaining oil, gas, sulfur, and other products from oil shale. The process comprises drying, pyrolysis, gasification, combustion, and

**Table 18 Typical shale oil properties by Petrosix**

| | |
|---|---|
| Gravity, API | 19.60 |
| Sulfur, wt. % | 1.06 |
| Nitrogen, wt. % | 0.86 |
| Pour point, °F | 25.0 |

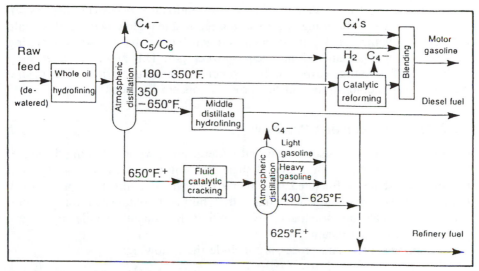

**Figure 25** Shale oil hydrofining process. (Reproduced with permission from *Chemical and Engineering News*, January 9, 1978, American Chemical Society, Washington, DC.)

cooling of pyrobituminous shale or similar rocks in a single passage of the shale continuously in a moving bed. The charge and discharge of the oil shale are intermittent, and the maximum temperature of the bed is maintained in the range 1050 to 1200°C or even higher. The shale is essentially completely freed from the organic matter, fixed carbon, and sulfur, resulting in a clean solid residue that can be disposed of without harming the ecology, according to the inventor's claim. The advantage of this process is its retorting efficiency, as the process operates at fairly high temperatures. However, the main concern is in the energy efficiency.

No pilot plant research data have been published for this process. No large-scale demonstration of the process has been done.

**The carbon dioxide retorting process.** Lee and Joshi [43] developed a retorting process that differs from other retorting processes in its chemistry because of use of a different sweep gas, carbon dioxide. They claim kerogen is swept out far better in the carbon dioxide medium and, therefore, most rich oil shales can be retorted with significantly higher yields than the Fischer assay when carbon dioxide is used as a process gas. Other advantages of the processes include suppression of the dolomite and calcite decomposition reaction, due to the partial pressure of $CO_2$ in the system.

The carbon dioxide retorting process can be used with various retort designs with little or no system configuration change. The preliminary experimental data show that the $CO_2$ retorting process substantially enhances the oil yields from

Colorado and Australian shales, over the conventional process (Fischer assay). However, if the $CO_2$ retorting process is applied to a lean, low-permeability shale like Ohio Devonian shale, the swollen kerogen blocks the poreways, resulting in poorer oil yields than with nitrogen as sweep gas. This process is interesting in the sense that kerogen swells in both subcritical and supercritical $CO_2$ media. No large-scale demonstration of this process concept has been done.

## 7.5.2 In Situ Retorting Processes

Oil shale retorting can also be achieved underground, without mining the shale. Such a process is called an in situ oil shale retorting process. In a typical in situ process, the shale is fractured by either explosive or hydrostatic pressure. A portion of the oil shale organic matter is then burned to obtain heat for retorting. The retorted shale oil is extracted by pumping in a manner similar to the extraction of crude petroleum.

Advantages of in situ processing include the following:

1. Oil can be recovered from deep deposits of an oil shale formation.
2. Mining costs can be avoided or minimized.
3. There is no solid waste disposal problem, because all operations are conducted through well bores. Therefore, the process is environmentally desirable.
4. Shale oil can be extracted from leaner shale, such as deposits containing less than 15 gal/ton of oil.
5. The process is ultimately more economical because of elimination or reduction of costs of mining, transportation, and crushing.

Disadvantages of in situ processing are:

1. It is difficult to control combustion because of insufficient permeability within the shale formation.
2. Drilling costs are high.
3. Recovery efficiencies are generally low.
4. It is difficult to establish the required porosity.
5. There is concern about possible contamination of aquifers.

The in situ technology for production of shale oil from shales optimizes recovery process economics while minimizing environmental impact. This is why considerable emphasis has also been placed on in situ processes.

In situ retorting processes can be roughly classified into two types: modified in situ (MIS) and true in situ (TIS). Modified in situ retorting involves partial mining of the oil shale deposit to create a void space and rubblizing the rest into this space to increase the overall permeability of the shale. The underground rubblized shale is then ignited, using an external or internal fuel source. The MIS process is the brainchild of Occidental Petroleum. In situ retorting is similar to modified in situ, but in this process no mining is done. The shale deposits are rubblized to increase the permeability and then the underground burning is begun.

A review of the oil shale literature indicates that all in situ oil shale processes can be classified into the following categories [44]:

1. Subsurface chimney

    A. Hot gases
    B. Hot fluids
    C. Chemical extraction

II. Natural fractures

    A. Unmodified
    B. Enlargements by leaching

III. Physical induction

    A. No subsurface voids

Other ways of classifying the in situ oil shale retorting processes are:

I. Formation of retort cavities

    A. Horizontal sill pillar
    B. Columnar voids
    C. Slot-shaped columnar voids
    D. Multiple zone design
    E. Multiple horizontal units
    F. Multiple adjacent production zones
    G. Multiple gallery-type retort zones
    H. Spaced-apart upright retort chambers
    I. Permeability control of rubble pile
    J. Formation of rich and lean zones
    K. Successive rubblization and combustion
    L. Thermomechanical fracturing
    M. Water leaching and explosive fracturing
    N. Inlet gas means
    O. Fluid communication
    P. Cementation to minimize plastic flow

Q. Near-surface cavity preparation
R. Dielectric heating

II. Retorting techniques

A. Ignition techniques
B. Multistage operation
C. Steam leaching and combustion
D. Pressure swing recovery
E. Multistratum reservoir
F. Production well throttling
G. Combined combustion techniques
H. Laser retorting
I. Low-heat fans for frontal advance units
J. Gas introduction and blockage
K. Water injection
L. Oil collection system
M. Handling system for feed and products
N. Uniform gas flow
O. Postretorting flow
P. Sound monitoring
Q. Underground weir separator
R. Emulsion breaking technique
S. Off-gas recycling
T. Prevention of off-gas leakage

III. Others

A. Molecular sulfur and benzene recovery
B. Hydrogen sulfide and carbon dioxide treatment
C. Hot-fluid injection into solvent-leached shale
D. Steam treatment and extended soak period
E. Steam-driven excavating unit
F. Anaerobic microorganisms
G. Hot aqueous akaline liquids and fluid circulation
H. Plasma arc

Tables 19 and 20 show the process patents on the processes categorized above. It should be clearly noted that these tables are not a comprehensive list of processes developed or invented for in situ retorting processes for oil shale.

During the past 20 years, an increasing amount of research has been focused on the production of oil from oil shale in situ or underground.

Most oil shales, in general, have little natural permeability, and this makes it difficult to recover oil from them in situ. Basically, two techniques have been

**Table 19  Classification of in situ retorting process**

| Method | Company name | Patents |
|---|---|---|
| I. Subsurface chimney | | |
|   A. Hot gases | Phillips Petroleum | U.S. 3,490,529 |
| | | U.S. 3,548,939 |
| | | U.S. 3,618,633 |
| | Atlantic Richfield | U.S. 3,586,377 |
| | McDonnell Douglas | U.S. 3,596,993 |
| | Continental Oil | U.S. 3,480,082 |
| | Mobil Oil | U.S. 3,542,131 |
|   B. Hot fluids | Shell Oil | U.S. 3,759,328 |
| | | U.S. 3,572,838 |
| | | U.S. 3,565,171 |
| | | U.S. 3,593,789 |
| | Cities Service Oil | U.S. 3,601,193 |
| | Garrett Corp. | U.S. 3,661,423 |
|   C. Chemical extraction | Shell Oil | U.S. 3,593,790 |
| | | U.S. 3,666,014 |
| II. Natural fractures | | |
|   A. Unmodified | Shell Oil | U.S. 3,501,201 |
| | | U.S. 3,500,913 |
| | | U.S. 3,513,913 |
| | | U.S. 3,513,914 |
| | Marathon Oil | |
| | | U.S. 3,730,270 |
| | Resources R & D | South African |
| | | 6,908,904 |
|   B. Enlarged by leaching | Shell Oil | U.S. 3,481,398 |
| | | U.S. 3,759,328 |
| | | U.S. 3,759,574 |
| III. Physical induction, no | Woods R & D | Neth. Appl. |
|   subsurface voids | Corp. | 6,905,815 |

(Reprinted with permission from Yen, T. F., *Science and Technology of Oil Shale*, p. 12, Ann Arbor Science Publishers, Ann Arbor, MI, 1976.)

proposed for creating some permeability in such formations. With the modified in situ processes, part of the shale bed is removed to create a void that will allow fracturing the rest of the bed with explosives. With the true in situ technique, no mining or other void-producing preparation is required; the permeability is produced with well bores from the surface.

**Sinclair Oil and Gas Company process.** In 1953 Sinclair Oil and Gas Company performed one of the early experiments in in situ oil shale retorting. Their process concept was similar to that shown in Figure 26. From their study, they found that communication between wells could be established through induced or natural fracture systems, that wells could be ignited successfully, and that combustion could be established and maintained in the oil shale bed. They also realized that high pressures were required to maintain injection rates during the

## Table 20 Classification of in situ oil shale retorting processes

| Method | Company name | Patents |
|---|---|---|
| I. Formation of retort cavities | | |
| A. Horizontal still pillar | Occidental Oil Shale | U.S. 4,118,070 |
| | | U.S. 4,118,071 |
| | | U.S. 4,045,085 |
| | | U.S. 4,025,115 |
| B. Columnar voids | Occidental Oil Shale | U.S. 4,043,595 |
| C. Slot-shaped voids | Occidental Oil Shale | U.S. 4,043,596 |
| D. Multiple zone design | Occidental Oil Shale | U.S. 4,043,597 |
| | | U.S. 4,043,598 |
| E. Multiple horizontal units | Occidental Oil Shale | U.S. 4,106,814 |
| F. Multiple adjacent production zones | Atlantic Richfield | U.S. 3,917,346 |
| | | U.S. 3,917,347 |
| | | U.S. 3,917,348 |
| G. Multiple gallery-type retort zones | Mobil | U.S. 3,950,029 |
| H. Spaced-apart upright retort chambers | Atlantic Richfield | U.S. 3,917,344 |
| I. Permeability control in rubble pile | Occidental Oil Shale | U.S. 3,951,456 |
| | Geokinetics | U.S. 3,980,339 |
| J. Formation of rich and lean zones | U.S. ERDA | U.S. 4,017,119 |
| K. Successive rubblizing and combustion | Gulf R & D | U.S. 4,015,664 |
| | Exxon Prod. Res. | U.S. 4,091,869 |
| L. Thermochemical fracturing | TRW | U.S. 4,083,604 |
| M. Water leaching and explosive | Shell Oil | U.S. 3,957,307 |
| N. Inlet gas means | Occidental Oil Shale Inc. | U.S. 4,047,760 |
| O. Fluid Communication | Standard Oil (Indiana) | U.S. 4,120,355 |
| | Standard Oil | U.S. 4,131,416 |
| P. Cementation to minimize | U.S. D.O.E. | U.S. 4,096,912 |
| Q. Near-surface cavity preparation | AZS Corp. | U.S. 4,063,780 |
| | Lekas | U.S. 4,037,657 |
| R. Dielectric heating | Synfuel (Indiana) | U.S. 4,401,162 |
| II. Retorting Techniques | | |
| A. Ignition techniques | Occidental Petrol. Corp. | U.S. 3,952,801 |
| | | U.S. 4,027,917 |
| | | U.S. 4,005,752 |
| B. Multistage operation | Occidental Oil Shale Inc. | U.S. 4,072,350 |
| C. Steam leaching and combustion | TRW, Inc. | U.S. 4,065,183 |
| D. Pressure swing recovery system | TRW, Inc. | U.S. 4,059,308 |
| E. Multistratum reservoir | | U.S. 3,978,920 |
| F. Production well throttling to control combustion drive | Cities Service Co. | U.S. 3,999,606 |
| G. Combined combustion tech. | Marathon Oil | U.S. 4,084,640 |
| H. Laser retorting | U.S. NASA | U.S. 4,061,190 |
| I. Low-head fans | Mobil Oil | U.S. 4,018,280 |
| J. Buffer Zone | Occidental Oil Shale, Inc. | U.S. 4,126,180 |
| K. Gas introduction and blockage | Occidental Oil Shale, Inc. | U.S. 4,119,345 |
| L. Water injection | Occidental Oil Shale, Inc. | U.S. 4,089,375 |
| | | U.S. 4,036,299 |
| M. Oil collection system | Occidental Petrol. Corp. | U.S. 4,007,963 |

## Table 20 Continued

| Method | Company name | Patents |
|---|---|---|
| N. Handling system for feed and products | Occidental Petrol. Corp. | U.S. 4,014,575 |
| O. Uniform gas flow | Occidental Petrol. Corp. | U.S. 3,941,421 |
| P. Postretorting tech. | Occidental Oil Shale | U.S. 4,105,072 |
| Q. Sound monitoring | Occidental Oil Shale | U.S. 4,082,145 |
| R. Underground weir separator | Gulf Oil & Standard Oil (Indiana) | U.S. 4,119,349 |
| S. Emulsion breaking tech. | Occidental Oil Shale | U.S. 4,109,718 |
| T. Off-gas recycling | Occidental Petrol. Corp. | U.S. 3,994,343 |
| U. Prevention of off-gas leakage | Occidental Oil Shale | U.S. 4,076,312 |
| III. Others | | |
| A. Molecular sulfur and benzene recovery | Marathon | U.S. 3,929,193 |
| B. Hydrogen sulfide and carbon dioxide treatment | Cities Service R & D | U.S. 3,915,234 |
| C. Hot fluid injection into solvent-leached shale | | U.S. 3,739,851 |
| | | U.S. 3,741,306 |
| | | U.S. 3,759,574 |
| D. Steam treatment | Cities Service R & D | U.S. 3,882,941 |
| E. Steam-driven excavator | | U.S. 3,941,423 |
| F. Anaerobic microoganisms | IGT | U.S. 4,085,972 |
| G. Hot aqueous alkaline liquids | Shell Oil | U.S. 4,026,360 |
| H. Plasma arc | Tech. Appl. Serv. Corp. | U.S. 4,067,390 |

heating period. These tests were made near an outcrop in the southern part of the Piceance Creek Basin [45]. Additional tests were made several years later at a depth of about 365 m in the north-central part of the Piceance Creek Basin with only limited success. The partial success was believed to be due to inability to obtain the required surface area for the heat transfer. However, their experiments established the basic technology for in situ retorting of oil shale and suggested further study areas.

**Equity Oil Company process.** Equity Oil Company of Salt Lake City [46] studied an in situ process that is somewhat different from the Sinclair process. This process involves injecting hot natural gas into the shale bed to retort the shale. One injection well and four producing wells were drilled into the oil shale formation in an area of the Piceance Creek Basin. The natural gas was compressed to about 85 atm, heated to approximately 480°C, and delivered through insulated tubing to the retorting zone. Based on the experimental results and a mathematical model developed from them, it was concluded that this technique was feasible and potentially an economic method for extracting shale oil. However, the process economics are undoubtedly strongly dependent on the cost of natural gas and the amount required for makeup of natural gas.

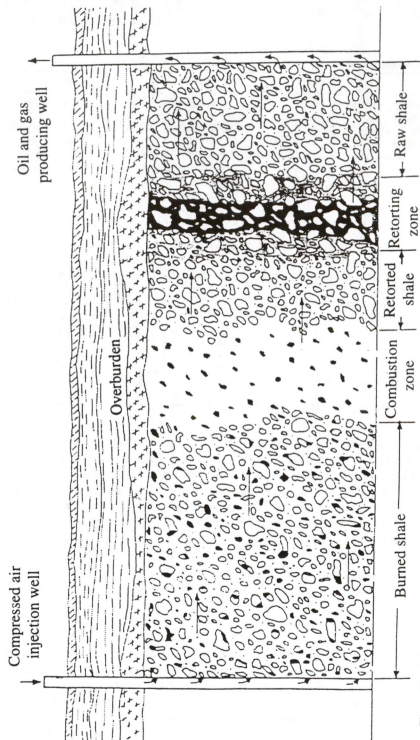

**Figure 26** Schematic diagram of in situ oil shale retorting process. (Reproduced with permission from T.F. Yen and G.V. Chilingarian, *Oil Shale*, Elseveir, Amsterdam, 1976.)

**Occidental Petroleum process.** Occidental Petroleum developed a modified in situ process in which conventional explosives are used to expand solid blocks of shale into a vertical mined-out cavity, creating underground chimneys of fractured shale. Figure 27 shows a design schematic of this retort. In order to improve the fluid communication, about 10 to 25% of the shale in the chimney is removed. After this removal, air is blown down through the remaining crushed shale, and the top is ignited with a burner that can be fueled with shale oil or off-gas from other retorts. On ignition, the burner is withdrawn and steam is mixed with the inlet air to control the process [40]. The liquid and gaseous products flow to the bottom of the chimney, leaving the char in the shale behind as the main source of fuel for the slowly advancing flame front.

Occidental Petroleum did a series of field tests on this process at Logan's Wash in Debeque, Colorado, with three retorts, 30 ft across and 72 ft deep, each containing 6000 to 10,000 tons of oil shale. Based on their experimental success, full-scale production of 57,000 barrels per day was planned.

**LETC process (or LERC process).** The former Laramie Energy Technology Center (LETC) of the U.S. Department of Energy sponsored several field projects to demonstrate the technical and economic feasibility of shale oil recovery by in situ technology.

LETC initiated their study of in situ retorting in the early 1960s with laboratory tests, simulated pilot plant tests on 10-ton and 150-ton retorts, and field tests at Rock Spring, Wyoming. The test results demonstrated that it was possible to move a self-sustaining combustion zone through an oil shale formation and to produce shale oil.

The underground shale bed is prepared for the LETC process by first boring injection and production wells into the shale and then increasing the permeability of the formation by conventional fracturing techniques. Based on the LETC tests, the sequential use of hydraulic fracturing and explosives was the best. Once the formation has been fractured, hot gases are forced into it to heat the area around the injection point. As the desired temperature is reached and air is substituted for the hot gas, combustion begins and becomes self-sustaining across a front that gradually moves through the bed. As retorting progresses, oil and gas products are pumped out through the production well. A schematic [47] of the LETC process is shown in Figure 28.

**Dow Chemical Company process.** The Dow Chemical Company, under contract with the U.S. Department of Energy, conducted a 4-year research program to test the feasibility of deep, in situ, recovery of gas with a low heat content from Michigan Antrim Shale [48].

The Antrim shale is a part of the eastern and midwestern oil shale deposits, formed some 260 million years ago during the Devonian and Mississippian ages. These oil shales underlie an area of 400,000 square miles ($1.04 \times 10^6$ m$^2$). In

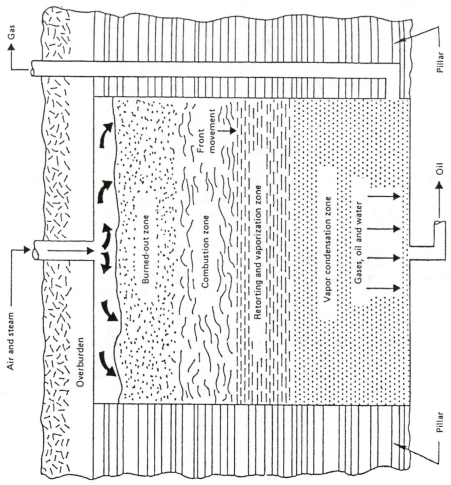

**Figure 27** Modified in situ retort of Occidental Petroleum. (Source: Reference 19.)

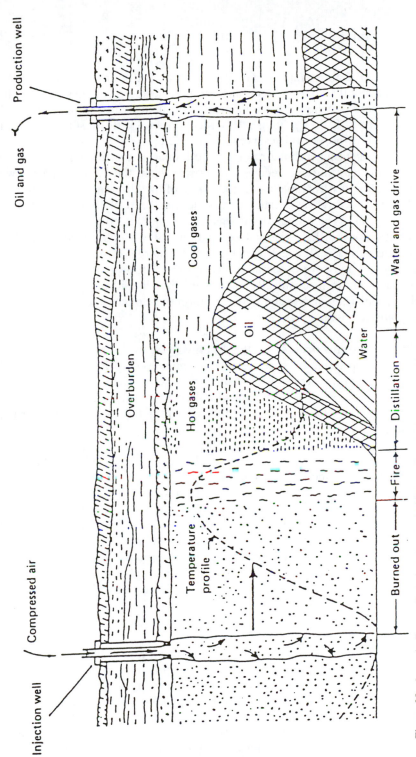

**Figure 28** Laramie Energy Technical Center (LETC) in situ retorting process. (Source: Reference 20.)

Michigan, the oil shale is approximately 61 m thick and is in a basin at depths ranging from about 0.8 km to outcroppings in three northern counties. The Michigan Antrim shale is believed to contain an equivalent hydrocarbon volume of 2500 billion barrels. Even applying a 10% recovery factor, this resource is about nine times the U.S. proven oil reserve.

Extensive fracturing (rubblizing) of the oil shale is considered essential to adequate in situ retorting and recovery of energy from the Antrim shale. Two wells were explosively fractured using 19,000 kg of metalized ammonium nitrate slurry. Their test facility was located at 75 miles northeast of Detroit, Michigan, over 1 acre of field. The process used was true in situ retorting.

Combustion of the shale was started using a 440-V electric heater (52 kW) and a propane burner (250,000 Btu/hr). Special features of the process include shale gasification and severe operating conditions. The tests also showed that explosive fracturing in mechanically underreamed wells did not produce extensive rubblization. They also tested hydrofracturing, chemical underreaming, and explosive underreaming.

**Talley Energy Systems process.** Talley Energy Systems, Inc. carried out a U.S. DOE-industry cooperative oil shale project 11 miles west of Rock Springs, Wyoming. The shale in the area is part of the Green River shale formation, which is some 50 million years old. This is also a process based on true in situ processing. This process uses explosive fracturing, additional hydraulic fracturing, and no mining.

**Geokinetics process.** Geokinetics, Inc. developed an in situ process that may be best described by modified horizontal technology. Explosive fracturing is used, and the process can be used even for shallow thin-seam recovery. This was one of the U.S. DOE-industry cooperative oil shale projects and was located 61 miles northwest of Grand Junction, Colorado. The geological formation for this process study was Green River Formation, Parachute Creek member, and Mahogany zone, the same as for the Occidental Oil process. The only difference of significance between the two is that the Geokinetics process is modified horizontal, whereas the Occident Oil process is modified vertical. Therefore, the Geokinetics process is good for shallow thin-seam recovery, and the Occidental Oil process is better suited for deep thick-seam recovery.

**Osborne in situ process.** The Osborne in situ process was developed by Osborne in 1983; a U.S. Patent [49] describing the process has been assigned to Synfuel (an Indiana limited partnership). The process is unique, and improvement of oil recovery is achieved by forming generally horizontal electrodes by injection of molten metal into preheated or unheated fractures of formation. A nonconductive spacing material is positioned in the casing of the bore hole between the electrodes. A fracture horizontally intermediate between the metallic electrodes is propped with a nonconductive granular material. Unterminated standing waves from a radio frequency (RF) generator are passed between the electrodes to heat the oil shale formation. The hydrocarbons in the formation are vaporized and recovered at the surface by transport through the intermediate fracture and tubing. By this method, radial metallic electrodes can be formed at various depths throughout a subterranean oil shale formation to devolatilize the hydrocarbons contained in the oil shale formation [49].

One advantage of this process is that uniform heating of the rock formation can be achieved by using RF electrical energy that corresponds to the dielectric absorption characteristics of the rock formation. An example of such techniques is described in U.S. Patents 4,140,180 and 4,144,935, in which a plurality of vertical conductors are inserted into the rock formation and bound a particular volume of the formation. A frequency of electrical excitation is selected to attain relatively uniform heating of the rock formation. The energy efficiency of the process is good; however, the economics of the process depend strongly on the cost of the electrodes and RF generation. Other merits of the process include relative ease of controlling the retort size.

The difficulty with this process is that it is neceessary to implant an electrode in the subterranean rock formation at a precise distance. A schematic for this process is shown in Figure 29.

**True and modified in situ retorting.** In situ retorting of oil shale is often classified as true in situ (TIS) and modified in situ (MIS). In this section, these two terms are further clarified.

True in situ retorting involves drilling wells and fracturing oil shale rock to increase its permeability. A hot gas mixture is used to heat the oil shale rubble. Forced air then helps burn the oil shale. A flame front forms that gradually moves through the bed, and the oil and gas produced are drawn from the production well to the surface.

In modified underground retorting, a blocked-out area is mined to remove approximately 10 to 25% of the oil shale. Vertical or horizontal wells are drilled through the remaining portion and are detonated. The voids produced help to fracture and rubblize the oil shale. This is a modification of the true in situ conversion process and was first developed by Occidental Oil.

## 7.5.3 Shale Oil Refining and Upgrading

As the demand for light hydrocarbon fractions constantly increases, there is much interest in developing economical methods for recovering liquid hydrocarbons from oil shale on a commercial scale. However, the recovered hydrocarbons are not yet economically competetive with petroleum crudes recovered by more conventional methods.

Furthermore, the value of hydrocarbons recovered from oil shale is diminished by the presence of undesirable contaminants. The major contaminants of concern are sulfurous, nitrogenous, and metallic compounds that have detrimen-

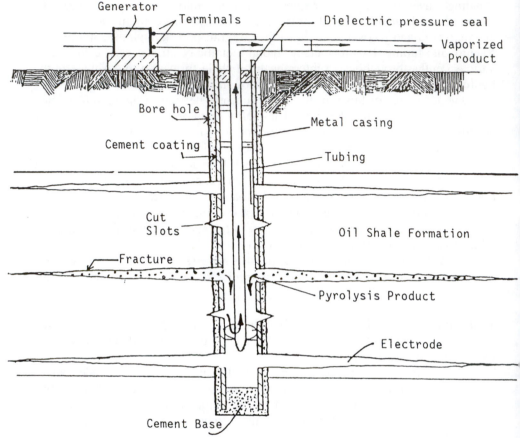

**Figure 29** Vertical sectional view of a borehole penetrating a subterranean oil shale formation in completed condition for recovery of hydrocarbons from the shale. Osborne's process. (Based on U.S. Patent 4,401,162.)

tal effects on various catalysts used in the subsequent refining processes. These contaminants are also undesirable because of their disagreeable odor, corrosive characteristics, and combustion products that further cause environmental problems.

Accordingly, there is great interest in developing more efficient methods for converting the heavier hydrocarbon fractions obtained in the form of shale oil into lighter materials. The conventional processes include catalytic cracking, thermal cracking, and coking.

Heavier hydrocarbon fractions and refractory materials can be converted to lighter materials by hydrocracking. Hydrocracking processes are most commonly employed on liquefied coals or heavy residual or distillate oils for the production of substantial yields of low-boiling saturated products and to some extent of intermediates, which are used as domestic fuels, and still heavier cuts, which find uses as lubricants. These destructive hydrogenation processes or hydrocracking processes may be operated on a strictly thermal basis or in the presence of a catalyst. Thermodynamically, larger hydrocarbon molecules are broken into lighter species when subjected to heat. The H/C ratio of such molecules is lower than that of saturated hydrocarbons and abundantly supplied hydrogen improves this ratio by saturating reactions, thus producing liquid species. These two steps may occur simultaneously.

However, the application of hydrocracking has been hampered by the presence of certain contaminants in such hydrocarbons. The presence of sulfur- and nitrogen-containing and organometallic compounds in crude shale oils and various refined petroleum products has long been considered undesirable.

**Thermal cracking process.** Gulf Research & Development [50, 51] developed a process for the noncatalytic thermal cracking of shale oil in the presence of a gaseous diluent and an entrained stream of inert heat carrier solids. The cracking process is directed toward the recovery of gaseous olefins as the primarily desired cracked product, in preference to gasoline range liquids. With this process, it is claimed that at least 15 to 20% of the feed shale oil is converted to ethylene, which is the most prevalent gaseous product. Most of the feed shale oil is converted to other gaseous and liquid products. Other important gaseous products include propylene, 1,3-butadiene, ethane, and other $C_4$ compounds. Hydrogen is also recovered as a valuable nonhydrocarbon gaseous product. Liquid products can make up 40 to 50 wt. % or more of the total product. Recovered liquid products include benzene, toluene, xylene, gasoline boiling range liquids, and light and heavy oils.

Coke is a solid product of the process and is produced by polymerization of unsaturated materials. Most of the formed coke is removed from the process as a deposit on the entrained inert heat carrier solids.

The thermal cracking reactor does not require a gaseous hydrogen feed. In the reactor, entrained solids flow concurrently through the thermal riser at an average riser temperature of 700 to 1400°C. The preferred high L/D ratio is in the range of a high 4:1 to 40:1, or 5:1 to 20:1 preferably.

**Moving bed hydroprocessing reactor.** This process was developed by and assigned to Universal Oil Products Company [52]. The process was meant to be for crude oil derived from oil shale or tar sands containing large amounts of highly abrasive particulate matter, such as rock dust and ash. The hydroprocessing takes place in a dual-function moving bed reactor that simultaneously removes particulate matter by the filter action of the catalyst bed. The effluent from the moving bed reactor is then separated and further hydroprocessed in fixed bed reactors with fresh hydrogen added to the heavier hydrocarbon fraction to promote sulfurization.

The preferred way to treat the shale oil involves using a moving bed reactor followed by a fractionation step to divide the wide-boiling-range crude oil produced from the shale oil into two separate fractions. The lighter fraction is hydrotreated to remove of residual metals, sulfur, and nitrogen, and the heavier fraction is cracked in a second fixed bed reactor normally operated under high-severity conditions.

**Fluidized bed hydroretort process.** This process was developed by Cities Service Company [53] in 1978. The process eliminates the retorting stage of conventional shale upgrading, by directly subjecting crushed oil shale to a hydro-retorting treatment in an upflow, fluidized bed reactor, such as that used for the hydrocracking of heavy petroleum residues. This process is a single-stage retorting and upgrading process. Therefore, the process involves (1) crushing oil shale; (2) mixing the crushed oil shale with a hydrocarbon liquid in order to provide a pumpable slurry; (3) introducing the slurry along with a hydrogen-containing gas into an upflow, fluidized bed reactor at a superficial fluid velocity sufficient to move the mixture upward through the reactor; (4) hydroretorting the oil shale; (5) removing the reaction mixture from the reactor; and (6) separating the reactor effluent into several components [54].

The mineral carbonate decomposition is minimized, because the process operating temperature is lower than those used in retorting. Therefore, the gas product of this process has a greater heating value than that of other conventional methods. In addition, because of the exothermic nature of the hydroretorting reactions, less energy input is required per barrel of product obtained. Furthermore, there is practically no upper or lower limit on the grade of oil shale that can be treated.

**Hydrocracking process.** Hydrocracking is essentially a cracking process in which higher molecular weight hydrocarbons pyrolyze to lower molecular weight paraffins and olefins in the presence of hydrogen. The hydrogen saturates the olefins formed during the cracking process. Hydrocracking is used to process low-value stocks with a high heavy metal content. It is also suitable for highly aromatic feeds that cannot be processed easily by conventional catalytic cracking. Shale oils are not highly aromatic, whereas coal liquids are highly aromatic.

Middle-distillate (often called mid-distillate) hydrocracking is carried out with a noble metal catalyst. The average reactor temperature is 480°C, and the average pressure is around 130–140 atm.

# REFERENCES

1. Tisot, P. R., and W. I. R. Murphy. *Chem. Eng. Prog. Symp. Ser.* 61(54):25, 1965.
2. Tisot, P. R., *J. Chem. Eng. Data*, 12(3):405, 1967.

3. Thumann, A. *The Emerging Synthetic Fuel Industry.* Fairmont Press, 1981.
4. Nottenburg, R., K. Rajeshwar, R. Rosenvold, and J. Dubow, *Fuel* 58:144, 1979.
5. Nottenburg, R., K. Rajeshwar, R. Rosenvold, and J. Dubow, *Fuel.* 57:789, 1978.
6. Tihen, S. S., H. C. Carpenter, and H. W. Sohns. Thermal conductivity and thermal diffusivity of Green River oil shale. Conf. Thermal Conductivity Proc. 7th, NBS Special Publ. 302, p. 529, September 1968. Lesser, H., G. Bruce, and H. Stone, *Colo. Sch. Mines* 62(3):111, 1967.
7. Lee, S. *Oil Shale Technology.* Boca Raton, FL: CRC Press, 1991.
8. Barnes, A. L., and R. T. Ellington. *ibid.,* 63(4), 827, 1968.
9. Prats, M., and S. M. O'Brien, *J. Pet. Tech.* 97, 1975.
10. Sladek, T. Ph.D. thesis, Colorado School of Mines, Golden, CO, 1970.
11. Dubow, J., R. Nottenburg, K. Rajeshwar, and Y. Wang. The effects of moisture and organic content on the thermophysical properties of Green River oil shale. 11th Oil Shale Symposium Proc., p. 350, Colorado School of Mines, Golden, 1978.
12. Rajeshwar, K., R. Nottenburg, and J. Dubow, *J. Mater. Sci.* 14:2025–2052, 1979.
13. Smith, J. W. *U.S. Bur. Mines Rept. Inv.* 7248, 1969.
14. Skrynnikova, G. N., E. S. Avdonina, M. M. Golyand, and L. Ya. Akhmedova, *Tr. Vsesoyuz Nauch-Issledovatel Inst. Pererabotke Slantsev* 7:80, 1959.
15. McKee, R. H., and E. E. Lyder, *J. Ind. Eng. Chem.* 13:613, 1921.
16. Wang, Y., K. Rajeshwar, R. Rosenvold, and J. Dubow, *Thermochim Acta* 30:141, 1979.
17. Wise, R. L., R. C. Miller, and H. W. Sohns. *U.S. Bur. Mines Rept. Inv.* 7482, 1971.
18. Shaw, R. J. *U.S. Bur. Mines Rept. Inv.* 4151, 1947.
19. Sohns, H. W., L. E. Mitchell, R. J. Cox, W. I. Burnet, and W. I. R. Murphy. *Ind. Eng. Chem.* 43:33, 1951.
20. Cook, W. E. *Q. Colo. Sch. Mines* 65(4):133, 1970.
21. Johnson, P. R., N. B. Young, and W. A. Robb, *Fuel* 54:249, 1975.
22. Allred, V. D. *Colo. Sch. Mines Q.* 59(3):47, 1964.
23. Branch, M. C. *Prog. Energy Combust. Sci.* 5:193, 1979.
24. Joshi, R. M.S. thesis, University of Akron, Akron, OH, 1983.
25. Eggertsen, F. T., S. Groennings, and J. J. Holst. *Anal. Chem.* 32(8):904, 1960.
26. Johnson, D. R., N. B. Young, and J. W. Smith, LERC/R1, Laramie Energy Research Center, Laramie, WY, June 1977.
27. Dyni, J. R., W. Mountjoy, P. L. Hauff, and P. D. Blackman, *U.S. Geol. Surv. Prof. Pap. 750B,* 1971.
28. Loughman, F. C., and G. T. See. *Am. Mineral.* 52:1216, 1967.
29. Agroskin, A. A., and I. G. Petrenko. *Zavodskaya Lab.* 14:807, 1948.
30. Scott, J. H., R. D. Carroll, and D. R. Cunningham. *Geophys. Res.* 72:5101, 1967.
31. Nottenburg, R., K. Rajeshwar, M. Freeman, M., and J. Dubow, *Thermochim. Acta* 31:39, 1979.
32. Rajeshwar, K., R. Nottenburg, J. Dubow, and R. Rosenvold, *Thermochim. Acta* 27:357, 1978.
33. Yen, Y. F. Structural investigations on Green River oil shale. In *Science and Technology of Oil Shale*, Ed. Y. F. Yen. Ann Arbor, MI: Ann Arbor Science Publishers, 1976.
34. Cook, E. W., *Fuel* 53:16, 1976.
35. ASTM, D2887-73 (Reapp. 78), p. 799, 1979.
36. Taylor, R. B. Oil shale commercialization: The risks and the potential. *Chem. Eng.*, September 7, 1981.
37. Matzick, A., R. O. Dannenburg, J. R. Ruark, J. E. Phillips, J. D. Lankford, and B. Guthrie. *U.S. Bur. Mines Bull. 635,* 99, 1966.
38. Whitcombe, J. A., and R. G. Vawter. The TOSCO-II oil shale process. In *Science and Technology of Oil Shale*, ed. Y. F. Yen. Chapter 4. Ann Arbor, MI: Ann Arbor Science Publishers, 1976.
39. National Research Council (U.S.), Panel on R & D needs in refining of coal and shale liquids. In *Refining Synthetic Liquids from Coal and Shale*, chapter 5, pp. 78–135. Washington, DC; National Academy Press, 1980.
40. Rammler, R. W. *Q. Colo. Sch. Mines* 65(4):141–168, 1970.

41. Matar, S. *Synfuels: Hydrocarbons of the Future.* Tulsa, OK: Pennwell Publishing, 1982.
42. Barcellos, E. D. U.S. Patent 4,060,479, November 29, 1977.
43. Lee, S., and R. Joshi. U.S. Patent, 4,502,942, March 5, 1985.
44. Yen, T. F. Oil shales of United States—a review. In *Science and Technolgy of Oil Shale,* ed. T. F. Yen, pp. 1–17, Ann Arbor, MI: Ann Arbor Science Publishers, 1976.
45. Dinneen, G. U. Retorting technology of oil shale. In *Oil Shale,* ed. Teh Fu Yen and G. V. Chilingarran, chapter 9, pp. 181–197, Amsterdam: Elsevier, 1976.
46. Dougan, P. M., F. S. Reynolds, and P. J. Root. The potential for in situ retorting of oil shale in the Piceance Creek Basin of northwestern Colorado. *Q. Colo. Sch. Mines* 65(4):57–72, 1970.
47. Sladek, T. A. Recent trends in oil shale—Part 2: Mining and shale oil extraction processes. *Colorado Sch. of Mines Mineral Ind. Bull.* 18(1):1–20, 1975.
48. McNamara, P. H., and J. P. Humphrey. Hydrocarbons from eastern oil shale. *Chem. Eng. Prog.* 88: September 1979.
49. Osborne, J. U.S. Patent 4,401,162, August 30, 1983.
50. Wynne, F. E. Jr. U.S. Patent, 4,057,490, November 8, 1977.
51. McKinney, J. D., R. T. Sebulsky, and F. E. Wynne, Jr. U.S. Patent 4,080,285, March 21, 1978.
52. Anderson, R. F. U.S. Patent 3,910,834, October 7, 1975.
53. Gregoli, A. A. U.S. Patent 4,075,081, February 21, 1978.
54. Ranney, M. W. *Oil Shale and Tar Sands Technology—Recent Developments,* p. 238. Noyes Data Corporation, New Jersey, 1979.
55. Johnson, W. F., D. K. Walton, H. H. Keller, and E. J. Couch. *Q. Colo. Sch. Mines* 70(3):237, 1975.
56. Shih, S.-M., and H. Y. Sohn. *Fuel* 57:662, 1978.
57. Gregg, M. L., J. H. Campbell, and J. R. Taylor. *Fuel* 60:179, 1981.
58. Wang, Y. M.S. thesis, University of Akron, Akron, OH, 1982.
59. Wang, Y., J. Dubow, K. Rajeshwar, and R. Nottenburg. *Thermochim. Acta* 28:23, 1979.
60. Huggins, C. W., and T. E. Green. *Am. Mineral.* 58:548, 1973.

# PROBLEMS

1. What is the difference between oil shale and shale oil?
2. Discuss the difference in geological formation between coal and oil shale.
3. In fossil fuels, what does the C/H ratio indicate?
4. Is the dolomite calcination reaction endothermic or exothermic?
5. Is the kerogen pyrolysis reaction endothermic or exothermic?
6. In the pyrolysis of oil shale, why is the total energy required more than the energy needed to pyrolyze kerogen?
7. By analyzing the mineral matter compositions only, how can we tell the difference between raw and burned shales?
8. In a fixed bed retorter, it took 20 minutes to pyrolyze 75% of the kerogen at 540°C. If the kerogen decomposition reaction follows global first-order kinetics, how long does it take to pyrolyze 50% of the kerogen of the same oil shale?
9. In a preheated environment at 500°C, a round shale particle of 1 cm diameter is introduced. The oil shale particle was originally at room temperature (25°C). How long does it take the center of the shale particle to "feel" the exterior "hot" temperature?

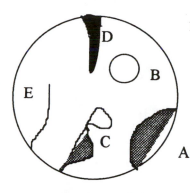

10. This is a drawing of a porous granular particle. You may call it a standard rock sample.
Find the following:
   (a) Superficial pore
   (b) Capillary pore
   (c) Closed pore
   (d) Internal pores
   (e) Total open pores
   (f) Saturable pores

11. Upon heating, calcite ($CaCO_3$) liberates carbon dioxide ($CO_2$). The partial
pressure of this carbon dioxide is called the dissociation pressure, which is
a function of temperature.

$$CaCO_3(s) = CaO(s) + CO_2(g)$$

Based on the Gibbs free energy of formation from the literature, calculate
the dissociation pressure of calcite as a function of temperature.
12. What is supercritical $CO_2$? Discuss the properties of supercritical carbon
dioxide.
13. Convert the following units:
   (a) $20°API = (\quad) g/cm^3$
   (b) 28 gal/ton Fischer assay $= (\quad)$ L/ton
   (c) 10 W/°C-m $= (\quad)$ cal/cm K

# EIGHT

## TAR SANDS

## 8.1 WHAT ARE TAR SANDS?

Tar sands are sands and other rock materials saturated with crude bitumen, water, and gas. Tar sand deposits are found commonly at or near the earth's surface and normally entrapped in large sedimentary basins. Oil sands and bituminous sands are of the same nature, and these terms normally are treated as alternative names to tar sands.

Unlike oil shales, tar sands are found predominantly in extremely large quantities in a few areas. Approximately 98% of the world's heavy oil is found in seven large tar deposits [1]. The Athabasca deposit in western Canada is the world's largest self-centered one.

### 8.1.1 Tar Sand Reserves

Canada has the largest confirmed tar sand reserves in the world. The Athabasca deposits alone extend over 100 miles, and an equivalent of 600 billion barrels of oil is estimated to be present. Considering that the total in-place reserves of tar sands deposits and heavy oils in Canada are estimated to be 900 billion barrels of oil, the Athabasca deposit alone accounts for two-thirds of the total Canadian reserve. Unfortunately, the present technology allows only one-third or one-half of these reserves to be ultimately recoverable [1]. Tar sand reserves in eastern Venezuela are estimated to be as high as 1.05 trillion barrels of oil. Two U.S. reserves are estimated at 30 billion barrels of oil, and the U.S. deposits are concentrated mainly in the state of Utah.

## 8.1.2 Definitions

Tar sands are defined as sedimentary rocks (consolidated or unconsolidated) that contain bitumen or other heavy petroleum that cannot be recovered by conventional petroleum recovery methods [2]. This condition is applicable to oils having a gravity less than 12° API.

The quartzose tar sands are of fluviatile and lacustrine origins with the highest tar contents being present in the clean, well-sorted fluviatile sands [2]. The maximum heavy oil content is 18–20% by mass of saturated sand [3].

The descriptive term tar sand has been defined under the Alberta Oil and Gas Conservation Act as sand having a "highly viscous crude hydrocarbon material not recoverable in its natural state through a well by ordinary production methods" [4]. Strictly speaking, the term tar sand is scientifically inappropriate, because tar along with pitch is a substance resulting from the destructive distillation of organic matter, and such an origin for tar in tar sands is rarely implied [9]. In this regard, a distinction between tar sands and sands containing heavy crude oil is not easily drawn [4]. Moreover, distinction between tar sands and sands contaminated with spilled oil is not automatic.

## 8.1.3 Recoverability of Synthetic Crude

The economic feasibility of producing a synthetic crude from a particular tar sand deposit is an exponential function of the size of the accumulation [5]. The only two deposits that have sufficient size and for which the technology is reasonably advanced are those in Canada and Venezuela. In order to extract and upgrade the tar, a substantial capital investment is required, and the deposit size becomes a crucial factor in determining the economic feasibility. The reserves of recoverable syncrude from tar sands cannot be estimated until studies of the geology, process engineering, and mechanics of in situ operation are completed. However, there is little doubt that tar sands provide an excellent source of liquid fuels and can be commercially exploited with currently available technology.

## 8.1.4 Major Tar Sand Deposits

Major tar sand deposits in the world are shown in Figure 1 [4]. An excellent article on this subject was published by Phizackerley and Scott [4]. The age, extent, thickness, bitumen saturation, oil characteristics, and overburden thickness of 20 major world tar sand deposits are summarized in their work [4].

The major tar sands are limited to eight countries: Canada, the United States, Albania, Venezuela, Trinidad, Rumania, the former USSR, and Malagasy. Sta-

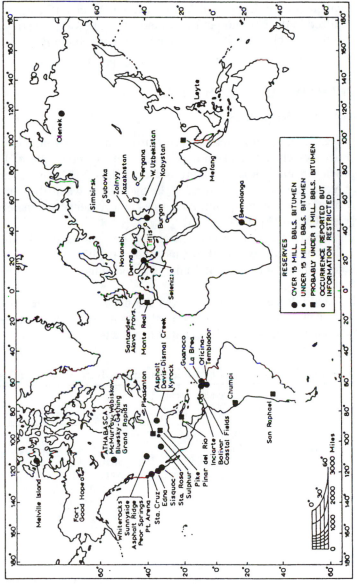

**Figure 1** Location map of the major tar sands of the world. (Source: Reference 9).

tistical information for all these areas and areas that are less well known, except for the Athabasca deposit, is usually incomplete or roughly estimated. Therefore, direct comparison of the deposits may not have much scientific significance. As has been realized in the gas and petroleum industries, new discoveries and better statistics are foreseeable only if the tar sands are commercially exploited.

## 8.2 PROPERTIES OF TAR SANDS

### 8.2.1 Elemental Properties

Tar sand differs greatly from regular sand. It has a very strong oil scent. The material properties change substantially with ambient temperature. During the summer, when the ambient temperature is higher than 20°C, the material is soft and sticky, generating a strong oil odor. In the winter when the ambient temperature is below the freezing point, it changes to a cement-hard material resembling carborundum, SiC [1].

Tar sand deposits are fairly permeable and porous. Tar sands are typically composed of approximately 84–88% sand and clay, 4% water, and 8–12% bitumen. Table 1 shows the elemental analyses and mineral compositions of tar sand solids [1].

The tar sand bitumen slowly dissolves the natural rubber in mining equipment, such as conveyor belts and tires, in the summer. In the cold weather, tar sands are extremely abrasive and wear down bucket-wheel teeth rapidly [1].

**Table 1 Elemental analyses and mineral composition of tar sand solids**

| Elemental analysis | | Minerals present | |
|---|---|---|---|
| Element | Percent by wt | Main constituents | Composition |
| Si | 50.0 | Quartz | $SiO_2$ |
| Al | 1.5 | Chert | $SiO_2$, limestone matrix |
| Mg | 0.006 | Chalcedony | $SiO_2$, as flint/agate |
| Ca | 0.015 | Kaolinite | $Al_2O_3 \cdot 2SiO_2 \cdot 1 - 2H_2O$ |
| Fe | 0.07 | Illite | $K_{1-1.5}Al_4Si_{7-6.5}Al_{1-1.5}O_{20}(OH)_4$ |
| Ti | 0.5 | Feldspars | Sodium/potassium aluminum silicates |
| Zr | Trace | Dolomite | $CaCO_3 \cdot MgCO_3$ |
| Mn | 0.005 | Calcite | $CaCO_3$ |
| Cu | 0.0025 | Micas | $H_2K(Mg, Fe)_3(Al, Fe)Li_{0-1}(SiO_4)_3$ |
| C | 0.46 | | |
| H | 0.40 | Minor constituents | |
| N | 0.08 | Tourmaline | Complex Al borosilicate |
| S | 0.14 | Iron minerals | Sulfides, carbonates, oxides |
| | | Titanium Minerals | Oxides and mixed oxides |
| | | Staurolite | $2FeO \cdot 5Al_2O_3 \cdot 4SiO_2 \cdot H_2O$ |
| | | Epidote | $4CaO \cdot 3(Al, Fe)_2O_3 \cdot 6SiO_2 \cdot H_2O$ |

*Source:* Reference 1.

## 8.2.2 Properties of Tar Sand Bitumen

Bitumen is a semisolid, viscous, dense material that varies in consistency. It is slightly heavier than water and does not flow at room temperature. When heated, bitumen flows and will float on water. An important property of Athabasca and Wabesca tar sands is the presence of a thin water film that surrounds each grain of sand, separating the bitumen from the sand. This property facilitates the hot water extraction process.

Bitumen contains large amounts of asphaltene that consist mainly of polynuclear aromatics linked by alkyl chains. The H/C ratio (by mole) is low, and contents of nitrogen, oxygen, and heavy metals are high [1]. Table 2 lists the properties of Athabasca bitumen by Considine [6]. Table 3 gives similar information for Athabasca tar sands and their bitumens by Mungen and Nicholls [7]. The sulfur content as shown in Table 2 is quite high, making the subsequent upgrading process more difficult.

## 8.2.3 Geochemistry of Tar Sands

The total amount of organic carbon in living matter of the earth is estimated as $2 \times 10^{11}$ tons [2]. Considering the carbon distribution on the entire earth, this amount is quite insignificant. Table 4 shows the carbon distribution on the earth from Skirrow [8]. As shown, carbonates are overwhelmingly predominant ma-

### Table 2 Properties of Athabasca bitumen (after Considine)

| | |
|---|---|
| Gravity at 60°F (15.6°C) | 60°API |
| UOP characterization factor | 11.18 |
| Pour point | +50°F (10°C) |
| Specific heat | 0.35 cal/(g)(°C) |
| Calorific value | 17,900 Btu/lb |
| Viscosity at 60°F (15.6°C) | 3000–3,000,000 poise |
| Carbon/hydrogen ratio | 8.1 |
| Components, % | |
|   Asphaltenes | 20.0 |
|   Resins | 25.0 |
|   Oils | 55.0 |
| Ultimate analysis, % | |
|   Carbon | 83.6 |
|   Hydrogen | 10.3 |
|   Sulfur | 5.5 |
|   Nitrogen | 0.4 |
|   Oxygen | 0.2 |
| Heavy metals, ppm | |
|   Nickel | 100 |
|   Vanadium | 250 |
|   Copper | 5 |

*Source:* References 1 and 6.

**Table 3 Properties of Athabasca oil sands and their bitumens**

| | |
|---|---|
| Sands | |
| Grain size | −200 mesh |
| Porosity | 30–40% |
| Permeability of | |
|   Most tar zones | 200–300 mD |
|   Clean sand tar zones | Several darcies |
|   Silt zones | A few millidarcies |
| Saturation | 0–90% bitumen (remainder water with little or no gas saturation) |
| Bitumen content | 0–18 wt. % basis |
| Bitumens | |
| Specific gravity | 1.08 |
| Viscosity at | |
|   200°F | 1000 cP |
|   50°F | 2,000,000–5,000,000 cP |
| Hydrogen/carbon ratio | 1.44 |

*Source:* Reference 7.

terials containing carbon, about 2000 times more than coal and petroleum combined.

The petroleum formation process is generally accepted as the thermal transformation of complex soluble (bitumen) and insoluble (kerogen) organic substances dispersed in source sediments [9]. The petroleum generation reaction rates also increase with increasing temperature and, consequently, with increasing burial depth of the generative sediments [9]. Figure 2 shows the data on bitumen compositions in sedimentary sequences from various localities [10]. This diagram of Tissot et al. [10] is illustrative and informative in understanding petroleum formation in sedimentary rocks.

According to Chilingarian and Yen [2], the principal model of petroleum formation can be represented as in Figure 3.

### 8.2.4 Classification of Bituminous Materials

Bitumens extracted from different bituminous substances vary significantly in physical and chemical properties. Much confusion exists in the use and interpretation of the terms "bitumens," "asphaltic bitumen," and "native asphalts." Efforts have been made to develop a uniform nomenclature for bituminous materials [12]. Figure 4 shows a modified version of the classification scheme proposed by Abraham [11, 12].

The bitumens in Figure 4 can be separated into fractions on the basis of their solubility in various organic solvents. A complete fractionation scheme is presented in Figure 5 [11]: (1) the oily constituents are propane soluble, (2) the

**Table 4  Carbon distribution on the earth**

| Source | $C_{org}$(g/cm$^2$ of each surface) |
|---|---|
| Carbonates | 2340 |
| Shales and sandstones | 633 |
| Seawater (HCO$_3^-$ + CO$_3^{2-}$) | 7.5 |
| Coal and petroleum | 1.1 |
| Living matter and dissolved organic carbon | 0.6 |
| Atmosphere | 0.1 |

*Source:* Reference 8.

resins are *n*-pentane soluble but propane insoluble, (3) asphaltenes are benzene soluble but *n*-pentane insoluble, and (4) carbenes are carbon disulfide (CS$_2$) soluble but benzene insoluble. The sequential progression from oils to resins to asphaltenes to carbenes is related to increasing the molecular weight of the macromolecules.

Various analytical instruments have been used to characterize diverse forms of bitumens from different source rocks. Parameters of interest have been molecular weight, chemical structure, aromaticity, color intensity, hydrogen distribution, heteroatom sites, oxidation rate, refractive index, density, and elemental

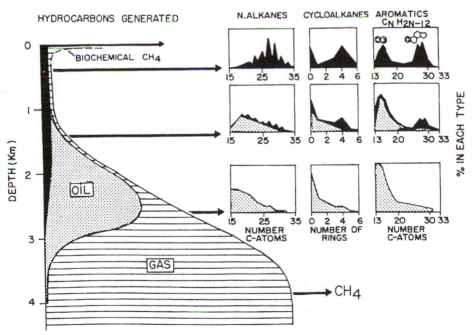

**Figure 2**  Generalized scheme of hydrocarbon generation. Depth scale represented is based on examples of mesozoic and paleozoic source rocks. (Source: Reference 10.)

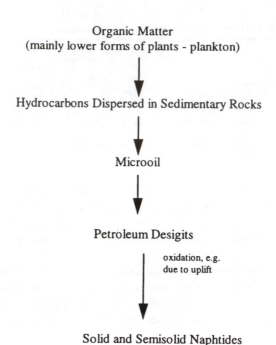

**Figure 3**   Petroleum formation model. (Source: Reference 2).

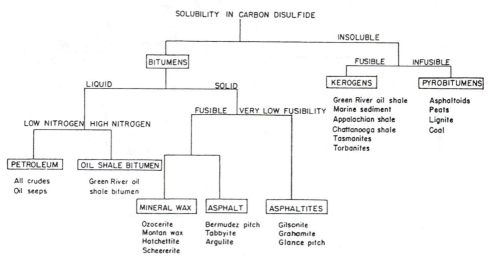

**Figure 4**   Terminology and classification of bitumens and related materials. (Source: References 11 and 12.)

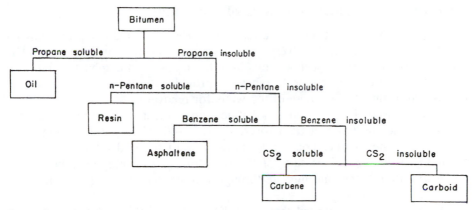

**Figure 5**   Fractionation scheme for bitumens. (Source: Reference 11.)

composition. Analytical methods frequently used include X-ray diffraction (XRD), electron spin resonance (ESR), nuclear magnetic resonance (NMR), Fourier transform infrared spectrometry (FTIR), gas chromatography–mass spectrometry (GC/MS), gel permeation chromatography (GPC), elemental analysis (CHNS), and vapor pressure osmometry (VPO).

## 8.3  TAR SAND PROCESSES

The existence of bituminous sands along the Athabasca River was first reported in 1788 by Peter Pond, a fur trader [13]. Little interest was shown until Ells of the Canadian Department of Mines and Technical Surveys commenced his work on mapping and exploration of the deposits in 1913. However, interest was still insignificant until the possibility of an oil shortage during World War II was threatened. Several pilot plants were constructed and the hot water separation process was demonstrated under Clark of the Research Council of Alberta [13]. Blair continued the research on the production of a synthetic crude from the Athabasca tar sands and concluded that the production is economically feasible [13].

The Alberta Oil and Gas Conservation Board estimated that the Athabasca deposit contains 626 billion barrels of oil in place [13]. Of this total, 250 to 400 billion barrels are considered recoverable raw tar sand oil, from which 180 to 300 billion barrels of upgraded synthetic crude are expected to be produced [13]. Considering the conventional petroleum reserves of the world, this amount is enormous, providing some optimism for the future liquid fuel supply.

In order to extract oil from tar sands, several principal routes are considered: in situ thermal cracking, ex situ thermal cracking, and hot water separation process.

### 8.3.1 In Situ Thermal Processing of Tar Sands

Elkins [14] reported that oil can be recovered from the Athabasca tar sands by in situ combustion, which is very similar to oil shale in situ processing. The process involves drilling several wells into the deposit and combusting bitumen in the presence of injected air. Combustion provides the necessary heat and oil that is driven into one or more of the wells for recovery.

Salomonsson [15, 13] proposed that heating elements be inserted to heat the formation and the bitumen be pyrolyzed at 250–400°C to produce lighter hydrocarbons. The hydrocarbon vapors move toward the collection zone, going through the denser, colder parts of the formation. Heavier hydrocarbons condense to a liquid state and migrate through a more permeable charred tar sand structure.

Watson [16, 13] reported that cracking of the Athabasca bitumen and recovery of a light oil could be achieved by injection of hot inert gases into the sand until a temperature of 370–650°C is attained. Carpenter [17, 13] claimed that a heat transfer medium could be used to heat the bitumen. He also claimed that the heat transfer medium contained in a wellbore in the formation could be heated electronically.

Davis [18, 13] reported that the economic feasibility of using electrical energy for devolatilization is not bright and that the process could be initiated using electrical energy to provide passages for compressed air injection. The thermal energy needed for subsequent stages would be provided via partial combustion of the bitumen. Natland [19] proposed in situ thermal treatment of the bitumen involving peaceful use of nuclear energy; a 9-kiloton nuclear device would be detonated at a depth of 1250 ft at a remote undeveloped area. Even though the process is innovative and has several merits, the use of nuclear energy to create a large cavity of some 70 m diameter has not been appealing to policymakers and citizens.

### 8.3.2 Ex Situ Thermal Processing of Tar Sands

Ex situ techniques, unlike in situ processes, involve mining of tar sands and sometimes transportation to processing sites. Because of the treatment above ground, there are more technological options than in situ processing. If the treatment method is the only matter of interest, practically all thermal and separative processes can be applied. However, the most important factor for choice of an ex situ process is the process economics.

**Fluidized bed technology.** In general, fluidized bed technology involves separation of the bitumen-sand mixture and simultaneous thermal cracking of bitumen. In this process, a tar sand bed is fluidized by a hot mixture containing about 85% nitrogen and light petroleum gas. Gishler [20] and co-workers published several articles on the fluidized bed treatment of tar sands. In their process, raw tar sands are fed into a coker maintained at approximately 500°C, where the tar sand is heated by contact with a fluidized bed of clean sand and volatile portions of the bitumen distill from the tar sand as a heavy synthetic crude oil [13]. Residual portions are converted to coke. Regeneration of clean sand is achieved in a second unit at approximately 760°C, where coke is combusted and hot sand is recirculated to the coker, the first unit. Reaction off-gases are recirculated to fluidize the clean sand in the coker.

Rammler [21] applied the Lurgi-Ruhrgas process to the production of synthetic crude oil from the Athabasca tar sands. It also involves a fluidized bed technique using fine-grained heat carriers. Tests using 45 tons of raw tar sand showed that the process is suitable for syncrude recovery from the tar sand.

Nathan and co-workers [22] claimed that the maximum recovery of a synthetic crude oil can be produced by mixing tar sand and fluid bed material at 120–300°C prior to introduction into a fluid-coking zone maintained at temperatures above 450°C.

Fluidized bed technology can be more readily applied to tar sands than oil shales, because tar sands are basically granular whereas oil shales are hard sedimentary rocks.

**Moving bed technology.** Bennett [23] claimed that the retorting system for tar sands can be refined by using a horizontally moving bed of the sands that are compacted and perforated before entry into the retort. This provides vertical flow channels allowing a controlled distribution of heat and burning within the retort. The process is claimed to produce excellent conversion of the bitumen to oil and allows efficient reuse of the heat generated during the burning.

In general, moving bed technology has definitive advantages in handling solids, compared with fixed bed technology.

## 8.3.3 Hot Water Separation

The hot water extraction process was first developed by Clark, who worked for the Research Council of Alberta [24]. The principle behind the process was a simple one: slurrying the tar sand with hot water, introducing this mixture into greater quantities of hot water, removing sand at the bottom, and skimming off the oil froth at the top. In many senses, it is similar to Ells' flotation cell technology [24].

Fitzsimmins was one of the first to try to put the hot water process into commercial use. He operated at Bitumount, about 50 miles downstream from Fort McMurray, in 1923. His product was used in western Canada for several years as a waterproofing material for roofs [24]. Abasand [25] built a 250 ton/day plant on the banks of the Horse River southwest of Fort McMurray in 1936. In 5 years, the plant capacity was increased to 400 ton/day. The plant burned to the ground in a fire. Bad luck stopped this valuable operation. Several pieces of choice land acquired by Abasand are currently being mined by Great Canadian Oil Sands (GCOS).

**GCOS process.** Great Canadian Oil Sands started construction of its 45,000 bbl synthetic crude plant in 1963 and started up the plant in September 1967. Figure 6 shows a four-stage process: mining, material handling, extraction, and upgrading. The product of this operation is a pipeline quality, low-sulfur, 38° API synthetic crude [24]. Spragins [24] pointed out that the final product from tar sand production is normally referred to as synthetic crude; however, this term is a misnomer, since it is by no means synthetic.

In the slurry drums, hot water and steam are combined with mechanical energy to transform the tar-sand ore into a frothy pulp. The main body of pulp is then pumped to extraction vessels filled with hot water. Oily froth on top is skimmed off. The middle stream is pumped to a secondary recovery cell where additional oil can be recovered. Makeup water is added to keep the cells in balance.

In the upgrading process, the diluted centrifuge product is heated, the diluent recovered, and the remaining oil passed to six delayed-coking drums, each 26 ft in diameter and 95 ft high. The overhead products from these drums are separated into three streams: naphtha, kerosene, and gas oil.

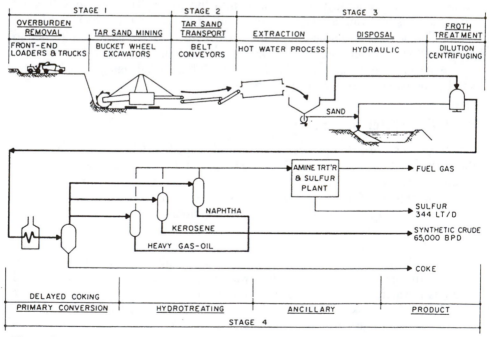

**Figure 6**  Schematic illustration of the GCOS plant. (Source: Reference 24.)

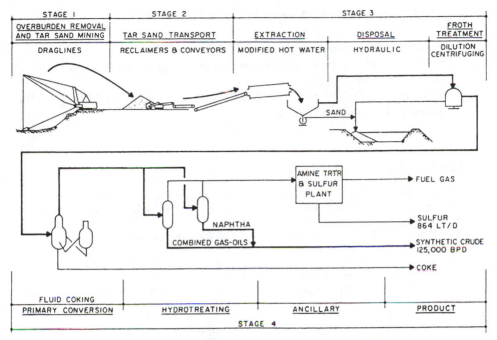

**Figure 7** Schematic illustration of the syncrude plant. (Source: Reference 24.)

**Syncrude project.** The syncrude plant, shown in Figure 7, has been designed to produce 125,000 bbl/day of synthetic crude oil. The process is similar to the GCOS project. The tar sands are pulped in tumblers with steam and hot water and placed in extraction vessels for primary recovery, followed by secondary recovery of oil from a middling stream; final cleanup of the oily froth takes place in a dilution centrifuge process [24].

The upgrading step is somewhat different from that in the GCOS project. Fluid coking is used in the primary conversion of the heavy oils, from which two overhead streams are taken off: naphtha and gas oil. The products are hydrotreated and blended, producing a low-sulfur synthetic crude oil of approximately 33° API.

## 8.4 OUTLOOK

Utilization of tar sands is a great idea for several reasons:

1. The oil reserves in tar sands are enormous.
2. Only minimal efforts are needed in mixing.
3. Because of their high permeability, in situ operation is facilitated more readily.
4. The extraction processes are simple but effective.

However, full utilization of tar sands has been hampered for several reasons:

1. The petroleum crude price has been stable.
2. Problems involving process water treatment and spent sand disposal are of potential environmental concern.
3. Unlike the case of oil shales, serious efforts have been made only on the Athabasca tar sands in Canada.

# REFERENCES

1. Matar, S. *Synfuels: Hydrocarbons of the Future.* Tulsa, OK: Penn Well Books, 1982.
2. Chilingarian, G. V., and T. F. Yen. Bitumens, Asphalts, and Tar Sands, Chapter 1. Amsterdam: Elsevier, 1978.
3. Conybeare, C. E. B. *Geomorphology of Oil and Gas Fields in Sandstone Bodies.* Developments in Petroleum Science, 4. Amsterdam: Elsevier, 1976.
4. Phizackerley, P. H., and L. O. Scott. Major tar sand deposits in the world. In *Bitumens, Asphalts, and Tar Sands*, eds. G.V. Chilingarian, and T. F. Yen. Amsterdam: Elsevier, 1978.
5. Doscher, T. M. An overview of the potential of tar sands. In *Bitumens, Asphalts, and Tar Sands*, eds. G. V. Chilingarian, and T. F. Yen. Amsterdam: Elsevier, 1978.
6. Considine, D. M. *Energy Technology Handbook.* New York: McGraw-Hill, 1977.
7. Mungen, R., and J. H. Nicholls. Recovery of oil from Athabasca oil sands and from heavy oil deposits of northern Alberta by in-situ methods. *Proc. 9th Word Pet. Congr. Panel Discussion* 22(2):11, 1975.
8. Skirrow, G. The dissolved gases—carbon dioxide. In *Chemical Oceanography*, eds. J. P. Riley, and G. Skirrow. New York: Academic Press, 1965.
9. Silverman, S. R. Geochemistry and origin of natural heavy-oil deposits. In *Bitumens, Asphalts, and Tar Sands*, eds. G. V. Chilingarian and T. F. Yen. Amsterdam: Elsevier, 1978.
10. Tissot, B., B. Curand, J. Espitalie, and A. Combaz. Influence of nature and diagenesis of organic matter in formation of petroleum. *Bull. Am Assoc. Pet. Geol.* 58(3):499–506, 1974.
11. Wen, C. S., G. V. Chilingarian, and T. F. Yen, Properties and structure of bitumens. In *Bitumens, Asphalts, and Tar Sands*, eds. G. V. Chilingarian, and T. F. Yen, Amsterdam: Elsevier, 1978.
12. Abraham, H. *Asphalts and Allied Substances*, vol. 1, 6th ed. Princeton, NJ: Van Nostrand, 1960.
13. Speight, J. G. Thermal cracking of Athabasca bitumen. In *Bitumens, Asphalts, and Tar Sands*, eds. G. V. Chilingarian, and T. F. Yen. Amsterdam, Elsevier, 1978.
14. Elkins, L. E. Oil production from bituminous sands. U.S. Patent 2,734,579, February 14, 1956.
15. Salomonsson, G. J. W. Underground gasification of precarbonized fuel deposits. U.S. Patent 2,914,309, November 24, 1959.
16. Watson, K. M. Oil recovery by subsurface thermal processing. Can. Patent 621,230, May 30, 1961.
17. Carpenter, C. A. Methods and apparatus for heating oil sands. U.S. Patent 3,113,622, December 10, 1963.
18. Davis, C. M. *Proc. Athabasca Oil Sands Conf.*, Research Council of Alberta, Edmonton, Alberta, p. 141, 1964.
19. Natland, M. L. Project oil sand. In *Athabasca Oil Sands*, ed. M. A. Carrigy, 45,143,155. Research Council of Alberta, 1963.
20. Gishler, P. E. The fluidization technique applied to direct distillation of oil from Alberta bituminous sands. *Can. J. Res.* F28:62–70, 1950.
21. Rammler, R. W. Production of synthetic crude oil from oil sand by application of the Lurgi Ruhrgas process. *Can. J. Chem. Eng.* 48:552–560, 1970.

22. Nathan, M. F., G. T. Skaperdas, and G. C. Grubb. Fluid coking of tar sands. U.S. Patent 3,320,152, May 16, 1967.
23. Bennett, J. D. Method of damping vibration and article. U.S. Patent 3,623,972, November 30, 1971.
24. Spragins, F. K. Athabasca tar sands: Occurrence and commercial projects. In *Bitumens, Asphalts, and Tar Sands*, eds. G. V. Chilingarian and T. F. Yen. Amsterdam: Elsevier, 1978.
25. Lynett, S. Digging for oil. *Imp. Oil Rev.* 56(4):21, 1973.

# PROBLEMS

1. Discuss the differences in geological formation between oil shale and tar sands.
2. Oil shales are found all over the entire world, whereas tar sands seem to be distributed in a more scattered manner. What is the reason behind this?
3. What is the major difference in properties between bitumen and pyrolytic bitumen?
4. Is hot water extraction effective on oil shale? If not, explain why not from the process engineering standpoint.
5. What are the alternative names for tar sands?
6. Tar sands are effectively processed in a fluidized bed mode, whereas oil shales are not. Why is this so?
7. How can you tell the difference between the bituminous sand and recently contaminated sand?
8. If you have to add some chemical additives to hot water extraction process, what do you recommend? Acidic, caustic, or interfacial agent?
9. If you had to carry out hot water extraction experiments on tar sands in your laboratory, what would your water temperature be? List your reasons for the choice.
10. For your laboratory experiment for hot water extraction, what would an ideal apparatus be?
11. If Athabasca tar sands are heated to 200°C, what kinds of changes in the bitumen properties would be expected?
    (a) Molecular weight
    (b) Viscosity
    (c) Density
    (d) Sulfur content
    (e) Aromaticity
    (f) Percent of heteroatoms
    (g) Moisture content

# UTILIZATION OF GEOTHERMAL ENERGY

## 9.1 GEOTHERMAL ENERGY

### 9.1.1 History of Geothermal Energy

Ancient people and even primitive people of today have regarded the depths of the earth with horror, as the seat of hell and of malignant gods, due to natural phenomena like earthquakes and volcanic eruptions—a result of the heat energy present in the earth's crust. Not so long ago, geothermal energy was regarded as merely an interesting freak of nature, a tourist attraction in the form of geysers, fumaroles, and pools of boiling mud. Its practical side was then more or less confined to its alleged curing properties for various human ailments and to the availability in somewhat remote places of "natural" hot baths. Those days have passed, and underground heat has become an undisputed commercial competitor with other forms of energy for many applications. Fortunately, we have come to view it as a seat of hope.

Although we are rapidly becoming familiar with the conditions of outer space, we have, until recently, been surprisingly ignorant of what is going on only a mile or two beneath our feet. As long ago as 1904, a British engineer, Sir Charles Parsons, sought to remedy some of this ignorance by advocating the sinking of a shaft 12 miles deep into the earth, where he expected to encounter temperatures of the order of 600°C. However, this proposal, referred to by its author as "The Hellfire Exploration Project" [1], was never implemented.

It is not surprising that humans have sought to make use of this heat. In ancient times the Romans and in modern times the Icelanders, Japanese, Turks, and others have used it for baths and for space heating. The Maoris in New Zealand have exploited natural heat for their domestic needs. One of the more interesting sights in their country is a Maori village near Rotura, in North Island, where one may see a fisherman catch a trout and drop it into a nearby pool of boiling water to cook it. A few yards away may be seen his wife administering a geothermal bath to the baby, while his daughter is doing the household laundry and the potatoes are cooking over a fumarole.

"The gates of hell," now known as The Geysers, were first acknowledged by William Bell Elliott in 1847. Elliott, while exploring and surveying the mountainous terrain of northern California, observed a valley of steam jets and discovered what is probably the largest reservoir of geothermal steam in the world [2].

The Larderello field in Tuscany, Italy first began to produce electricity in 1904 and developed over the next 10 years to a capacity of 250 kW. Pioneering in the geothermal production of electricity, the Italians had tried operating reciprocating engines with this steam in the 1890s and today are the world leaders with respect to geothermal energy utilization.

The United States has been rather slow to exploit its resources of geothermal energy. It was not until the early 1920s that the United States examined the possibilities for commercial usage of geothermal steam. Eight steam wells were drilled, but the competition from hydroelectric power was too keen to promote further development at that time.

In Japan in 1919, Beppu was the first site for experimental geothermal work and these experiments led to a pilot plant in 1924 producing 1 kW of electricity. Somewhat earlier than this the Japanese began to use geoheat to heat their greenhouses.

Iceland was also an early pioneer in geoheat. Municipal heating was effected by hot thermal waters in the 1930s and is still the major source of heating today.

## 9.1.2  What Is Geothermal Energy?

The natural heat of the earth is geothermal energy. It includes all the heat contained in about 260 billion cubic miles of rocks and metallic alloys at or near their melting temperatures, constituting the entire volume of earth except for a relatively thin, comparatively cool outer surface. As a supply of heat, it is second only to another natural source, solar energy. And like solar energy, geothermal energy is widely distributed over the earth's surface in amounts too small to be converted directly into an useful source of energy [3]. Available geothermal energy is concentrated in underground reservoirs, usually as steam, high-temperature water, or hot rock.

The more widely known and accepted forms of geothermal energy are geysers, boiling pools of mud, fumaroles, and hot springs. However, a greater potential exists in regions not yet valuable for their energy possibilities—hot dry rocks.

### 9.1.3 Need for Geothermal Energy

Technology has produced a standard of living for most of us that we expect and accept as our own. We fancy the ability to create light, build skyscrapers, land on the moon, and so on, using energy and resources.

New, clean energy sources have become necessary not only because of the depletion of fossil fuels but also because oil and coal consumption results in detrimental by-products.

The "new technologies" of energy production, which include geothermal energy, solar energy, crude oil from shale, and hydroelectric power, were predicted to meet only 3% of total energy needs in 1985. Although research and development on nuclear power were heavily financed by the U.S. government and capable of supplying 50% of U.S. electrical needs, nuclear did not prove to be an attractive source [4]. It had multiple problems with respect to plant safety and government regulations. In addition to the detrimental effects of radioactive wastes and thermal pollution, we should realize that the availability of nuclear fuel is not infinite.

The rate of consumption of oil by the world and especially countries like the United States, Russia, England, and Japan is alarming. The gap between the production and consumption rates goes on widening. In order to maintain the life-styles to which we have become accustomed, alternate means of energy production must be developed. In fact, the proven worldwide resources will be depleted within 50 years [5]. Also, these new energy sources must be environmentally sound to alleviate the problems associated with fossil fuels and their effect on society.

At a depth of about 6 miles from the surface, the temperature of the earth is greater than 100°C. Thus the total amount of geothermal energy in storage exceeds by several orders of magnitude the total thermal energy in all the nuclear and fossil fuel resources of this planet. Solar energy is the only comparable resource. Therefore, our energy priorities must be shifted to incorporate a resource such as geothermal energy.

### 9.1.4 Occurrence of Geothermal Energy

The occurrence of geothermal heat (also known as geoheat) can be explained by one of the following theories [4]:

1. The theory that 6 million years ago the earth was a hot molten mass of rock and this mass has been cooling through the epochs of time, with the outer crust forming as a result of faster cooling.

2. The theory that the earth is a giant furnace. The decay of radioactive materials within the earth provides a constant heat source.

Some combination of both theories is widely accepted.

The interior of the earth consists of a molten fluid of extremely high temperature called magma. This magma is cooling and/or expelling heat to the earth's surface according to the second law of thermodynamics. The flow of heat is from a hot sink (the earth's core) toward a cold sink. The cold sink consists of earth's crust, surface, and atmosphere. This is a very slow process.

**Geothermal energy sources: Characteristics and classification.** Figure 1 shows the geological setting of a geothermal energy source. The molten magma has been pushed near the earth's surface by the stress and strain force from within the planet. Heat moves by means of convection through the crystalline rock layer to the porous rock layer, which has been saturated with water that has descended through fissures in the earth's surface. The semipermeable rock traps the heat in the rock layers below it, isolating and collecting large quantities of heat.

As the water at this depth is subjected to high pressure and temperature, it remains liquid. It then rises to the surface through fissures as a result of its density change and, in effect, it vents the system, thereby reducing the pressure of the system. When the pressure decreases, the water is allowed to boil, turn into steam, and rise to the surface through fissures or wells.

The immense geothermal energy of the earth cannot be properly defined. Heat of the rocks above a depth of 10 km has been estimated to be $3 \times 10^{26}$ calories.

However, with current technology, only a fraction of this heat is available as a resource recoverable to be utilized. A mere 0.03% or $10^{23}$ calories of this energy is considered hot enough and near enough to the earth's surface to be recoverable; the rest of the energy is too dispersed throughout the crust. Geothermal resources can be classified as shown in Table 1 [6].

Although not all resources can be used to produce electricity, they are useful for many industrial, agricultural, and domestic needs. U.S. Geological Survey research found that full utilization of geothermal resources will be dependent on achieving the following [7]:

1. Advances in technology to make available a low-temperature reservoir for electrical generation.
2. Advance in drilling technology to make available geothermal reservoirs at depths greater than 3 km, which is at present the limit of geothermal drilling, as economics now dictates.

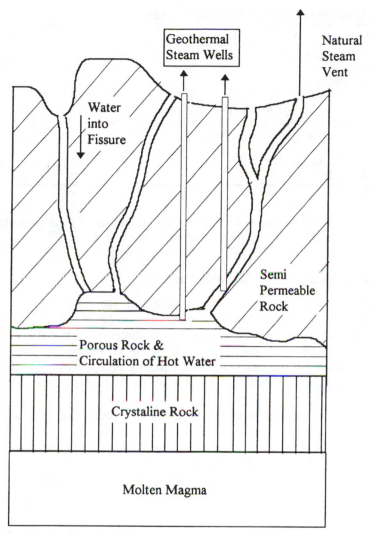

**Figure 1** Geological setting of a geothermal energy source.

3. Techniques of artificial simulation to increase the productivity of the existing geothermal reservoirs.
4. Expansion of low-grade and low-temperature geothermal resources to include space heating, product processing, agriculture, desalination, etc.

As a concluding remark, the potential of this resource is great, availability is great, and economics are competitive with or better than those of other sources

**Table 1 Classification of geothermal resources**

| Resource type | Temperature (~ °C) |
|---|---|
| Convective hydrothermal resource | |
|   Vapor dominated | 240 |
|   Hot water dominated | 30–350 |
| Other hydrothermal resources | |
|   Sedimentary basins/regional aquifers (hot fluid in sedimentary rocks) | 30–150 |
|   Geopressured (hot fluid under pressure that is greater than hydrostatic) | 90–200 |
|   Radiogenic (heat generated by radioactive decay) | 30–150 |
| Hot rock resources | |
|   Part still molten (magma) | >600 |
|   Solidified (hot, dry rock) | 90–650 |

of energy. Geothermalists look forward to geothermal energy contributing a great deal more to the total energy requirements of the world.

## 9.1.5 Advantages of Geothermal Energy

Geothermal resources are continuous sources of energy regardless of climate or weather conditions. This fact makes geothermal energy especially attractive as a source for base load electricity generation or direct-use applications requiring constant heat. Geothermal power plants compete economically with coal, oil, and nuclear plants in meeting base load capacity needs, with significant environmental advantages. Specific advantages of geothermal systems include:

1. Indigenous energy: Geothermal energy helps reduce dependence on imported oil, gas, coal, or nuclear fuels.
2. Clean energy: Use of geothermal energy helps reduce combustion-related emissions from conventional fuels.
3. Diversity of use: Geothermal energy has three common economic uses—electricity generation, the direct use of heat, and as a source of heat for ground-coupled heat pumps.
4. Long-term resource potential: With optimum development strategies, geothermal energy can provide a significant portion of a nation's long-term energy needs.
5. Flexible system sizing: Power generation projects range in capacity from a 200-kW system in China to 1200 MW at The Geysers in California.
6. Power plant longevity: Geothermal power plants are designed for a life span of 20 to 30 years. With proper resource management strategies that minimize reservoir depletion, life spans can exceed design periods.
7. High availability: Availability is defined as the percentage of time that a system is capable of producing electricity. Availability of 95 to 99% is

typical for modern geothermal plants, compared with 85% for coal and 80% for nuclear plants [8].

8. Combined use: Geothermal energy can be used simultaneously for power generation and direct-use applications. The combined use of geothermal energy results in higher thermal efficiencies and associated cost savings.
9. Low operating and maintenance costs: The annual operation and maintenance costs of a geothermal electric system are typically 5 to 8% of the capital cost.

## 9.2 STATISTICAL INFORMATION ON GEOTHERMAL ENERGY

### 9.2.1 Geothermal Energy (World)

Generating electric power from geothermal resources began in the early twentieth century. Italy became the pioneer of the international geothermal-electric community with the installation of the first dry steam geothermal power plant in 1913 at Larderello. Today the total installed capacity of geothermal power stations throughout the world is slightly over 6000 MW [9].

The United States remains the biggest producer of electricity from geothermal energy, as shown in Table 2. The table also presents projections for the year 2000 [10]. The developing countries represent 35% of the figures for 1995, but if the predictions are correct, their portion will increase to 48% in the year 2000.

Geothermal developments in other nations, not mentioned above, are as follows:

**Table 2  Worldwide geothermal capacity [11–13]**

| Country | Capacity installed in 1995 (MW) | Projection for 2000 (MW) |
|---|---|---|
| United States | 2913 | 3299 |
| Philippines | 894 | 2561 |
| Mexico | 700 | 870 |
| Italy | 545 | 907.6 |
| New Zealand | 295 | 285.2 |
| Japan | 270 | 552.5 |
| Indonesia | 142 | 859.5 |
| El Salvador | 95 | 155 |
| Nicaragua | 70 | 70 |
| Iceland | 45 | 49.9 |
| Kenya | 45 | 109 |
| Others[a] | 65 | 260.7 |
| Total | 6079 | 9979.7 |

[a]Others include China, Turkey, Russia, France, Portugal, and Greece.

1. Canada. The first Canadian geothermal electricity project in the Meager Creek area is expected to take approximately 5 years, with the first phase to be on line by 1995–1996.
2. Costa Rica. The first 55-MW power plant began operating at Miravalles in early 1994. A second 55-MW plant is planned for start-up in 1996.
3. Ethiopia. There are significant geothermal resources in the Ethopian rift valley and the Afar depression. A 5-MW power plant is expected in the rift valley at Aluto-Langano.
4. France. It is a significant user of geothermal heat for district heating. Other direct uses include greenhouses, aquaculture, and balneology.
5. Guatemala. It is favorably located with respect to geothermal potential because it is the site of a triple junction of crustal plates. There are plans to install 20 MW by the year 2000.
6. Hungary. Hungary's low-temperature resources are located mostly within the Pannonian Basin, where thermal waters with temperatures of 30°C and above are used for direct-use applications including production of drinking water.

### 9.2.2 U.S. Energy Sources

Fossil fuels—oil, coal, and gas—provide 85% of all the energy used in the United States. Renewable energy sources supply just 8%, and most of that comes from hydropower and the burning of biomass (wood, wood wastes, agricultural wastes, and municipal wastes). A mere 4% comes from geothermal sources (Figure 2).

**Geothermal energy (U.S.).** It is estimated that in the United States alone there is a potential for generating 23,000 MW of electric power from recoverable hydrothermal geothermal energy [14]. Undiscovered reserves may add significantly to this total.

The direct steam cycle is typical of power plants at The Geysers in northern California, the largest geothermal field in the world. In Nevada, electrical generation began at the Brady's Hot Spring geothermal field in 1993, where a large vegetable drying plant was operating successfully on geothermal energy for 10 years. The major geothermal fields along with their capacities are shown in Table 3 [6].

**Cost projection for geothermal energy.** As mentioned earlier, geothermal energy is second only to another natural source, solar energy. This makes it a highly attractive form of energy. Solar energy was the dream technology of environmentalists in the 1970s. But solar thermal has major drawbacks: it requires intense sunshine, available only in the southwest of this country, and it takes up large amounts of space. As a result, the natural heat trapped in rocks

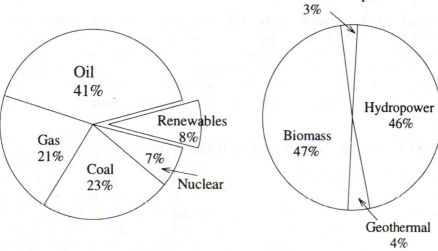

**Figure 2** U.S. energy sources, U.S. Department of Energy [13].

**Table 3 U.S. geothermal power plants [6]**

| Location | Capacity installed (MW) |
| --- | --- |
| The Geysers, CA | 2115 |
| East Mesa, CA | 119 |
| Salton Sea | 198 |
| Heber, CA | 94 |
| Mammoth, CA | 7 |
| Coso, CA | 225 |
| Amadee, CA | 2 |
| Wendel, CA | 0.6 |
| Puna, HI | 18 |
| Steamboat, NV | 31 |
| Beowave, NV | 17 |
| Brady, NV | 6 |
| Desert Peak, NV | 9 |
| Wabuska, NV | 1.2 |
| Soda Lake, NV | 3.6 |
| Stillwater, NV | 14 |
| Empire Farms, NV | 4.8 |
| Roosevelt, UT | 20 |
| Cove Fort, UT | 4.2 |
| Total | 2889.4 |

and fluids beneath the earth's surface has become a key energy resource in the United States. In addition, the projected cost of power generation from geothermal energy is lower than that from solar energy, as seen from the graphs in Figure 3 [12].

**Geothermal direct-use projects.** Although attractive in terms of fuel savings, widespread industrial use of geothermal energy has not occurred. Part of the difficulty is in situating commercial enterprises close to geothermal sites. Optimum industrial use would result if the geothermal resource occurred near or within a short distance of a suitable industrial process. Table 4 shows the direct use of geothermal heat in various projects in the United States [8].

## 9.3 SCIENTIFIC AND TECHNOLOGY UPDATES

### 9.3.1 Scientific Updates

The goal of the U.S. Department of Energy's Geothermal Program, as in the rest of the world, is to establish a scientific and technological base that assists in developing specific processes and products at competitive costs in response to increasing energy needs. This goal is being achieved in collaboration with the U.S. geothermal industry. The program strategy is to support research and development that will:

1. Help the geothermal industry meet critical near-term needs.
2. Provide technology base through longer range research for the continued growth of geothermal energy use.

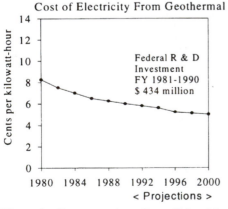
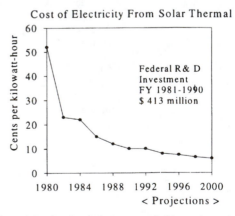

**Figure 3** Cost comparison. Cost of electricity from (a) solar thermal energy and (b) geothermal energy.

**Table 4 U.S. geothermal direct-use projects**

| Application | No. of sites | Thermal capacity (MW) | Annual energy (GWh) |
|---|---|---|---|
| Geothermal heat pumps | Most states | 2072 | 2402 |
| Space and district heating | 126 | 188 | 433 |
| Greenhouses | 39 | 66 | 166 |
| Aquaculture | 21 | 66 | 346 |
| Resorts/pools | 115 | 68 | 426 |
| Industrial processes | 13 | 43 | 216 |
| Totals | | 2503 | 3969 |

**U.S. government–funded research.** The following major activities are examples of U.S. government–funded research being conducted in accordance with the program strategy [8]:

1. Advanced techniques to detect and delineate hidden geothermal resources are being developed, including remote sensing techniques and improvements of various electric and acoustic methods.
2. Slim-hole drilling and coring, a cost-effective option for exploratory drilling, continues to be investigated. According to preliminary estimates, the cost of exploratory drilling can be reduced by as much as 50% if the acquisition of reliable resource data from slim holes proves to be technically feasible. This research includes developing slim-hole reservoir engineering techniques and logging tools.
3. Improved materials able to withstand the high temperature and corrosive nature of geothermal brines are being developed.
4. Methods for increasing the net brine effectiveness of geothermal power plants are being pursued, as are ways to reduce power plant costs.

**Research in national laboratories and schools.** Research activity going on in national laboratories and schools in the United States and certain parts of the world are cited below:

1. In New Mexico, scientists from Los Alamos National Laboratory are working on ways to heat water by pumping it through hot, dry rocks. The scheme is similar to the long-running project at the Camborne School of Mines in Cornwall, but the temperatures achieved in New Mexico are much higher [15–17].
2. Scientists at Sandia National Laboratories are moving ahead on an $8 million project to drill a 3.8-mile-deep well in the Long Valley Caldera, near Mammoth Lakes, California to evaluate the use of magma—molten subsurface rock—as a clean energy source [18].

3. A group of developing companies conducted a successful injection test at The Geysers geothermal field in California as part of their work to lessen the continuing pressure decline in the reservoir. Condenser water from nearby power plants was injected into the center of a low-pressure zone, and steam production from surrounding wells was monitored. The decline in production was arrested in the wells, and some wells showed an increase in steam flow. Another injection test is being planned in the same area [19].
4. Engineers at the Brookhaven National Laboratory, using geothermal funds, have developed a system for lining well casing and surface pipes with polymer cement to protect against corrosion. Brookhaven and Unocal have done experiments over the past 10 years at the Salton Sea Field in California using polymer-cement composites to protect pipes from corrosion by high-chloride brine [9].
5. Hot-rock researchers from the Camborne School of Mines drilled 2 km down in Camborne to reach 60°C rock, in a project sponsored by the U.K. Department of Energy [9].
6. The German Industrial HDR Group Erlangen, BRGM (Orleans, France), and Rio Tinto Zinc Consultants Ltd. (Bristol, U.K.) have formed European Hot Dry Rock Industries, a consortium to provide hot dry rock power for the European Community [9].
7. Another project of note is being conducted by Stanford University and Leningrad's Mining Institute. The institutions are collaborating on a project to create geyserless geothermal power. The researchers are also studying the feasibility of setting fire to underground coal seams or spent oil deposits. Using an igniting agent such as ammonium nitrate and adding water should produce boiler-quality steam, they believe [9].

## 9.3.2 Technology Updates

The development of a successful geothermal energy project relies on a variety of specialized technologies. Cost-effective use of each technology requires its own special expert scientists and engineers. The various technologies and innovations made in the past few years are mentioned below.

**Exploration technology.** Exploration is a key to the discovery of new geothermal resources. It identifies geothermal resources, estimates resource potential, and establishes resource size, depth, and potential production. It relies on surface measurements of subsurface geological, geochemical, and geophysical conditions to develop a conceptual model of the system. Geothermal exploration of unmapped regions typically proceeds in two basic phases: reconnaissance and detailed exploration.

*Reconnaissance, or delineation, of one or more geothermal provinces.* During this phase regional geology and fracture systems are studied. These include young volcanic fractures, tectonically active fault zones (as deduced from seismic information), and overt or subtle geothermal manifestations. Information is collected from various sources, including:

Regional geologic and geophysical maps
Satellite imagery
Geochemical sampling of thermal and nonthermal waters and gases
Analysis of surface rocks and soils
Aerial photography
Measurements of thermal gradients in existing wells

*Detailed exploration for exploitable reservoirs within the provinces.* If the reconnaissance phase confirms that the province has geothermal potential and that specific sites in the province should be explored further, the second phase, detailed exploration, focuses on one or more of the individual prospects mentioned above. Important technology innovations in this area in the past few years include [13] improved algorithms for interpreting data from geophysical surveys, especially seismic and self-potential surveys, and emphases on slim-hole drilling as a less expensive method for drilling discovery and resource confirmation wells.

**Well drilling and testing technology.** Wells are used to measure subsurface temperatures and flow rates, measure other subsurface conditions, and produce and reinject the geothermal fluid. After each well is completed, production or injection tests are run. Geothermal wells producing fluids for low-temperature, direct-use applications are usually less than 300 m deep and can be drilled using conventional water-well rigs. In contrast, power generation systems require deep production and injection wells to produce and dispose of the geothermal fluids. The predominant method for drilling these deep wells is rotary drilling, adapted from the oil and gas industry.

Important technology innovations in this area in the past few years include [8]:

1. Improved drilling bits for medium and medium-hard rock
2. Emphasis on development of methods based on fast-acting cements to handle episodes of lost circulation
3. Improved high-temperature cements for casings, especially carbon dioxide–resistant cements

**Reservoir engineering and development.** Reservoir engineering uses information gathered from subsurface measurements and well testing to generate and refine a model of how the reservoir works. These models are used to optimize energy extraction and maximize the economic lifetime of the resource. Major design consideration determined by reservoir engineers include the locations, depths, flow rates, configurations, and numbers of production and injection wells. Important technology innovations in this area in the past few years include [8] improved methods for conducting and analyzing injection tracer tests and improved numerical simulation codes for fractured reservoirs.

**Electric power plants.** Electricity generation using geothermal fluids is a well-developed technology. Selection of specific power conversion processes depends on the nature of the resource. Power extraction from steam-dominated resources uses dry-steam plants. Flashed-steam or binary plants are used for liquid-dominated resources. The resource temperature dictates the conversion process to be used and is also a significant driver in the economics. Higher temperature resources produce lower cost electricity. Important technology innovations in this area in the past few years include [8]:

1. Modest cost reduction in both flash-cycle and binary-cycle power plants
2. Research on ammonia-based cycles to further reduce binary-cycle plant costs
3. Conversion of upstream energy through radial-flow or rotary separation turbine-generators

**Brine handling technology.** The chemical composition of geothermal fluids varies greatly from one reservoir to another. Similarly, the salinity of geothermal fluids varies greatly, ranging from nearly potable geothermal waters to highly saline brines. Variations in chemistry and salinity affect the design, maintenance, and longevity of wells and surface equipment. Special materials and process treatments are used to overcome such problems. Important technology innovations in this area in the past few years include [8]:

1. Use of scale-inhibiting chemicals to reduce carbonate scaling of flashing wells
2. Development of acidization approaches ("pH modification") to control silica scaling in power plants
3. Development of highly accurate computer codes to estimate and predict chemistry effects in geothermal systems
4. Continued development of polymer cement coating to reduce corrosion in heat exchangers and process piping

**Environmental technology.** Geothermal energy is one of the cleanest and safest means of generating electric power. It uses a natural heat source, requires no combustion, and causes minimal disruption to the environment. However, effects on water resources, air quality, and noise during geothermal development and operation must be understood and mitigated. Among these are emissions to air (particularly of hydrogen sulfide), land use, and disposal of solid wastes. Effects can vary greatly from site to site. U.S. development of appropriate environmental control technologies has facilitated sound geothermal projects in areas with strict environmental regulations.

**Development in direct-use systems.** Historically, geothermal resources have been used for various nonelectric, direct uses, including bathing, cooking, heating, and recreation. Today, low- and moderate-temperature resources ($<150°C$) are used on a much larger scale for applications including district heating and cooling, light industrial and agricultural use, and aquaculture. Most direct-use applications employ shallow geothermal waters with low heat contents and operate on smaller fluid volumes than for electric power generation. Cooler and shallower geothermal waters are also less mineralized, resulting in fewer chemistry-related problems. The reliability, economics, and environmental acceptability of direct-use systems have been demonstrated throughout the world. Conventional fuel price increases over the years have further helped direct-use applications become more cost competitive and attractive.

Important technology innovations in this area in the past few years include [8]:

1. Use of lower cost plastic and concrete pipes for some portion of the system
2. Application of geothermal fluids in heap leaching of precious metals
3. Modest increase in the overall cost effectiveness of geothermal heat pumps
4. Innovative approaches to the installation of the ground-loop part of geothermal heat pump systems

# 9.4  PAST, CURRENT, AND FUTURE

## 9.4.1  Current

Growing concerns in many developed countries about the global effects of increasing $CO_2$ and methane in the atmosphere and the concerns of oil-producing countries that they must displace the growing consumption of oil to sustain exports are working to enhance the role of geothermal resources worldwide. Hence the utilization of geothermal energy to generate electric power dominates all other applications.

The slow economy and low price of natural gas continued to restrain geothermal energy developments, not only in the United States but also throughout the world. Although some significant events in geothermal resource development took place in the United States, most of the geothermal activities occurred in other parts of the world until around 1993. The government is encouraging geothermal exploration and power generation.

Two legislative actions proved favorable for geothermal energy development in the United States [10]:

1. The Energy Policy Act of 1992 made permanent the investment tax credit for geothermal installations and authorized an incentive payment of $0.015 per kilowatt-hour to geothermal operators for electricity produced from fields other than dry-steam reservoirs.
2. The National Geologic Mapping Act of 1992 authorized a major increase (from $37.5 million to $55.5 million) in funding over the next 4 years to expand the availability of geologic maps, which have added to the discovery of new geothermal fields.

In addition, some other factors boosted the production of geothermal energy [19]:

1. The economics of geothermal energy became more favorable because of the increase in petroleum prices.
2. The implementation of the Clean Air Act Amendments of 1990 also provided an economic benefit because of the well-developed technology for control of gas emissions from geothermal power plants.
3. Amendment of the Public Utilities Regulatory Act removed the 80-MW limit from independent power plants selling electricity to utilities and is expected to help competitiveness of geothermal energy.

These moves by the government and the U.S. Department of Energy resulted in the United States becoming the highest generator of geothermal power. At present, it is estimated that the United States has installed capacity of slightly less than 3000 MW, which is almost 50% of the world capacity [8].

## 9.4.2 Future

The problems associated with energy needs are extremely complex. It has, however, become apparent that the world must develop all the energy supplies possible with present and future technologies. Among the sources of energy that are likely candidates for development in substantial quantities is geothermal energy. Power generation from hot, dry rock seems to be more promising than power generation from hydrothermal sources because of its massive quantity. However, geothermalists and engineers are exploiting both sources.

**Future of hydrothermal power.** Geothermal power plants have historically been designed without consideration of potential resource variations or changes. The next generation of geothermal power plants will be designed using long-term projections for resource production as the basis for cycle selection, optimization, and system design. Equipment and system design will be capable of flexible operation to respond, with minimal modification, to anticipated resource changes over the life of the facility.

**Future of hot dry rock systems.** Current hot, dry rock (HDR) technology seems to be competitive with modern coal-fired plants in regions with geothermal gradients exceeding 60°C/km [20]. Reasonable improvements in reservoir performance or reductions in drilling and completion costs may substantially lower the cost of HDR power. In areas with steep geothermal gradients, the use of HDR may develop a substantial cost advantage over coal. This advantage may increase over time, allowing the use of HDR to produce a significant portion of the future electricity of the world.

Many of the known resources can be developed with today's technology to generate electric power and for various direct uses including greenhouses or heating of buildings. For other reserves, technical breakthroughs are necessary before this naturally occurring energy source can be developed.

# 9.5 PROCESSES AND APPLICATIONS

## 9.5.1 Processes

The utilization of geothermal resource is unique compared with other energy sources, and many difficulties are associated with it. Each geothermal reservoir is different from the next with respect to geological formations; thickness; hardness; permeability of the bedrock; chemical, mineral, dissolved solids, and gaseous content of the subsurface fluids; temperatures; pressures; hot water; wet steam; and dry steam.

The common denominator for each geothermal reservoir is heat. Heat is supplied in different quantities and different forms. Whatever the form, the heat dissipates quickly upon arrival at the surface, and therefore hot fluids of a geothermal nature cannot be conveyed across the earth's surface for long distances without great energy losses and lower heat qualities. Depending on the application of the heat source, these transportation heat losses are magnified many times. Furthermore, piping and insulation costs have become unbearable. Another problem is that the known geothermal reserves of the world and especially the western United States are not readily available to the majority of population centers. To reach these metropolitan areas, the heat energy of the geothermal well is converted to electricity, which travels along transmission lines to its destination.

**Geothermal power plants.** As seen earlier, geothermal resources may be described as hydrothermal, hot dry rock, or geopressured. Hydrothermal resources contain hot water, steam, or a mixture of water and steam. Although research and development continue to look for ways to efficiently extract and use the energy contained in hot, dry rock and geopressured resources, virtually all current geothermal power plants operate on hydrothermal resources.

The characteristics of the hydrothermal resource determine the power cycle of the geothermal power plant. A resource that produces dry steam uses a direct steam cycle. A power plant for a liquid-dominated resource with a temperature above 330°F typically uses a flash steam cycle. For liquid-dominated resources with temperatures below 330°F, a binary cycle is the best choice for power generation. Power plants on liquid-dominated resources often benefit from combined cycles, using both flash and binary energy conversion cycles.

*Direct steam cycle* [11]. Figure 4 shows a direct steam cycle. The cycle diagram shows the major components that are used to generate electric power and are influenced by changing resource conditions.

In a direct steam cycle power plant, a geothermal turbine can operate with steam that is far from pure. Chemicals and compounds in solid, liquid, and gaseous phases are transported with the steam to the power plant. At the power plant the steam passes through a separator that removes water droplets and particles before it is delivered to the steam turbine. The turbines are of conventional design with special materials, such as 12Cr steel and precipitation-hardened stainless steel, to improve reliability in geothermal services.

The other components present in a cycle include:

- A condenser, used to condense turbine exhaust steam. Both direct contact and surface condensers have been used in direct steam geothermal power plants.
- Noncondensable gas removal system, to remove and compress the noncondensable gases. A typical system uses two stages of compression. The first stage is a steam jet ejector. The second stage is another steam jet ejector, a liquid ring vacuum pump, or a centrifugal compressor.
- Cooling tower. The cooling tower is a multicell wet mechanical draft design. Cooling is accomplished primarily by evaporation. Water that is lost from the cooling system to evaporation and drift is replaced by steam condensate from the condenser.

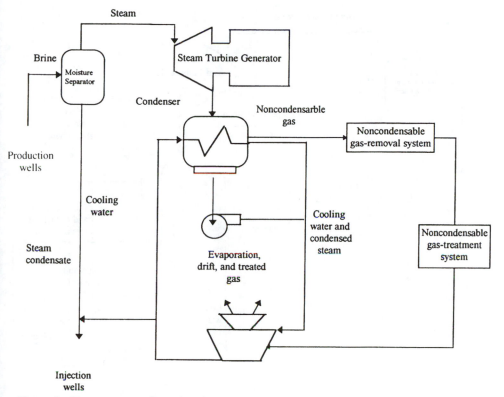

**Figure 4** Direct steam geothermal cycle.

• An injection well, through which excess water is returned to the geothermal resource.

The direct steam cycle is typical of power plants at The Geysers in northern California, the largest geothermal field in the world. The primary operator of The Geysers is Pacific Gas and Electric.

*Flash steam cycle* [11]. A flash steam cycle for a high-temperature liquid-dominated resource is shown in Figure 5. This dual-flash cycle is typical of most larger flash steam geothermal power plants. Single-flash cycles are frequently selected for smaller facilities.

Geothermal brine or a mixture of brine and steam is delivered to a flash vessel at the power plant by either natural circulation or pumps in the production wells. At the entrance to the flash vessel, the pressure is reduced to produce flash steam. The steam is delivered to the high-pressure inlet to the turbine. The

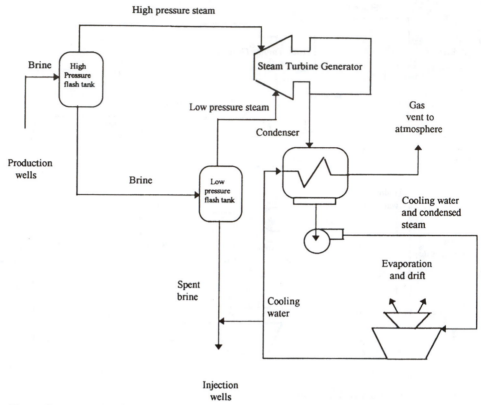

**Figure 5** Double-flash steam geothermal cycle.

remaining brine drains to another flash vessel, where the pressure is again reduced to produce low-pressure flash steam.

The other components present in cycle include:

1. Direct contact condenser as hydrogen sulfide is not produced in large quantities.
2. A cooling tower. This cycle too uses a multicell wet mechanical draft cooling tower. The water lost to evaporation and drift is replaced by steam condensate.
3. The excess water and spent brine from the flash vessels are injected back into the geothermal resource in an injection well.

***Binary cycle*** [9]. Binary geothermal plants have been in service for less than 10 years. A binary cycle is the economic choice for hydrothermal resources with temperatures below approximately 330°F. A binary cycle uses a secondary heat transfer fluid instead of steam in the power generation equipment. A typical binary cycle is shown in Figure 6.

The cycle shown in the figure uses isobutane as the binary heat transfer fluid. Heat from geothermal brine vaporizes the binary fluid in the brine heat exchanger. The spent brine is returned to the resource in injection wells, and the binary fluid vapor drives a turbine generator. The turbine exhaust vapor is delivered to an air-cooled condenser, where the vapor is condensed. Liquid binary fluid drains to an accumulator vessel before being pumped back to the brine heat exchangers to repeat the cycle. The brine heat exchangers are typically shell-and-tube units fabricated from carbon steel.

***Hot dry rock (dry geothermal sources) systems*** [4, 20, 21]. Since the vast majority of the geothermal heat resources of the world are located in dry, hot rock (HDR) sources rather than the water system, it is only natural for this energy source to receive more attention from geothermalists. The more accessible HDR resources in the United States alone would provide an estimated 650,000 quads

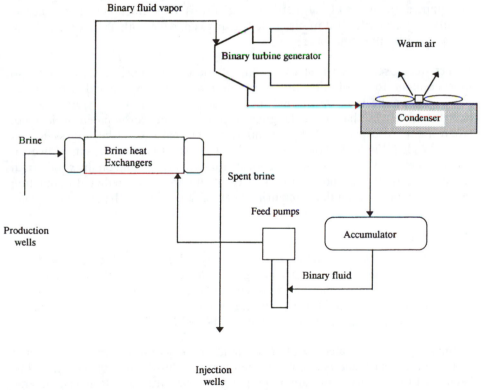

**Figure 6**   Binary geothermal cycle.

of heat, one quad being the amount of energy contained in 171.5 million barrels of oil. As the U.S. consumption is 84 quads, whoever figures out how to economically tap even a fraction of the potential in HDR could earn a place in history.

Hot dry rock is a deeply buried crystal rock at a usefully high temperature. Current engineering designs plan to tap its heat by drilling a wellbore, fracturing or stimulating preexisting joints around the wellbore, and directionally drilling another wellbore through the fracture network. Cold water then flows down one wellbore, pushes through the fractured rock, warms, returns up the other wellbore, and drives a power plant. The major technical uncertainty is establishing the fracture network between the two wellbores. If adequate connectivity can be established and a sufficiently large fracture surface area can be exposed between the two wellbores, HDR can be a competitive source of energy.

Figure 7 is a schematic diagram of the experimental Los Alamos system in New Mexico. Water at 65°C and 1000 psia is pumped into a hydraulic fracture network approximately 3000 ft in diameter and circulated at 7500 ft, where temperatures range between 260 and 320°C. The pressure is around 4100 psia. The water is then pumped out of the ground, and when it reaches the surface its temperature is 230°C at 1250 psia. For this experiment the hot water is circulated through a 100-m air-cooled heat exchanger with the extracted heat dissipated into the atmosphere.

**Direct heat use.** Direct heat use is one of the oldest, most versatile, and also most common forms of utilization of geothermal energy. Space and district heating is one of the best known and most widespread form of utilization. Space and district heating has made the greatest progress and development in Iceland, where the total capacity of the operating geothermal district heating system is 800 MW [12]. Figure 8 shows an example of a district heating system. However, each system has to be adapted to the local situation, depending on the geothermal resource available, the population density of the area and predicted population growth, the types of buildings requiring heating, and, above all, the local climate.

### 9.5.2 Applications

Geothermal energy is a highly versatile entity whose uses through direct and indirect means and their corresponding by-products are countless. In fact, the vast applications of geothermal resource are bounded in many respects by the conceptual limitations of humans. The various applications of geothermal energy in addition to generating electricity are mentioned below.

**Balneology.** Water heated by the hot magma rock within the earth rises to the surface of the earth and is known as a hot spring. This thermal phenomenon has been used for centuries for bathing purposes by the Etruscans, Romans, Greeks, Turks, Mexicans, Japanese, and no doubt others.

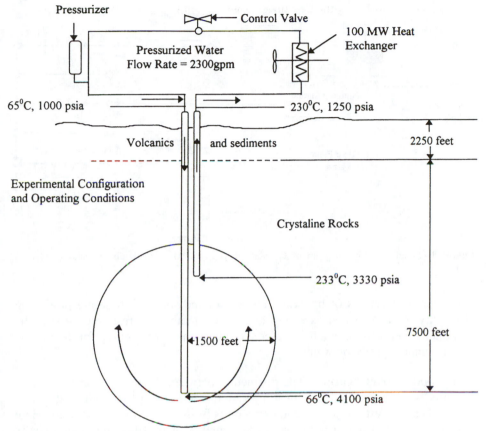

**Figure 7**   Schematic diagram of experimental work at Los Alamos.

The hot spring water not only is soothing to tired limbs but also is alleged to possess healing and prophylactic properties when applied externally, as in bathing, or internally when taken orally or used for douches. Thus was born the balneology industry—the oldest application of earth's heat.

**Fresh water.** Water as available for drinking is estimated as less than 2% of the earth's supply. The oceans, atmosphere, and rocks or rock formations (and polluted resources) contain the remaining 98%. This supply remains constant, although the consumption goes on increasing in proportion to world population growth and also advance in standards of living. From the standpoint of water shortage, all the systems recognized to date (desalination, recycling, and/or transportation over long distances) consume enormous amounts of energy and have also proved to be uneconomical. Geothermal resources, on the other hand,

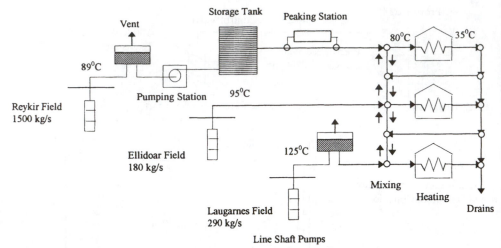

**Figure 8** Simplified flow diagram of Hitaveita Reykjavikur (Reykjavik district heating system) [25].

contain vast reservoirs of hot water and steam and some of these are producing electricity and fresh water as a by-product. The geothermal resource satisfies the two main criteria for alleviating water shortages: an energy source for distillation and an ample supply of water.

**Space and district heating.** The distinction between district and space heating systems is merely that space heating systems serve only one building, while district heating systems serve many structures from a common set of wells. Here it should be noted that a high load density makes district heating economically feasible, because the cost of the distribution network transporting hot water to consumers is shared.

Geothermal district heating pumps are capital intensive in the early stages. The principal costs are the initial investments for production and injection wells, downhole and circulation pumps, heat exchangers, and pipelines, as well as the distribution network. However, operating costs are comparatively low. Geothermal district heating systems offer significant life-cycle cost savings to consumers, as much as 30 to 50% of the cost of using natural gas or oil.

At present, a very successful district heating system exists in San Bernardino, California. The water production system, consisting of two wells, yields an average flow of 5200 L/min at 54°C water. The system currently serves 33 buildings including government centers, a prison, a new blood bank facility, and other private buildings.

**Industrial process heat.** Industrial processes can be heat intensive and commonly use either steam or superheated water with temperatures of 150°C or above. This makes industrial processes the highest temperature users of geothermal direct-heat applications. However, lower temperatures can suffice in some cases, especially for some drying applications.

Two of the largest industrial users of geothermal heat are a diatomaceous earth drying plant in Iceland and a paper and pulp processing plant in New Zealand.

**Greenhouses.** Greenhouse heating is one of the most common worldwide applications of geothermal energy. Fruits, vegetables, flowers, and ornamental plants are successfully grown year-round in geothermally heated greenhouses using a low temperature (<38°C). Even in the coldest weather, the temperatures necessary for optimal plant growth remain stable. Geothermal energy can extend short growing seasons and significantly reduce fuel costs.

In California, a 650-m$^2$ greenhouse has been built as part of an agricultural park in Lake County. A geothermal well 150 m deep supplies 67°C water. The well is still capable of supplying heat for an additional 1800–3700 m$^2$ of greenhouses.

**Aquaculture.** Aquaculture is the raising of freshwater or marine organisms in a controlled environment. Geothermally heated water produces excellent yields of high-quality fish and crustaceans under accelerated growth conditions. Further geothermal aquaculture permits breeding in the winter, allowing fish farmers to harvest their products when product availability is low and market prices are high.

California has six geothermal aquaculture operations. The largest, the Hot Creek Hatchery near Mammoth, uses water from four springs with temperatures of 11 to 16°C.

**Geothermal heat pumps.** The geothermal heat pump (GHP) uses earth as a heat source for heating or as a sink for cooling. A water and antifreeze mixture circulates through a pipe buried in the ground and transfers thermal energy to a heat exchanger in the heat pump. The heat exchanger works through a water-to-refrigerant loop. In a typical reversible heat pump, the ground loop heat exchanger rejects heat from the condenser or delivers heat to the evaporator, depending on the mode of operation [22–25].

Geothermal heat pumps offer a distinct advantage over the use of air as a source or sink because the ground is at a more favorable temperature. Compared to air, the ground is warmer in winter and cooler in summer. Therefore, GHPs demonstrate better performance than airsource heat pumps. GHPs also reduce electricity consumption by approximately 30% compared with air source heat pumps. Aided by utility-sponsored campaigns, GHPs are becoming increasingly popular throughout the world. The U.S. GHP industry is expanding at a rate of 10 to 20% annually.

The industrial and other potentials of geothermal energy suggest that great economic advantages could be gained from dual or multipurpose plants combining power production with one or more other applications. Such plants would enable the costs of exploration drilling and certain other items to be shared between two or more end products.

Although power may remain the dominant reason for exploiting geothermal fields, it is important that planners should not be blind to these other wide possibilities.

## REFERENCES

1. Christopher, H., and H. Armsted. *Geothermal Energy.* p. 11. New York: Wiley, 1978.
2. Hadfield, P. *New Scientist* February, 17:58, 1990.
3. Geothermal Resource Group. *Geothermal Resource and Technology in the United States,* 1, Washington, DC: National Academy of Sciences, 1979.
4. Chermisinoff, P. N., and A. C. Morresi. *Geothermal Energy Technology Assessment.* Westport, CT: Technomic Publishing, 1970.
5. Around the world, alternative energy is sought. *New York Times*, p. 23. January 26, 1975.
6. Wright, P. M. *Am. Asso. Pet. Geol. Bull.* 23(12):366, 1989.
7. Griffin, R. D. *CQ Researcher* 2(25):575–588, 1992.
8. U. S. Department of Energy. United States Geothermal Energy—Equipment and Services for Worldwide Application. DOE/EE-0044, 1994.
9. Lynn, M., and M. J. Reed, *Energy Sources.* 14:443, 1992.
10. Reed, M. J. *Geotimes* 38:12, 1993.
11. Phair, K. A. *Mech. Eng.* September: 76, 1994.
12. Dickson, M. H., and M. Fanelli. *Energy Sources* 6:349, 1994.
13. Brown, P. R. L. *Geotimes* 40(2):35, 1995.
14. U. S. Department of Energy. Geothermal Energy: 1992 Program Overview, 1992.
15. Joyce, C. *New Scientist* February 4:58, 1989.
16. Anderson, I. *New Scientist,* 111:22, July 24, 1986.
17. Feature article. *Mech. Eng.,* 113:10, January 1991.
18. Feature article. *Power Eng.* 93:50, October 1989.
19. Reed, M. J. *Geotimes* 36(2):16, 1991.
20. Harden, J. *Energy* 17(8):777, 1992.
21. Tenenbaum, D. *Technol. Rev.* 98:38, January 1995.
22. Milora, S. L., and J. W. Tester. *Geothermal Energy as a Source of Electric Power— Thermodynamic and Economic Design Criteria.* Cambridge, MA: The MIT Press, 1976.
23. Berman, E. R. *Geothermal Energy.* Park Ridge, NJ: Noyes Data Corporation, 1975.
24. U.S. Department of Energy. Geothermal Progress Monitor, DOE/EE - 0043, 1994.
25. Gudmundsson, J. S. Direct uses of geothermal energy in 1984. *Int. Symposium of geothermal energy.* Geothermal Resources Council, pp. 19–29, 1984.

## PROBLEMS

1. A small town of 500 households is considering installing a geothermal electric power plant to supply all the electrical energy needs of the town. Based on the following predesign information and data, estimate the required capacity of the power plant in megawatts.

(a) The average monthly electricity usage per household is 260 kWh.
(b) The peak-time surge usage is 70% higher than the year-round average.
(c) The efficiency of conversion of geothermal energy is 92%.
(d) The town is expected to grow in population to 600 households in 20 years.
(e) The design safety factor (or overdesign flexibility, design contingency) is taken as 25%.

In drawing your conclusion, you may have to make several assumptions.

2. Explain the second law of thermodynamics in relation to the conversion of geothermal energy to electrical energy.
3. Which has a higher energy content per kilogram: (a) hot dry rock at 600°C and (b) supersaturated steam ($T_{superheat}$ = 100°C) at 600°C?
4. Heat of the rocks above a depth of 10 km for the entire earth has been estimated to be $3 \times 10^{26}$ calories. Verify this value by showing all computations and assumptions made.
5. Discuss the environmental benefits of geothermal energy utilization from standpoints of air pollution, water pollution, and soil contamination.
6. If we use the geothermal energy very heavily, is it going to cool the earth? Is it going to be a major concern, something similar to global warming?

# BIOMASS CONVERSION

## 10.1 INTRODUCTION

Energy conservation is one of the major means of reducing fossil fuel emissions [1]. One aspect of reducing the emissions is increased usage of biomass as an alternative energy source. Biomass is defined as different materials of biological origin that can be used as a primary source of energy [2]. By this definition, energy from biomass has existed for centuries. The burning, or incineration, of wastes such as wood products was first used to produce warmth. It has been estimated that, in the late 1700s, approximately two-thirds of the volume of wood removed from the American forest was for energy [3]. As wood is one of the only renewable energy sources, its use continued to grow. During the 1800s, single households consumed 70 to 145 m³ of wood annually for heating and cooking [4, 5]. A small percentage of the rural communities in the United States still use biomass for these purposes. Other developed countries such as Finland use the direct combustion of wood for a percentage of their total energy use [6]. Finland and the United States are not the only countries that use biomass to supplement their total energy usage. Table 1 shows the percentages of biomass consumption by several "developed" countries [7].

On a larger scale, biomass is currently the primary fuel in the residential sector in almost every developing country. For instance, it accounts for over 90% of total household use in the poorer countries of Africa and Central Amer-

**Table 1  Biomass consumption in developed countries**

| Country | Energy from biomass (%) |
|---------|-------------------------|
| Austria | 4.0 |
| Belgium | 0.2 |
| Canada | 3.0 |
| Denmark | 1.0 |
| Ireland | 13.0 |
| New Zealand | 0.4 |
| Norway | 4.0 |
| Sweden | 13.0 |
| Switzerland | 1.6 |

ica [8] and 35% in Latin America and Asia [9]. The biomass resource may be in the form of wood, charcoal, crop waste, or animal waste. In these countries its most critical function is for cooking, with the other principal uses being lighting and heating. Although direct combustion of charcoal and animal waste is used extensively in developing countries, this usage of biomass is not considered suitable or efficient for direct energy applications [10]. Three broad categories of biomass feedstocks have been deemed suitable for energy production applications:

1. Vegetable oils
2. Pure carbohydrates, such as sugar and starch
3. Heterogeneous "woody" materials, collectively termed lignocellulose

The annual world production of biomass for these three categories is estimated as 146 billion metric tons. Approximately 80% of this amount is attributed to uncontrolled plant growth. Trees and farm crop wastes can produce 10–20 tons of dry biomass per acre per year. Certain genera of algae and grass can produce up to 50 metric tons of biomass a year, whose heating value is 5000–8000 Btu/lb [11]. Compared to coal, fuel from biomass has essentially no sulfur (0.1–0.2%) or ash content [12, 13]. In addition, it does not add any significant net $CO_2$ to the atmosphere [14].

Meeting U.S. demands for oil and gas by the direct combustion of lignocellulose materials would require that 6–8% of the land area of the 48 states be cultivated solely for biomass production [15]. Although the direct combustion of biomass is not an efficient or economical alternative, the conversion of biomass feedstocks into a gaseous or liquid fuel is feasible. Biomass could be used to replace up to 50–60 million metric tons of petroleum and natural gas currently consumed in the manufacture of primary chemicals in the United States annually. Among the sources of biomass that could be used for chemical production are grains and sugar crops for ethanol manufacture, oil seeds, animal by-products, and manure and sewage for methane generation.

Regardless of whether the biomass feedstock is lignocellulose or crop waste, several factors must be addressed when considering a large-scale biomass program [16]:

1. Short-term and long-term land availability
2. Productivities, species, mixtures
3. Environmental sustainability
4. Social factors
5. Economic feasibility
6. Ancillary benefits
7. Disadvantages and perceived problems

Although item 1 applies primarily to lignocellulose materials, all of the remaining factors apply to each of the biomass feedstocks. Many of the disadvantages and perceived problems diminish if biomass energy is viewed as a long-term entrepreneurial opportunity.

Each of the aforementioned biomass feedstocks can be converted into a viable fuel by three primary routes. As indicated in Figure 1, these routes are categorized as thermal, biological, and extractive. Extraction processes, which supply oils for food and industrial uses, have been in commercial use for over 25 years. Therefore, extraction processes will not be addressed at this time. The

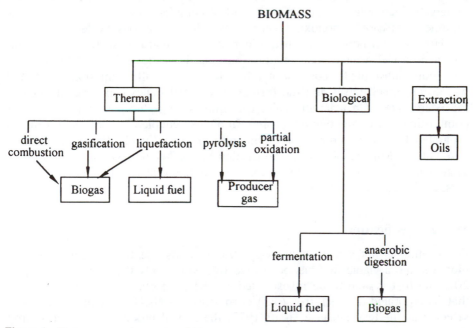

**Figure 1**    Various conversion routes of biomass.

following sections discuss the thermal and biological methods for the conversion of biomass to fuel.

## 10.2 THERMAL CONVERSION

Five thermal approaches are commonly used to convert biomass into an alternative fuel: direct combustion, gasification, liquefaction, pyrolysis, and partial oxidation. Liquefaction has the easiest capability for producing either a liquid or gaseous product. The other processes mainly produce a gaseous fuel.

### 10.2.1 Direct Combustion

In the past, indoor combustion of biomass fuels in unvented cooking and heating spaces caused considerable health problems to the direct user, primarily the women and children of developing countries [17]. Biomass fuels, when used improperly in this manner, release considerable amounts of toxic gases into the unvented area. These gases are typically carbon monoxide, nitrous oxide, hydrocarbons, organics, aldehydes, and trace amounts of aromatics and ketones. As the moisture content of the wood increases and as other biomass fuels of lower energy content (such as animal and crop waste) are used, the emissions rise. The woody components of biomass burn much more efficiently during complete combustion. In the earthen kilns of developing countries, the wood undergoes incomplete combustion instead of complete combustion. This causes the release of carbon monoxide, carbon dioxide, and nitrous oxide.

However, all hope is not lost. When direct combustion of biomass is conducted in a well-vented area, biomass-burning domestic stoves and boilers can be a sound substitute for conventional fossil fuel [17]. Sulfur emissions (0.05–0.2 wt. %) are much lower and the formation of particulates can be controlled at the source [18]. On a larger scale, the biomass is reduced to fine pieces for combustion in a close-coupled turbine. In a close-coupled system, the turbine is separated from the combustion chamber by a filter. Energy Performance Systems, Inc. of Minneapolis has demonstrated 87% efficiency with a close-coupled system using lignocellulose material. The company states that the process is feasible for 25–400-MW plants.

### 10.2.2 Gasification

Gasification is not a new technology; however, its use for the conversion of biomass into a viable fuel has been investigated for only the past 20 years [19, 20]. The first system to be investigated on the pilot scale was a fluidized bed that incorporated dry ash–free (daf) corn stover as the feed. Corn stover was selected as the feed because, as of 1977, the annual production of "corn crop wastes" exceeded 300 million tons [20]. The corn stover, if treated properly, has the potential to be converted into an energy source that would supply up to 2% of U.S. energy needs. The pilot-scale system, shown in Figure 2, has a 45.5-kg bed capacity. Fluidizing gas and heat for the gasification were supplied by the combustion of propane in the absence of air. The particulates and char are removed using a high-temperature cyclone. A venturi scrubber was then used to separate the volatile material into noncondensable gas, a tar-oil fraction, and an aqueous waste fraction. Raman et al. [21] conducted a series of tests with temperatures ranging from 840 to 1020 K. The optimal gas production was obtained using a feed rate of 27 kg/hr and 930 K. At these conditions, $0.25 \times 10^6$ Btu/hr of gas was produced. This is enough to operate a 25-hp internal combustion engine operating at 25% efficiency [21].

Another of the extensively studied gasification systems for biomass conversion is Sweden's VEGA gasification system. VEGA is a biomass fuel–based IGCC system that combines heat and power (CHP) for a district heating system. The pilot plant system is 140-MW fuel with an output of 60 MW of electricity and 65 MW of heat [22]. As indicated in Figure 3, the moisture of the entering biomass feedstock is removed via a biofuel dryer to decrease gaseous emissions. The dried biomass is then converted into a biofuel in a combined cycle gasifier. The resulting gas is cooled before entering the heat recovery boiler and distribution to the district heating. In 1989 the Swedish Environmental Protection Agency estimated that the VEGA system has the potential to produce 146–149 TWh/year of fuel from biomass [22].

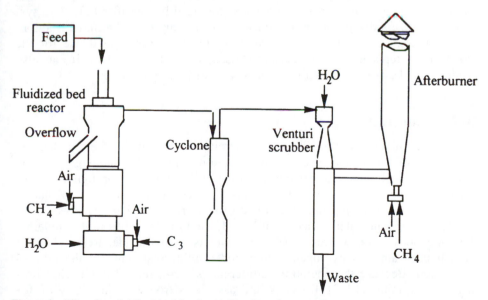

**Figure 2**  Pilot plant fluidized bed for the gasification of corn stover.

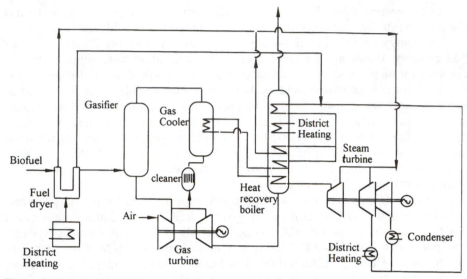

**Figure 3** Schematic of VEGA process for the gasification of biomass.

Sweden is not the only country making great advances in the gasification of biomass. During the past 15 years, Canadian developments in biomass gasification for the production of medium- and high-energy gases has received worldwide technical acclaim. Their Biosyn process is a 10 ton/hr system that comprises a pressurized air- or oxygen-fed fluidized bed gasifier [23]. The system can utilize a diverse array of feedstocks including whole biomass, fractionated biomass, peat, and municipal solid waste. The primary end use for the biogas is in replacing the oil currently used in industrial boilers. It can also produce synthesis gas for methanol or low-energy gas production.

## 10.2.3 Liquefaction

During the middle to late 1980s, commercial interest in the thermochemical conversion of biomass focused on liquefaction. Unlike the initial gasification studies, the preliminary liquefaction studies utilized woody or lignocellulose material as the feedstock. Woody biomass was considered superior to that of corn stover due to its potentially lower cost and greater availability [24].

The first documented "successful" production of ethanol from the liquefaction of woody material was at McGill University in Canada. Researchers at McGill used aqueous hydrogen iodide under mild conditions for the ethanol production. Because the operating temperatures were mild (125°C), char production was minimal [23]. Boocock and associates from the University of Toronto have also contributed to the understanding of biomass liquefaction. Their research determined that the size of the wood chips used for liquefaction was directly related to the amount of ethanol produced. As chip size increased, the product yield increased [23]. This led to research on the combined process of liquefaction, fermentation, and distillation.

The pilot plant process, shown in Figure 4, was designed on the basis of a feed rate of wood chips of 579,270 Mg/yr. The mixed wood feedstock comprises cellulose, xylose, and lignin. Liquefaction converts the wood chips into a liquid and a solid fraction. The liquid fraction, consisting of xylose, is passed through a neutralization unit before entering the fermentation step. The solid fraction, cellulose and lignin material, is sent directly to the fermentation unit. After leaving the fermentation unit, the lignin components are removed for the generation of process heat and electricity. The remaining material is then sent to the distillation unit. From the pilot plant results, the yield of ethanol from cellulose and hemicellulose feedstocks is approximately 110 gallons of ethanol per ton of wood [25].

## 10.2.4 Pyrolysis

Biomass can be converted into gas, liquid, and char via pyrolysis. The exact proportion of the end products is dependent on the pyrolysis process (e.g., temperature, pressure) used [26]. During the early 1980s, the National Renewable Energy Laboratory (NREL) investigated the use of ablative pyrolysis for biomass conversion. The technology behind the biomass pyrolysis is identical to that used in the petroleum industry.

The first biocrude was produced on a scale of 30 kg/hr [18]. At operating conditions of 500°C and a 1-second residence time, the biocrude had the same oxygen and energy content as its "natural crude" counterpart. Since then, Canadian researchers have converted woody biomass into fuel via pyrolysis in a

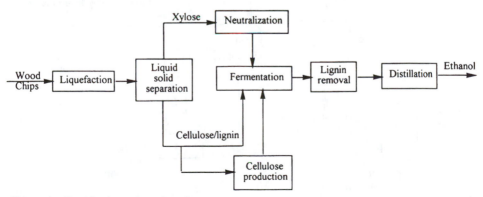

**Figure 4** Combined wood-to-ethanol process.

200 kg/hr pilot plant [27]. The fuel oil substitute was produced (on 1000 ton/day dry basis) at approximately \$3.4/GJ. At the time, the cost for light fuel oil was 4.0–4.6 \$/GJ [28], indicating that the pyrolyzed biomass fuel was a more economical alternative. However, skepticism about high transportation costs of the biomass to the pyrolyzer outweighing potential profits limited research funding for the next 2 years. To circumvent this problem, the Energy Resources Company (ERCO) in Massachusetts developed a mobile pyrolysis prototype for the Environmental Protection Agency.

ERCO's unit was designed to accept biomass with a 10% moisture content at a rate of 100 tons/day. At this rate, the system had a minimal net energy efficiency of 70% and produced gaseous, liquid, and char end products. The system, which is initially started using an outside fuel, is completely self-sufficient shortly after start-up. This was achieved by implementing a cogeneration system to convert the pyrolysis gas into the electricity required for operation. A small fraction of the pyrolysis gas is also used to dry the entering feedstock to the required 10% moisture. A simplified version of ERCO's mobile unit is shown in Figure 5.

The end products are pyrolytic oil and pyrolytic char. Both are more economical to transport than the original biomass feedstock. The average heating values are 10,000 and 12,000 Btu/lb for the pyrolytic oil and char, respectively [29]. The pyrolysis gas, which has a nominal heating value of 150 Btu/scf, is

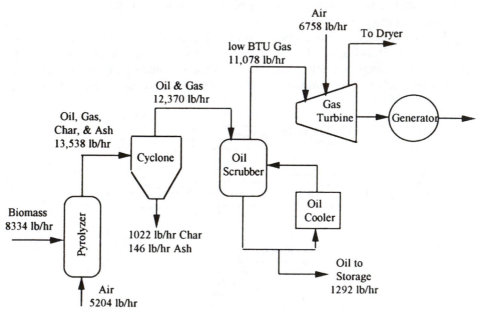

**Figure 5** Simplified schematic and material balance for ERCO's mobile pyrolysis unit [29].

not considered an end product because it is directly used in the cogeneration system. The mobility, self-sufficiency, and profitability of the system lifted some of the hesitancy about funding research on the pyrolysis of biomass. In addition, ERCO's success led to investigation of dual or cogeneration systems.

The most widely studied dual gasification-pyrolysis system is the Hydrocarb process [29–31]. The Hydrocarb process focuses on the configuration of gasi-fication and pyrolysis systems to convert a mixed biomass and natural gas feed-stock into methanol, gasoline, and char. The process combines three basic steps: (1) a hydropyrolyzer in which the biomass is gasified with a recycled hydrogen-rich gas to form a methane-rich gas, (2) a methane pyrolyzer in which methane is decomposed to carbon and hydrogen, and (3) a methanol synthesis reactor in which carbon monoxide is catalytically combined with hydrogen to form methanol [32]. Preliminary studies maintained the gasifier at 800°C, the methanol converter at 260°C, and the pyrolysis reactor at 1100°C. When process step 3 had a system pressure of 50 atm at the corresponding temperatures, the equilibrium compositions shown in Figure 6 were obtained. For every 100 kg of biomass and 18 kg of methane fed to the gasifier, approximately 67 kg of methanol and 40 kg of char were produced [33].

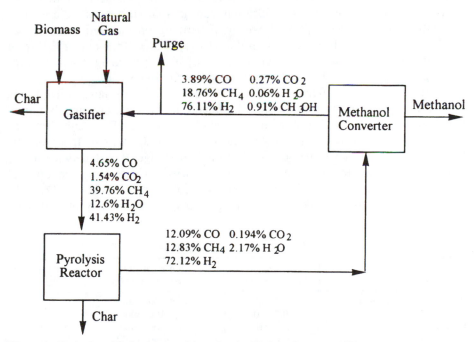

**Figure 6**  Typical equilibrium compositions for the Hydrocarb process [33].

One advantage of this process is that the char produced is essentially "pure" carbon. In other words, it is free of sulfur, ash, and nitrogen. Therefore, it can be used as a clean fuel by the industrial sector. Second, the conversion of methane to methanol in the presence of biomass decreases the carbon dioxide emissions typically associated with the gasification of methane. Another advantage is the potential to replace the feedstocks with other materials. The biomass component can be replaced with other carbonaceous materials, such as municipal solid waste. Methane could be replaced with coal. Although feedstock replacement is still in preliminary stages, researchers at Brookhaven National Laboratory anticipate similar results.

## 10.3 BIOLOGICAL CONVERSION: ANAEROBIC DIGESTION

Although the process of anaerobic digestion has been well known for the past 80 years, researchers have been reinvestigating the process for use as a potential fuel source [9, 10]. Specifically, the focus is on hastening the natural process of biomass conversion to a gaseous fuel referred to as biogas. Since 1977, universities and research institutes have been conducting experiments to ascertain the optimal conditions (feedstock, temperature, pressure, etc.) for the most efficient operations. Most of the biomass feedstocks studied have produced a biogas rich in methane. This medium- to high-Btu gas can, in some instances, be upgraded to a substitute natural gas [34]. However, depending on the feedstock, nonnegligible amounts of sulfur are also produced [35]. Table 2 contains a listing of different feedstocks, initial sulfur content, final sulfur percentage, and the projected power generation for studies conducted in Germany.

The first anaerobic digester studies, conducted at Penn State, utilized cow manure for biogas production. During the anaerobic process, organically bound materials are mineralized to methane and carbon dioxide. The Penn State digester was operated for a total of 450 days to treat 1200 tons of manure. The digester was started up using activated municipal sludge, then fed twice a day with manure. Digester retention times were varied to determine the optimal biogas production (Table 3). The biogas produced was approximately 60% methane, 32–34% carbon dioxide, 6–8% nitrogen, and trace amounts of hydrogen

**Table 2  Comparison of different biogas feedstocks [10, 35]**

| Feedstock | Power generation (MW/ton of biomass) | Original sulfur content (mg/m³) | Final percent sulfur |
|---|---|---|---|
| Liquid and solid manure | 0.2–0.5 | 300–500 | 0.5 |
| Organic waste | 0.5–2.0 | 100–300 | 0.3 |
| Wood chips | 5–50 | 300–1000 | 0.3 |
| Sewage sludge | 100–500 | 300–500 | 0.6 |

**Table 3 Capacity tests for 100 m³ digester [36]**

| Manure input (kg/day) | Retention time (days) | Total biogas production (m³/m³) | |
|---|---|---|---|
| | | m³/day | digester/day |
| 346 | 35 | 67 | 0.67 |
| 554 | 21 | 129 | 1.29 |
| 1030 | 11 | 202 | 2.02 |

sulfide [36]. Based on a methane content of 60%, the biogas had a total energy generation of 44 kW. This preliminary success led to the investigation of other feedstocks for anaerobic digesters.

Ghosh and Klass [37] conducted a series of anaerobic digestion experiments using a mixed biomass-waste feedstock. The mixed waste contained various ratios of *Eichhornia crassipies* (water hyacinth), *Cynodon dactylon* (Bermuda grass), sewage sludge, and municipal solid waste (MSW). All of the experiments were conducted in cylindrical Plexiglas digesters. The optimal blend was determined to be 32.3:32.3:32.3:3.1 of hyacinth, grass, sludge, and MSW. Once the optimal waste blend had been determined, studies were conducted in which the temperature and pH of the digester were varied. The final results are given in Table 4.

At first glance, the results indicate that varying the temperature and pH had very little effect on the amount of methane produced. However, the thermophilic digester had a higher pH and approximately three times the loading rate of its mesophilic counterpart. This indicates that additional experiments are required to determine the optimal operating conditions. Regardless of the operating conditions used, the anaerobic digestion of biomass is a promising alternative for supplementing the world's energy needs.

# 10.4 FUEL ETHANOL FERMENTATION

## 10.4.1 Introduction

Developing countries with characteristically weak economies and precarious infrastructures have been seriously hit by the energy crisis. To fully tap the potential of fossil-based fuels and other new renewable sources requires huge capital outlays, which these countries do not have. The trend has thus been toward small-scale utilization. One area in which the developing countries can have quick success is in supplementing their fossil fuel supplies with fuels derived from food crops such as sugarcane, cassava, maize, and sorghum. Focused primarily on the petroleum products as a primary source of energy fuels, ethanol has attracted attention all over the world as an alternative source to petrol or in

**Table 4 Mesophilic (35°C) and thermophilic (55°C) digestion of mixed biomass waste [37]**

|  | Mesophilic, pH not measured | Digester feed, pH 8–8.4 | Thermophilic, pH 9.0 | Digester feed, pH 10 |
|---|---|---|---|---|
| Operating conditions |  |  |  |  |
| Loading, lb/ft³ d | 0.1 | 0.1 | 0.4 | 0.43 |
| Detention time, d | 12 | 12 | 6.0 | 5.5 |
| Gas production rate, vol/vol/d | 0.61 | 0.67 | 2.85 | 2.91 |
| CH₄ content, mol % | 59.5 | 58.1 | 56.4 | 56.0 |
| CH₄ yield, scf/lb | 4.10 | 3.92 | 4.00 | 3.68 |
| % CH₄ collected | 46.1 | 44.1 | 45.0 | 41.4 |

blend with petrol to reduce the petrol consumption. In Brazil all cars are run on either a 20% mixture of ethanol with gasoline or pure ethanol. Arising from the crises of the 1970s, the program, called Proalcohol in Brazil, changed the profile of transportation fuels. Brazil produces about 4 billion gallons of ethanol annually [38]. The United States produces just over 900 million gallons.

Regarding the atmospheric concentrations of so-called greenhouse gases, the National Research Council (NRC), responding to a request from the Congress and with funding from the U.S Department of Energy, emphasizes substantially more research on renewable energy sources, improved methods of employing fossil fuels, energy conservation, and energy-efficient technologies [39]. The NRC committee on alternative energy research and development strategies emphasizes the importance of conservation or more efficient use of fossil fuels as the best near-term solution to reducing emissions of $CO_2$ and other greenhouse gases. The committee is also gaining better understanding of processes involving biomass production and conversion that will lessen U.S dependence on foreign oil.

The United States does not suffer from a lack of energy resources; it has plenty of coal reserves but it needs transportation fuels (liquid fuels). Gaddy [40] and his colleagues are working on the biological production of liquid fuels from biomass and coal. They have microorganisms that can produce ethanol from biomass, convert natural gas into ethanol, and convert syngas from coal gasification into liquid fuels. These microorganisms, Gaddy says, are very energy efficient. They work at ordinary temperature and pressure and offer significant advantages over chemical processes to produce liquid fuels from coal. Naee [41] focuses on using a renewable resource, lignin, to recover a nonrenewable resource, oil. Lignins are produced in large quantities, approximately 250 billion pounds per year in the United States as by-products of the paper and pulp industry. As a consequence, prices of some lignin products, such as lignosulfonates, are as low as 2–3 cents a pound.

**Ethanol.** Ethanol, $C_2H_5OH$, is one of the most significant synthetic oxygen-containing organic chemicals because of its unique combination of properties as a solvent, a fuel, a germicide, a beverage, and an antifreeze, and especially because of its versatility as an intermediate to other chemicals. Ethanol is one of the largest-volume chemicals used in industrial and consumer products. The main uses for ethanol are as an intermediate in the production of other chemicals and as a solvent. As a solvent, ethanol is second only to water. Ethanol is a key raw material in the manufacture of plastics, lacquers, polishes, plasticizers, perfumes, and cosmetics. The physical and chemical properties of ethanol are primarily dependent on the hydroxyl group, which imparts the polarity to the molecule and also gives rise to intermediate hydrogen bonding. In the liquid state hydrogen bonds are formed by the attraction of the hydroxyl hydrogen of one molecule and the hydroxyl oxygen of another molecule [42]. This makes liquid alcohol behave as though it were largely dimerized. Its association is confined to the liquid state; in the vapor state it is monomeric.

**Manufacture.** Industrial alcohol can be produced either (1) synthetically from ethylene, (2) as a by-product of certain industrial operations, or (3) by the fermentation of sugars, starch, or cellulose. There are two main processes for the synthesis of alcohol from ethylene. The earlier one (in the 1930s at Union Carbide) was an indirect hydration process, variously called the strong sulfuric acid–ethylene process, the ethyl sulfate process, the esterification hydrolysis process, and the sulfation hydrolysis process. The other synthetic process, designed to eliminate the use of sulfuric acid, is the direct hydration process. In addition to the direct hydration process, the sulfuric acid process, and fermentation routes, several other processes have been suggested [43–46]. None of these have been successfully implemented on the commercial scale.

**Fermentation ethanol.** Fermentation, one of the oldest chemical processes, is used to make a variety of useful products and chemicals. At present, however, many of the products such as ethanol are synthesized from petroleum feedstocks at lower costs. The future of the fermentation industry, therefore, depends on its ability to utilize the high efficiency and specificity of enzyme catalysis to synthesize complex products and on its ability to overcome variations in the quality and availability of the raw materials.

Ethanol can be derived by fermentation processes from any material that contains sugar. The raw materials used in the manufacture of ethanol via fermentation are classified as sugars, starches, and cellulosic materials [47]. Sugars can be converted to ethanol directly. Starches must first be hydrolyzed to fermentable sugars by the action of enzymes. Cellulose must likewise be converted to sugars, generally by the action of mineral acids. Once the simple sugars are formed, enzymes from yeasts can readily ferment them to ethanol.

**Sugars.** The most widely used form of sugar for ethanol fermentation is black-strap molasses, which contains about 30–40 wt. % sucrose, 15–20 wt. % invert sugars such as glucose and fructose, and 28–35 wt. % nonsugar solids. The direct fermentation of sugarcane juice, sugarbeet juice, beet molasses, fresh and dried fruits, sorghum, whey, and skim milk has been considered but none of these could compete economically with molasses. From the viewpoint of industrial ethanol production, sucrose-based substances such as sugarcane and sugarbeet juices have many advantages including their relative abundance and renewable nature. Molasses, the noncrystallizable residue that remains after sucrose purification, has additional advantages; it is a relatively inexpensive raw material, readily available, and already used for industrial ethanol production. Park and Baratti [48] have studied the batch fermentation kinetics of sugarbeet molasses by *Zymomonos mobilis*. This bacterium has several interesting properties that make it competitive with the yeasts, the most important being higher ethanol yields and specific productivity. However, when cultivated on molasses, *Z. mobilis* generally shows poor growth and low ethanol production in comparison with those in glucose media [72]. The low ethanol yield is explained by the formation of by-products such as levan and sorbitol. Other components of molasses such as organic salts, nitrates, or the phenolic compounds could also be inhibitory for growth. Park and Baratti found that in spite of good growth and prevention of leven formation, the ethanol yield and concentration were not sufficient for the development of an industrial process. Yeasts of the *Saccharomyces* genus are mainly used in industrial processes. However, there are continued efforts to develop mutant strains of *Z. mobilis* to avoid the costly addition of yeast extract [78, 79, 82].

**Starches.** The grains generally provide cheaper ethanol feedstocks and the conversion is less expensive because they can be stored more easily than most sugar crops, which often must be reduced to a syrup before storage. Furthermore, grain distillation produces a by-product that can be used for protein meal in animal feeds [49]. Fermentation of starch from grains is somewhat more complex than from sugars because the starch must first be converted to sugar and then to ethanol. Simplified equations for the conversion of starch to ethanol are:

$$C_6H_{10}O_5 + H_2O \xrightarrow[\text{fungal amylase}]{\text{enzyme}} C_6H_{12}O_6 \xrightarrow{\text{yeast}} 2C_2H_5OH + 2CO_2$$

As shown in Figure 7, in making grain alcohol, the distiller produces a sugar solution from feedstock, ferments the sugar to ethanol, and then separates the ethanol from water through distillation.

Among the disadvantages of grain are its fluctuations in price. Whenever the price of grain falls there is interest in the use of grain alcohol as an auto-

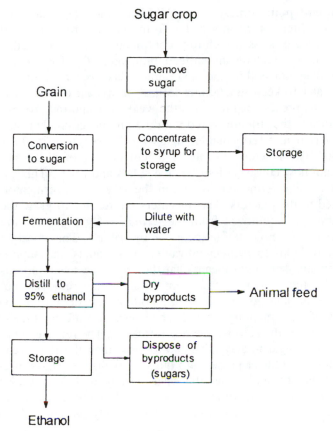

**Figure 7**   Synthesis of ethanol from grains and sugar crops [49].

motive fuel additive (gasohol). Ethanol in the petrol boosts the fuel's octane value without the use of dangerous lead additives.

European farms produce more food than they can consume. Farmers and oil companies have been lobbying [50] in the European parliament on a scheme to blend petrol with 5% bioethanol. They insisted that farmers should grow energy crops to make alcohol to replace petrol. In Western Europe and the United States, agricultural policies no longer concern themselves with the problem of producing enough food. Instead they struggle with storing the food of increasingly abundant harvests. They are paying high costs for these policies. It costs them as much to store as it would to promote ethanol as a fuel. The United States also has huge surpluses of cereal. The European Community has several advantages over its rivals, the United States and Japan. Europe has more land than Japan and, unlike the United States, no shortage of water. Furthermore, Europe now has the highest yields of grain in the world.

**Alcohol without pollution.** ALFA-LAVAL's ethanol fermentor in Sweden has brewed 20,000 liters of ethanol a day from surplus grain since 1983. All the products from the process, which runs continuously, are recovered and most are solids. The products include animal feed, bran, and $CO_2$. The company says that this process does not pollute the environment because it requires only a small amount of water to keep functioning [50]. A flowsheet of the system is shown in Figure 8. The process begins when the weak beer and the process water from the fermentor and the still are mixed with the ground grain. The starch of liquid feed is converted into fermentable sugars by the enzymes in the giant stirred tank reactors. The reactors operate at a temperature of about 60–90°C. The effluent stream now consists of suspended solids and fibers. This stream is continuously fed to the fermentor. While on the way to the fermentor it heats up the fresh feed to the reactors. The air enters the reactor to grow the yeast. The beer is continuously removed from the fermentor by a process that keeps the level of beer in the fermentor at about 7% by volume. This constant concentration of ethanol helps to prevent unwanted by-products and suppress bacteria. Keeping their number down means that the feedstock needs no pasteurization. The beer from the fermentor passes into a sieve that removes the fibers. The fibers are washed and go straight to the bottom part of the still, where steam removes any of the remaining ethanol. The beer enters a centrifuge that removes the yeast and passes the cells back to the fermentor. The recycled yeast consumes significantly less sugar to stay alive and to grow than the fresh yeast.

Newly fermented beer passes into the still. A 40% solution of ethanol exits at the top of the still and leaves the weak beer behind. The weak beer passes out of the side of the still, warms up the newly fermented one, and then moves to the start of the process. The weak beer is already pasteurized, which means

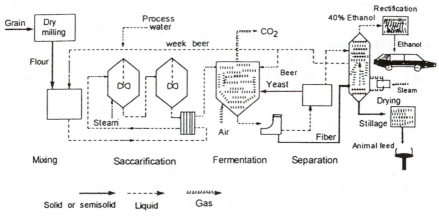

**Figure 8** Flowsheet of ALFA-LAVAL's ethanol fermentor [50].

it is better than the fresh water for mixing with the feedstock. Heat for the distillation comes from the steam that cleans the ethanol from the fibers. The fibers (stillage) leave the bottom of the still for a dryer that turns them into the animal feed. The dryer operates at 70°C, hot enough to dry the stillage and cool enough not to destroy the proteins in the fiber. The process is highly automated.

**Cellulosic materials.** Cellulose from wood, agricultural residue, and waste sulfite liquor from pulp and paper mills must first be converted to sugar before it can be fermented. Enormous amounts of carbohydrate-containing cellulosic wastes are generated every year. The technology for converting this material into ethanol is available, but the stoichiometry of the process is unfavorable. About two-thirds of the mass disappears during the conversion of cellulose to ethanol, most of it as $CO_2$ in the fermentation of glucose to ethanol. This amount of $CO_2$ leads to a disposal problem rather than to a raw material credit. Another problem is that the aqueous acid used to hydrolyze the cellulose in wood to glucose and other simple sugars destroys much of the sugar in the process. New ways to reduce the cost include the use of less corrosive acids and reduced hydrolysis time [51].

The conversion of cellulosic wastes is a two-step process: hydrolysis and fermentation. The two steps are sequential and in the order named. Whether the process is conducted on a batch or on a continuous basis, the two steps can be separated. Of the two steps in alcohol production, hydrolysis is more critical and rate determining because in it cellulose is transformed into glucose that is available to the microbes responsible for ethanol fermentation.

Among the several forms of cellulose hydrolysis (saccharification), the chemical and biological types are the most commonly applied. Because the actual hydrolysis is accomplished enzymatically, biological hydrolysis is designated as enzymatic hydrolysis. Although the use of enzymes avoids the corrosion problems and loss of fuel product associated with acid hydrolysis, enzymes have their own drawbacks. Enzymatic hydrolysis slows as the glucose product accumulates in the reactor vessel. The end product inhibition eventually halts the hydrolysis unless some way is found to draw off the glucose as it is formed.

In 1978 the Gulf Oil researchers [51] designed a commercial-scale plant producing $95 \times 10^6$ L/yr of ethanol by simultaneous enzymatic hydrolysis of cellulose and fermentation of resulting glucose as it is formed and overcoming the problem of product inhibition. The method consists of a pretreatment developed for this process that involves grinding and heating of the feedstock followed by hydrolysis with a mutant bacterium, also developed for this purpose. Mutated strains of the common soil mold *Trichoderma viride* can process 15 times as much glucose as natural strains. Simultaneous hydrolysis and fermentation reduces the time requirement for the separate hydrolysis step, reducing the cost and increasing the yield. Also, the process does not use acids, which would increase the equipment costs. The sugar yields from the cellulose are about 80% of what is theoretically achievable, but the small amount of hemicellulose in the sawdust is not being converted.

## 10.4.2 Agricultural Lignocellulosic Feedstocks

One reason the United States has depended so heavily up to now on natural gas and petroleum for energy and for the manufacture of most organic materials is that the liquids and gases are relatively easy to handle. Solid materials like wood, on the other hand, are difficult to collect, transport, and process into components that can make desired products.

Biomass is a relatively inexpensive material because it is made by the sun, resulting in enormous savings. In addition, the quantity of biomass available for conversion to fuel, chemicals, and other materials is virtually unlimited. Greater biomass utilization can also help ameliorate solid waste disposal problems. About 180 million tons of municipal waste is generated annually in the United States. About 50% of this is cellulosic and could be converted to useful chemicals and fuels [52]. For example, there is no reason for used diapers to go into landfills, which contain very high quality lignocellulosic fibers.

Although lignocellulose is inexpensive because it cannot be digested and does not compete as a food, its inability to be digested makes it difficult to convert to fermentable sugars. Furthermore, as shown in Figure 9, lignocellulose has a complex structure with three major components, which must be processed separately to make the best use of high efficiencies inherent in the biological process. A general schematic representation for the conversion of lignocellulose to ethanol is shown in Figure 10. The lignocellulose is pretreated to separate the xylose and sometimes the lignin from the crystalline cellulose. The xylose can then be fermented to ethanol and lignin processed to produce other liquid fuels. The crystalline cellulose, the largest (50%) and most difficult fraction, remains behind as a solid after the pretreatment and is sent to an enzymatic hydrolysis process that breaks the cellulose down into glucose. Enzymes, the biological catalysts, are specific and hence, the hydrolysis of cellulose to sugar does not break down the sugars. Enzymatic processes are capable of yields approaching 100%. The glucose is then fermented to ethanol and combined with the ethanol from xylose fermentation. This dilute beer is then concentrated to fuel-grade ethanol by distillation.

The hemicellulose fraction (25%) is composed primarily of xylan, which is easily converted to simple sugar xylose, but the xylose is difficult to convert to ethanol. Methods have been identified using new yeasts, bacteria, and processes combining enzymes and yeasts. Although none of these fermentations are yet ready for commercial use, considerable progress has been made.

Lignin, the third major component of lignocellulose (25%), is a large random phenolic polymer. In lignin processing, the polymer is broken down into frag-

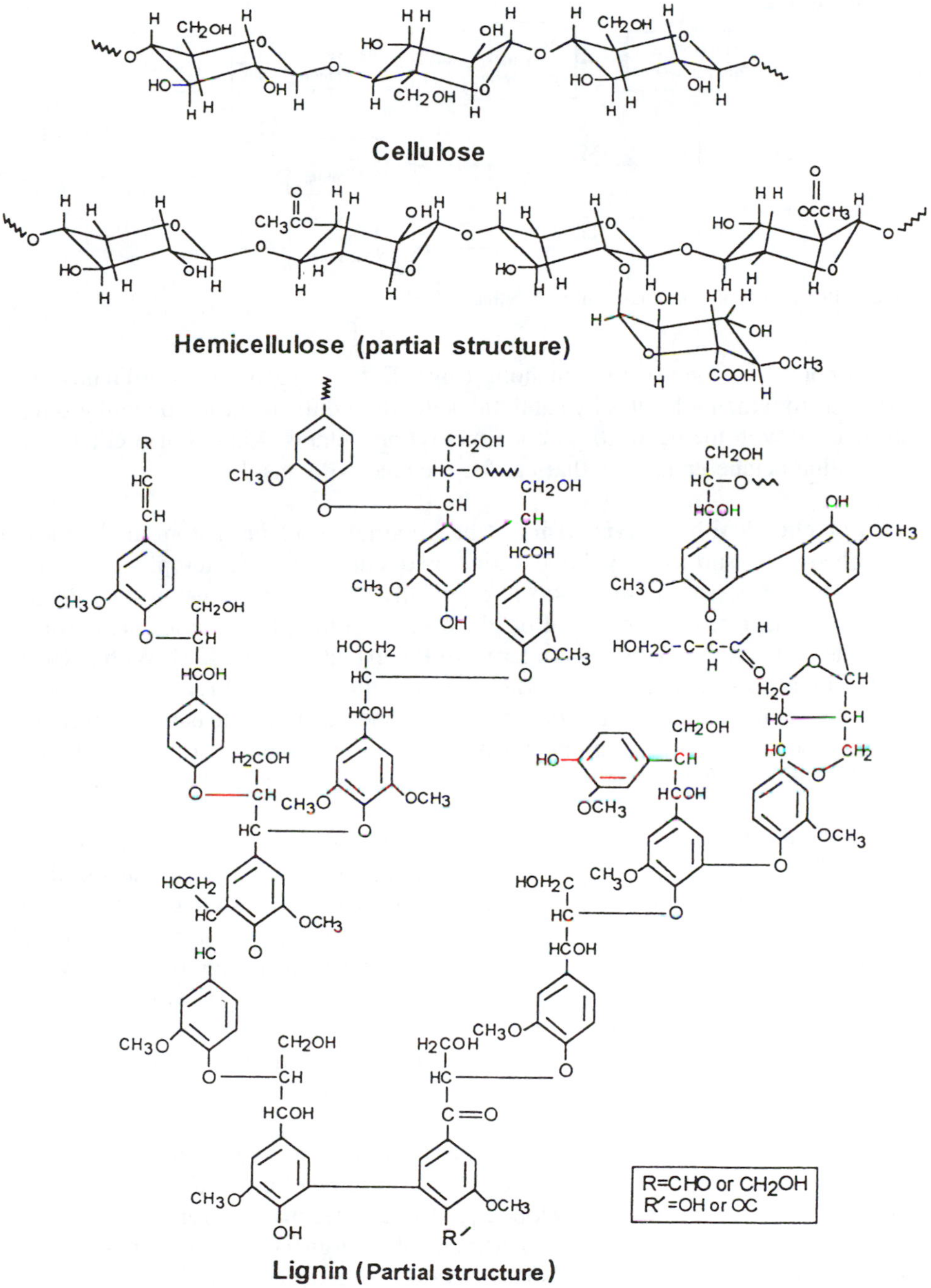

**Figure 9** Major polymeric components of plant materials [84].

Lignocellulose

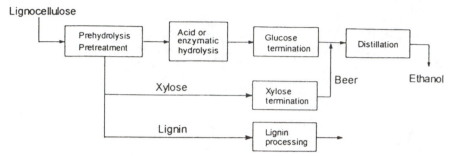

**Figure 10**  Conversion of lignocellulose to ethanol [53].

ments containing one or two phenolic rings. Extra oxygen and side chains are stripped from the molecules by catalytic methods and the resulting phenol groups are reacted with methanol to produce methyl aryl ethers. Methyl aryl ethers are high-value octane enhancers that can be blended with gasoline.

**Acid or chemical hydrolyis.**  Among the important specific factors in chemical hydrolysis are surface-to-volume ratio, acid concentration, temperature, and time. The surface-to-volume ratio is especially important in that it also determines the magnitude of the yield of glucose. Therefore, the smaller the particle size, the better is the hydrolysis in terms of the extent and rate [54]. With respect to the liquid-to-solids ratio, the higher the ratio, the faster the reaction. A trade-off must be made between the optimum ratio and economic feasibility, because an increase in the cost of equipment parallels an increase in ratio of the liquid to solids. For chemical hydrolysis, a ratio of 10:1 seems to be most suitable [54].

In a typical system for chemically hydrolyzing cellulosic wastes, the wastes are milled to micrometer-sized particles. The milled material is immersed in a weak acid (0.2 to 10%), the temperature of the suspension is elevated to 180 to 230°C, and a moderate pressure is applied. Eventually the hydrolyzable cellulose is transformed into sugar. The reaction has no effect on the lignin that may be present. The yield of glucose varies with the nature of the raw waste. For example, 84 to 86% of kraft paper or 38 to 53% of the weight of the ground refuse may be recovered as sugar. The sugar yield increases with the acid concentration as well as the elevation of temperature. A suitable concentration of acid ($H_2SO_4$) is about 0.5% of the charge.

A two-stage, low-temperature, ambient pressure acid hydrolysis process that utilizes separate unit operations to convert the hemicellulose and cellulose to fermentable sugars is being developed [55] and tested by the Tennessee Valley Authority (TVA) and the U.S. Department of Energy (DOE). Laboratory and bench-scale evaluations showed more than 90% recovery and conversion efficiencies of sugar from corn stover. Sugar product concentrations of more than

10% glucose and 10% xylose were achieved. The inhibitor levels in the sugar solutions never exceeded 0.02 g/100 ml, which is far below the level shown to inhibit fermentation. This experimental plant was designed and built in 1984. The acid hydrolysis plant provides fermentable sugars to a 38 L/hr fermentation and distillation facility built in 1980. The results of the studies are summarized as follows:

- Corn stover ground to 2.5 cm was adequate for the hydrolysis of hemicellulose.
- The time required for optimum hydrolysis in 10% acid at 100°C was 2 hours.
- Overall xylose yields of 86 and 93% were obtained in a bench-scale study at 1- and 3-hour reaction times.
- Recycled leachate, dilute acid, and prehydrolysis acid solutions were stable during storage for several days.
- Vacuum drying was adequate in the acid concentration step.
- Cellulose hydrolysis was successfully accomplished by cooking stover containing 66 to 78% acid for 6 hours at 100°C. Yields of cellulose conversion to glucose of 75 to 99% were obtained in the laboratory studies.
- Vinyl ester resin fiberglass-reinforced plastics were used for the construction of process vessels and pipings.

*Process description.* The process involves two-stage sulfuric acid hydrolysis, relatively low temperature, and a cellulose prehydrolysis treatment with concentrated acid. Figure 11 is a flow diagram of the TVA process. Corn stover is ground and mixed with the dilute sulfuric acid (about 10% by weight). The hemicellulose fraction of the stover is converted to pentose sugars by heating the solution to 100°C for 2 hours in the first hydrolysis reactor. Raw corn stover contains, on a dry basis, about 40% cellulose, 25% hemicellulose, and 25% lignin. Sulfuric acid for the hydrolysis reaction is provided by recycling the product stream from the second hydrolysis step, which contains the sulfuric acid and hexose sugars. The pentose and hexose sugars, which are primarily xylose and glucose, respectively, are leached from the reactor with warm water. The sugar-rich leachate is then neutralized with lime, filtered to remove precipitated material, and fermented to produce ethanol.

Residue stover from the first hydrolysis step (hemicellulose conversion) is dewatered and prepared for the second hydrolysis step (cellulose conversion) by soaking (prehydrolysis treatment step) in sulfuric acid (about 20–30% concentration) from 1 to 2 hours. The residue is then screened, mechanically dewatered, and vacuum dried to increase the acid concentration to 75–80% in the liquid phase before entering the cellulose reactor. The second hydrolysis reactor operates at 100°C and requires a time of 4 hours. The reactor product is filtered to remove solids (primarily lignin and unreacted cellulose). Because the second hydrolysis reactor product stream contains about 10% acid, it is used in the first hydrolysis step to supply the acid required for hemicellulose hydrolysis. Residue

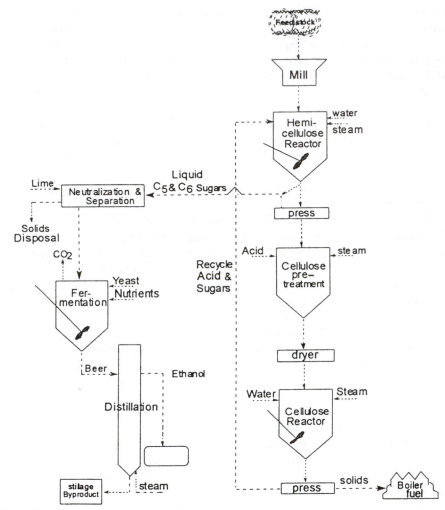

**Figure 11** Low-termperature, low-pressure, two-stage acid hydrolysis concept for conversion of nonwoody feedstocks to ethanol [55].

from the reactor is washed to recover the remaining sulfuric acid and the sugar not removed in the filtration step.

Lignin is the unreacted fraction of the feedstock, which can be burned as the boiler fuel. It has a heating value of about 5270 kcal/kg, comparable to that of subbituminous coal. Other products such as surfactants and adhesives can be made from lignin. Stillage can be used to produce several products, including methane. Preliminary research has shown that 30 liters of biogas containing 60% methane gas is produced from a liter of corn stover stillage. For each liter of ethanol produced, 10 liters of stillage is produced.

All process pipings, vessels, and reactors in contact with corrosive sulfuric acid are made of vinyl ester resin fiberglass-reinforced plastic. The dryer is made of carbon steel and lined with Kynar. Conveyor belts are made of acid-resistant material. Mild steel agitator shafts are coated with Kynar or Teflon. Heat exchangers are made with CPVC pipe shells and carpenter 20 stainless steel coils. Pumps are made with nonmetallic compound Teflon lining or carpenter 20 stainless steel. The two filter press units have plates made of polypropylene.

**Enzymatic hydrolysis.** As a fermentable carbohydrate, cellulose differs from other carbohydrates generally used as substrate for fermentation. Cellulose is insoluble and is polymerized as 1,4-β-glucosidic linkage. Cellulose is solubilized so that an entry can be made into cellular metabolic pathways. Solubilization is brought about by an enzymatic hydrolysis catalyzed by the cellulase system of certain bacteria and fungi.

*Enzyme system.* Each cellulolytic microbial group has an enzyme system unique to it. The enzyme capabilities range from those with which only soluble derivatives of cellulose can be hydrolyzed to those with which a cellulose complex can be disrupted. Based on the enzymatic capability the cellulase is characterized into two groups: $C_1$ enzyme or factor and $C_X$ enzyme or factor [54]. The $C_1$ factor is regarded as an "affinity" or prehydrolysis factor that transforms the cotton cellulose into the linear and hydroglucose chains. As such, it serves as an essential precursor to the action of $C_X$ factor. The $C_X$ (hydrolytic) factor breaks down the linear chains into the soluble carbohydrates, usually cellobiose and glucose.

Microbes rich in $C_1$ are more useful in the production of glucose from the cellulose. Moreover, because the $C_1$ phase proceeds more slowly than the subsequent step, it is the rate-determining step. Among the many microbes, *Trichoderma reesei* surpasses all others in the possession of $C_1$ complex. The site of action of cellulolytic enzymes is important in the design of hydrolytic systems ($C_X$ factor). If the enzyme is within cell mass, material to be reacted must diffuse into the cell mass. Therefore the enzymatic hydrolysis of cellulose usually takes place extracellularly, where enzyme is diffused from the cell mass into the external medium.

Another important factor in the enzymatic reaction is whether the enzyme is adaptive or constitutive. A constitutive enzyme is one that is present in the cell at all times. Adaptive enzymes are found only in the presence of a given substance, and the synthesis of the enzyme is triggered by an inducing agent. Most of the fungal cellulases are adaptive [47, 54].

Cellobiose is an inducing agent with respect to *T. reesei.* In fact, depending on the circumstances, cellobiose can be either an inhibitor or an inducing agent. It is inhibitory when its concentration exceeds 0.5 to 1.0%. Cellobiose is an intermediate product and generally is present in concentrations low enough to permit it to serve as a continuous inducer [56].

## 10.4.3 Enzymatic Processes

All enzymatic processes consist of four major steps that may be combined in a variety of ways as represented in Figure 12: pretreatment, enzyme production, hydrolysis, and fermentation.

**Pretreatment.** It has long been recognized that some form of treatment is necessary to achieve reasonable rates and yields in the enzymatic hydrolysis of biomass. This has generally been done to reduce the crystallinity of cellulose and to lessen the average polymerization of cellulose, the lignin-hemicellulose sheath that surrounds the cellulose, and the lack of available surface area for the enzymes to attack. Mechanical pretreatments such as intensive ball and roll milling have been investigated as means of increasing the surface area, but they require exorbitant amounts of energy. The efficiency of chemical processes can be understood by considering the interaction of the enzymes and the substrate. The hydrolysis of cellulose into sugars and other oligomers is a solid-phase reaction in which the enzymes must bind to the surface to catalyze the reaction. Cellulase enzymes are large proteins, with molecular weights ranging from 30,000 to 60,000, and are thought to be ellipsoidal with major and minor dimensions of 30 to 200Å. The internal surface area of wood is very large, but only 20% of the pore volume is accessible to cellulase-sized molecules. By breaking down the hemicellulose-lignin matrix, the hemicellulose or the lignin can be separated and the accessible volume greatly increased. This removal of material greatly enhances enzymatic digestibility.

The hemicellulose-lignin sheath can be disrupted by either acidic or basic catalysts. Basic catalysts simultaneously remove both lignin and hemicellulose, but this approach suffers from large consumption of base through neutralization by ash and acid groups in the hemicellulose. In recent years attention has been focused on the acidic catalysts. They can be mineral acids or organic acids generated in situ by the autohydrolysis of hemicellulose.

Various types of pretreatments are used for the biomass conversion. The pretreatments that have been studied in recent years are steam explosion, autohydrolysis, wet oxidation, organosolv, and rapid steam hydrolysis (RASH). The major objective of most pretreatments is to increase the susceptibility of cellulose

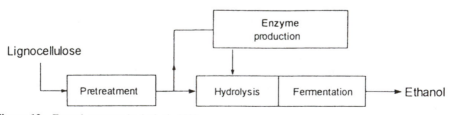

**Figure 12** Fungal enzyme hydrolysis [53].

and lignocellulose material to acid and enzymatic hydrolysis. Enzymatic hydrolysis is a sensitive indicator of lignin depolymerization and cellulose accessibility. Cellulose enzyme systems react very slowly with untreated material; however, if the lignin barrier around the plant cell is partially disrupted, rates of enzymatic hydrolysis are increased dramatically.

**Autohydrolysis steam explosion.** The process is represented as shown in Figure 13. Very high temperature processes may lead to significant pyrolysis, which produces inhibitory compounds. The ratio of the rate of hemicellulose hydrolysis to sugar degradation is greater at higher temperatures. Low-temperature processes have lower xylose yields and produce more degradation products than a well-controlled high-temperature process using small particles. In general, xylose yields in autohydrolysis are low (30–50%). An autohydrolysis system is used as the pretreatment in separate hydrolysis and fermentation (SHF). The reaction conditions are 200°C for 10 minutes, with a xylose yield of 35%.

Steam consumption in autohydrolysis is strongly dependent on the moisture content of the starting material. Wet feedstocks require considerably more energy because of the high heat capacity of water. An important advantage of autohydrolysis is that it breaks the lignin into relatively small fragments that can be easily solubilized in either base or organic solvents.

**Dilute acid prehydrolysis.** Lower temperature operation with reduced sugar degradation is achieved by adding a small amount of mineral acid to the pretreatment process. The acid increases reaction rates at a given temperature, and the ratio of the hydrolysis rate to the degradation rate is also increased.

A compromise between the reaction temperature and time exists for acid-catalyzed reactions. As for autohydrolysis, however, conditions explored range from several hours at 100°C to 10 seconds at 200°C with sulfuric acid concentration of 0.5 to 4.0%. Acid catalysts have also been used in steam explosion systems with similar results. Xylose yields generally range from 70 to 95%. However, sulfuric acid processes produce a lignin that is more condensed (52% of the lignin extractable in dilute NaOH) than that produced by autohydrolysis.

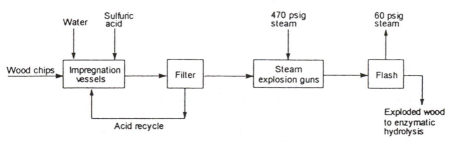

**Figure 13** Steam explosion pretreatment process flow diagram [53].

Sulfur dioxide has also been investigated as a catalyst to improve the efficiency of the pretreatment. Use of the excess water increases the energy consumption and decreases the concentration of xylose in the hydrolyzate, thus decreasing the concentration of ethanol that can be produced in the xylose fermentation step.

**Organosolv pretreatment.** In this process an organic solvent (ethanol or methanol) is added to the pretreatment reaction to dissolve and remove the lignin fraction. In the pretreatment reactor the internal lignin and hemicellulose bonds are broken and both fractions are solubilized, while the cellulose remains as a solid. After leaving the reactor, the organic fraction is removed by evaporation in the liquid phase; the lignin precipitates and can be removed by filtration or centrifugation. Thus, this process cleanly separates the feedstock into a solid cellulose residue, a solid lignin that has undergone a few condensation reactions, and a liquid stream containing the xylon (Figure 14).

Results have shown [57] that some reactions occur during the organosolv process that strongly affect the enzymatic rate. These reactions could be due to the physical or chemical changes in lignin or cellulose. In general, organosolv processes have higher xylose yields than the other processes because of the influence of organic solvent on hydrolysis kinetics. A major concern in these processes is complete recovery of the solvent.

**Combined RASH and organosolv pretreatment.** Attempts have been made to improve the overall process by combining the two individual pretreatments. Rughani and McGinnis (1989) [57] have studied the effect of a combined RASH-organosolv process on the rate of enzymatic hydrolysis and the yield of solubilized lignin and hemicellulose. A schematic diagram of the process is shown in Figure 15. For the organosolv pretreatment the steam generator is disconnected and the condensate valve closed. The rest of the reactor setup is similar to that in the RASH procedure. Organosolv processes at low temperature are generally ineffective in removing lignin; however, combining the two processes leads to increased solubilization of lignin and hemicellulose. RASH temperature is the major factor in maximizing the percentage of cellulose in the final product. The maximum yield of solubilized lignin was obtained at a temperature of 240°C for RASH and 160°C for the organosolv process.

**Enzyme production and inhibition.** The enzyme of interest is cellulase, needed for the hydrolysis of cellulose. Cellulase is a multicomponent enzyme system consisting of *endo*-β-1,4-glucanases, *exo*-β-1,4-glucan glucohydrolases, and *exo*-β-1,4-glucan cellobiohydrolase. Cellobiose is the dominant product of this system but is highly inhibitory to the enzymes and is not usable by most organisms. Cellobiase hydrolyzes cellobiose to glucose, which is much less inhibitory and highly fermentable. Many of the fungi produce this, and most of the work that

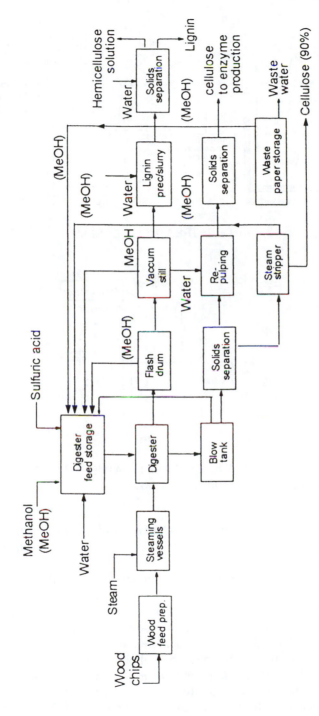

**Figure 14** Organsolv pretreatment process [53].

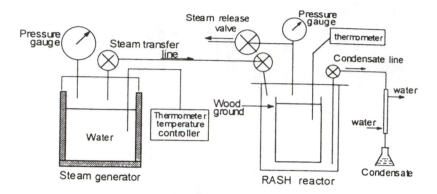

RASH PRETREATMENT SCHEME

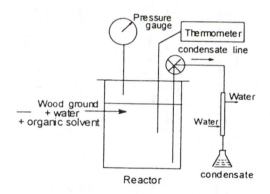

ORGANOSOLV PRETREATMENT SCHEME

**Figure 15**   RASH and organosolv pretreatment schemes [57].

is presently going on is on the *Trichoderma reesei* (*viride*). This cellulase is much less inhibited than other cellulases, which is a major advantage for industrial purposes [58].

The type of inhibition exhibited by cellulases is the subject of much confusion. Although most researchers favor competitive inhibition [59–64], some cellulases are noncompetitively [64–66] or uncompetitively inhibited [63]. *Trichoderma reesei* enzyme on substrates like solka floc, wheat straw, and bagasse is competitively inhibited by glucose and cellobiose. On the other hand, some enzyme is noncompetitively inhibited by cellobiose with other substrates such as rice straw and avicel. *Trichoderma viride* is noncompetitively inhibited by glucose in a cotton waste substrate [63].

Many mutants have been produced following *T. reesei*. The most prominent among these is Rut C-30, the first mutant with great β-glucosidase production

[53]. Other advantages of the strain are that it is hyperproducing and is carbolite repression resistant.

Cellulases from thermophilic bacteria have also been extensively examined. Among these *Clostridium thermocellum* is perhaps the most extensively characterized thermophilic, anaerobic, cellulose-degrading organism. The enzymes isolated from thermophilic bacteria may have superior thermostability and hence have longer half-lives at high temperatures. Although this is not always the case, cellulases isolated from *C. thermocellum* have high specific activities [67], especially against crystalline forms of cellulose that have proved resistant to other cellulase preparations.

Enzyme production with *T. reesei* is difficult because cellulase production discontinues in the presence of easily metabolizable substrates. Thus, most production work has been carried out on insoluble carbon sources such as steam-exploded biomass or solka floc. In such systems, the rate of growth and cellulase production is limited because the fungi must secrete the cellulase and carry out slow enzymatic hydrolysis of the solid to obtain the necessary carbon. Average productivities have been approximately 100 IU/L.h. [Hydrolytic activity of cellulose is generally expressed in terms of international filter unit (IU). This is a unit defined in terms of the amount of sugar produced per unit time from a strip of Whatman filter paper.] The filter paper unit is a measure of the combined activities of all three enzymes on the substrate. High productivities have been reported with *T. reesei* mutant in a fed-batch system using lactose as the carbon source and steam-exploded aspen as an inducer. Although lactose is not available in sufficient quantities to supply a large ethanol industry, this does suggest that it may be possible to develop strains that can produce cellulases with soluble carbon sources such as xylose and glucose.

Increases in productivity dramatically reduce the size and cost of the fermentors used to produce the enzyme. More rapid fermentations would also decrease the risk of contamination and might allow less expensive construction [80, 84]. Alternatively, using a soluble substrate may allow simplification of fermentor design or allow the design of a continuous enzyme production system.

**Cellulose hydrolysis**

*Cellulase adsorption.* The enzymatic hydrolysis of cellulose proceeds by the adsorption of cellulase enzyme on the lignacious residue as well as the cellulose fraction. The adsorption on the lignacious residue is also interesting from the viewpoint of recovering enzyme after the reaction and recycling it to use on the fresh substrate. Obviously, the recovery is reduced by adsorption of the enzyme on lignacious residue, an important consideration because a large fraction of the total operating cost is due to the production of enzyme. Since the capacity of lignacious residue to adsorb the enzyme is influenced by the pretreatment conditions, the pretreatment should be evaluated, in part, by how much enzyme adsorbs on the lignacious residue at the end of hydrolysis as well as its effect on the rate and extent of the hydrolysis reaction.

The adsorption of cellulase on cellulose and the lignacious residue have been investigated by Ooshima et al. [68] using cellulase from *T. reesei* and hardwood pretreated with dilute sulfuric acid with explosive decomposition. The cellulase was found to adsorb on lignacious residue as well as the cellulose during hydrolysis of the pretreated wood. A decrease in the enzyme recovery in the liquid phase with increased substrate concentration has been reported due to the adsorption on the lignacious residue. The enzyme adsorption capacity of the lignacious residue decreases as the pretreatment temperature is increased, whereas the capacity of the cellulose increases. The reduction of the enzyme adsorbed on the lignacious residue as the pretreatment temperature increases is important in increasing the ultimate recovery of the enzyme as well as enhancing the enzyme hydrolysis rate and extent.

An enzymatic hydrolysis process involving solid lignocellulosic materials can be designed in many ways. The common denominators are that the substrates and the enzyme are fed into the process, and the product stream (sugar solution) along with a solid residue leaves it at various points. The residue contains adsorbed enzymes, which are lost when the residue is removed from the system.

To ensure that the enzymatic hydrolysis process is economically efficient, a degree of enzyme recovery is essential. Both the soluble enzymes and the enzyme adsorbed to the substrate residue must be reutilized. It is expected that the loss of enzyme is influenced by the selection of the stages at which the enzymes in solution and adsorbed enzymes are recirculated and the point at which the residue is removed from the system. Vallander and Erikkson [69] defined an enzyme loss function, $L$, assuming that no loss occurs through filtration:

$$L = \frac{\text{amount of enzyme lost through removal of residue}}{\text{amount of enzyme at the start of hydrolysis}}$$

They developed a number of theoretical models and concluded that increased enzyme adsorption leads to increased enzyme loss. The enzyme loss decreases if the solid residue is removed late in the process. Both the adsorbed and dissolved enzymes should be reintroduced at the starting point of the process. This is particularly important for the dissolved enzymes. Washing of the entire residue is likely to result in significantly lower recovery of adsorbed enzymes than if a major part (60% or more) of the residue with adsorbed enzymes is recirculated. Uninterrupted hydrolysis over a given time period leads to a lower degree of saccharification than a process in which hydrolysate is withdrawn several times. Saccharification is also favored if the residue is removed at a late stage. Experimental investigations of the theoretical hydrolysis models have recovered more than 70% of the enzymes [69].

*Mechanism of hydrolysis.* The overall hydrolysis is based on the synergistic action of three distinct cellulase enzymes, depending on the concentration ratio and the adsorption ratio of the component enzymes: *endo*-β-gluconases, *exo*-β-gluconases, and β-glucosidases. *endo*-β-Gluconases attack the interior of the cellulose polymer in a random fashion [53], exposing new chain ends. Because this enzyme catalyzes a solid-phase reaction, it adsorbs strongly but reversibly to the crystalline cellulose (avicel). The strength of the adsorption is greater at lower temperatures. This enzyme is necessary for the hydrolysis of crystalline substrates. The hydrolysis of cellulose results in considerable accumulation of reducing sugars, mainly cellobiose, because the extracellular cellulase complex does not have cellobiose activity.

*exo*-β-Gluconases remove cellobiose units (two glucose units) from the non-reducing ends of cellulose chains. This is also a solid-phase reaction, and the *exo*-gluconases adsorb strongly on both crystalline and amorphous substrates. The picture is complicated because there are two distinct forms of both the *endo*- and *exo*-enzymes, each with a different type of synergism with the other members of the complex. As these enzymes continue to split off cellobiose units, the concentration of cellobiose in solution may increase. The action of *exo*-gluconases may be severely inhibited or even stopped by the accumulation of cellobiose.

The cellobiose is hydrolyzed to glucose by action of β-glucosidase. The effect of β-glucosidase on the ability of the cellulase complex to degrade avicel has been investigated by Kadam and Demain [70]. They determined the substrate specificity of the β-glucosidase and demonstrated that its addition to the cellulase complex enhances the hydrolysis of avicel specifically by removing the accumulated cellobiose. A thermostable β-glucosidase form, *C. thermocellum,* which is expressed in *Escherichia coli,* was used to determine the substrate specificity of the enzyme. The hydrolysis of cellobiose to glucose is a liquid-phase reaction and β-glucosidase absorbs either quickly or not at all on cellulosic substrates. Its action can be slow or halted by the action of glucose accumulated in the solution. The accumulation may also bring the entire hydrolysis to a halt as inhibition of the β-glucosidase results in a buildup of cellobiose, which in turn inhibits the action of *exo*-gluconases. The hydrolysis of the cellulosic materials depends on the presence of proper amounts of all three enzymes. If any enzyme is present in less than the required amount, the other enzymes will be inhibited or lack the necessary substrates to act on.

The hydrolysis rate increases with increasing temperature. However, because the catalytic activity of an enzyme is related to its shape, the deformation of the enzyme at high temperature can inactivate or destroy the enzyme. To strike a balance between increased activity and increased deactivation, it is preferable to run fungal enzymatic hydrolysis at approximately 40–50°C.

**Fermentation.** Cellulose hydrolysis and fermentation are being conducted in two processes depending on where the fermentation is carried out: separate hydrolysis and fermentation (SHF) or simultaneous saccharification and fermentation (SSF).

*Separate hydrolysis and fermentation.* In SHF, the hydrolysis is carried out in one vessel and the hydrolysate is then fermented in a second reactor. The most expensive items are the feedstock, enzyme production, hydrolysis, and utilities. The feedstock and utility costs are high because only 73% of the cellulose is converted to ethanol in 48 hours, while the remainder of the cellulose, hemicellulose and lignin, are burned. Enzyme production is expensive because of the large amount of enzyme used in an attempt to overcome the end product inhibition and because of its slow rate of production. Hydrolysis is expensive because of the large tanks and agitators. The most important parameters are the hydrolysis section yield, product, and required enzyme loading, which are all interrelated. Yields are higher in more dilute systems, where inhibition of enzymes by glucose and cellobiose is minimized. Increasing the enzyme loading can help to overcome inhibition and increase yield and concentration. Increased reaction times also result in higher yields and concentrations. Cellulase enzymes from different organisms can have markedly different performances. Figure 16 shows the effect of yield at constant solid and enzyme loading and shows the performance of different enzyme loadings. Enzyme loading beyond a particular point is of no use. It would be economical to operate at a minimum enzyme loading level. Other means could be to recycle the enzyme. As the cellulose is hydrolyzed, the *endo-* and *exo-*gluconase components are released back into the solution. Because of their affinity for cellulose, these enzymes can be recovered and reused by having the hydrolyzate make contact with fresh feed. The amount of recovery is limited because of β-glucosidase, which does not adsorb on the feed. Some of the enzyme remains attached to the lignin, and unreacted cellulose

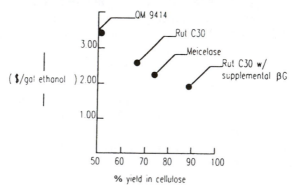

**Figure 16**   Effect of yield on selling price of ethanol [53] (βG, β-gluconase).

and enzyme are thermally denatured during hydrolysis. The major difficulty is in maintaining the sterility which would otherwise be contaminated. The power consumed in agitation is enormous and affects the economy of this process [53].

***Simultaneous saccharification and fermentation.*** This process is cheaper than SHF, and the hydrolysis and fermentation are carried out in the same vessel. In this process yeast ferments the glucose to ethanol as soon as the glucose is produced, preventing sugar accumulation and end product inhibition. Using the yeast *Candida brassicae* and the genecor enzyme, the yield is increased to 79% and the ethanol concentration produced is 3.7%.

Even in SSF the cellobiose (soluble sugar) inhibition occurs to an appreciable extent. The enzyme loading for SSF is only 7 IU/g cellulose, compared to 33 IU/g in SHF. The cost of energy and feedstock is somewhat reduced because of improved yield, and the increased ethanol concentration significantly reduces the cost of distillation and utilities. The cost of the SSF process is slightly less than the combined costs of hydrolysis and fermentation in the SHF process. The decreased reactor volume, because of the higher concentrations, offsets the increase in size because of the longer reaction times (7 days for SSF versus 2 days for hydrolysis and 2 days for fermentation). Experiments show that fermentation, not the enzymatic process, is the rate-controlling step. The hydrolysis is carried out at 37°C, and increasing the temperature increases the reaction rate, but this is limited by the yeast cell viability. The concentration of ethanol is also a limiting factor. (This was tested by connecting a flash unit to the SSF reactor and removing the ethanol periodically. This technique showed high productivities up to 44%.) Recycling the residual solids may also increase the process yield. However, the most important limitation in the enzyme recycling comes from the presence of lignin, which is inert to the enzymes. High recycling rates increase the fraction of lignin in the reactor and cause handling difficulties.

Two major types of enzyme recycling schemes have been proposed: those in which enzymes are recovered in the liquid phase and those in which enzymes are recovered by recycling unreacted solids [53]. Systems of the first type have been proposed for SHF processes that operate at 50°C. These systems are favored at such a high temperature because increasing temperature increases the proportion of enzyme that remains in the liquid phase. Conversely, as the temperature is decreased, the amount of enzyme adsorbed on the solid increases. At the lower temperatures encountered in SSF processes, solids recycling appears to be more effective.

***Comparison of SSF and SHF processes.*** SSF systems offer large advantages over SHF processes because of their reduction in end product inhibition of the cellulase enzyme complex. The SSF process has slightly increased yields (88 versus 73%) and greatly increased product concentrations (equivalent glucose concentration of 10 versus 4.4%). Most significant is the enzyme loading, which can be reduced from 33 to 7 IU/g cellulose; this cuts down the cost of ethanol appreciably. The approximate costs of two processes can be seen in Table 5.

**Xylose fermentation.** Because xylose accounts for 30–60% of the fermentable sugars in hardwood and herbaceous biomass, it becomes an important issue to ferment it to ethanol. The efficient fermentation of xylose and other hemicellulose constituents is essential for the development of an economically viable process to produce ethanol from biomass. Xylose fermentation using pentose yeasts has proved difficult due to the requirement for $O_2$ during ethanol production, acetate toxicity, and the production of xyletol as by-product. Other approaches to xylose fermentation include the conversion of xylose to xylulose using xylose isomerase before fermentation by *Saccharomyces cerevisiae* and the development of genetically engineered strains [76].

The method for integrating xylose fermentation into the overall process is shown in Figure 17. The liquid stream is neutralized to remove any mineral acids or organic acids liberated in the pretreatment process and then sent to xylose fermentation. Water is added before the fermentation, if necessary, so that organisms can make full use of the substrate without having the yield limited by end product inhibition. The dilute ethanol stream from xylose fermentation is then used to provide the dilution water for the cellulose-lignin mixture entering SSF. Thus, the water that enters during the pretreatment process is used in both the xylose fermentation and the SSF process.

The conversion of xylose to ethanol by recombinant *E. coli* has been investigated in pH-controlled batch fermentations [78]. Relatively high concentrations of ethanol (56 g/L) were produced from xylose with excellent efficiencies. In

**Table 5 Cost of production summary for SHF and SSF processing of lignocellulose to ethanol (25,000,000 gal/yr) [53]**

|  | SHF | SSF |
|---|---|---|
| Raw materials | | |
|   Wood (lb) | 82.01 | 68.4 |
|   Sulfuric acid (lb) | | 2.2 |
|   Lime (lb) | 1.8 | 1.2 |
|   Chemicals | 3.6 | 1.7 |
| Utilities | | |
|   Water (1000 gal) | 1.0 | 0.6 |
| Labor (hr) | 12.6 | 7.4 |
| Overhead and maintenance | 59.9 | 35.3 |
|  | 161.0 | 116.2 |
| Annual operating cost: capital charges (capital recovery factor = 0.13) (15% internal rate of return, 20 years straight line depreciation) | 105.4 | 62.2 |
| Ethanol selling price | 266.4 | 178.4 |

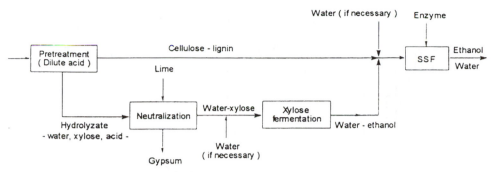

**Figure 17** Integration of xylose fermentation and SSF [53].

addition to xylose, all other sugar constituents of biomass can be efficiently converted to ethanol by recombinant *E. coli*. Neither oxygen nor strict maintenance of anaerobic conditions is required for ethanol production by *E. coli*. However, addition of base to prevent excessive acidification is essential. Although less base is needed to maintain low-pH conditions, poor ethanol yields and slower fermentations are observed below a pH of 6. Also, the addition of metal ions stimulates ethanol production. In general, xylose fermentation does not require precise temperature control provided the broth temperature is maintained between 25 and 40°C. Xylose concentrations as high as 140 g/L have been positively tested to evaluate the extent to which this sugar inhibits growth and fermentation. Higher concentrations considerably slow down growth and fermentation.

**Ethanol extraction during fermentation.** In spite of the considerable efforts given to fermentative alcohols, industrial applications have been delayed because of the high cost of production, which depends primarily on the energy input to the purification of dilute end products and on the low productivities of cultures. These two points are directly linked to inhibition phenomena.

Along with the conventional unit operations, liquid-liquid extraction with biocompatible organic solvents, distillation under vacuum, and selective adsorption on the solids have demonstrated the technical feasibility of the extractive fermentation concept. Lately, membrane separation processes, which decrease biocompatibility constraints, have been proposed. These include dialysis [71] and reverse osmosis [72]. More recently, the concept of supported liquid membranes is being reported. This method minimizes the amounts of organic solvents involved and permits simultaneous realization of the extraction and recovery phases. Enhanced volumetric productivity and high substrate conversion yields have been reported [73] using a porous Teflon sheet as the support (soaked with isotridecanol) for the extraction of ethanol during semicontinuous fermentation of *Saccharomyces bayanus*. This selective process results in ethanol purification

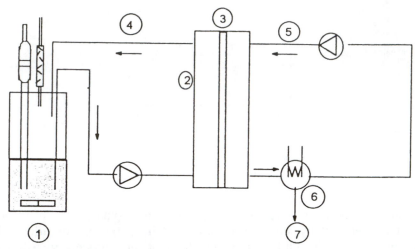

**Figure 18** Extractive fermentation system: (1) fermentor; (2) permeation cell; (3) supported liquid membrane; (4) extracted phase; (5) gaseous stripping phase; (6) cold trap; (7) condensed permeate [73].

and combines three operations, fermentation, extraction, and reextraction (stripping), as schematically represented in Figure 18.

## 10.4.4 Lignin Conversion

Lignin is produced in large quantities, approximately 250 billion pounds per year in the United States as by-products of the paper and pulp industry. Lignins are complex amorphous phenolic polymers, not sugar based, and hence cannot be fermented to ethanol. Lignin is a random polymer made up of phenylpropane units, where the phenol unit may be either a guaiacyl or a syringyl unit (Figure 19). These units are bonded together in many ways, the most common of which are α- or β-ether linkages. A variety of C—C linkages are also present but are less common (Figure 20). The distribution of linkage in lignin is random because lignin formation is a free radical reaction that is not under enzymatic control.

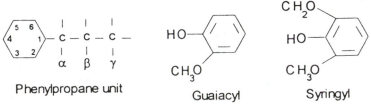

**Figure 19** Monomer units in lignin [53].

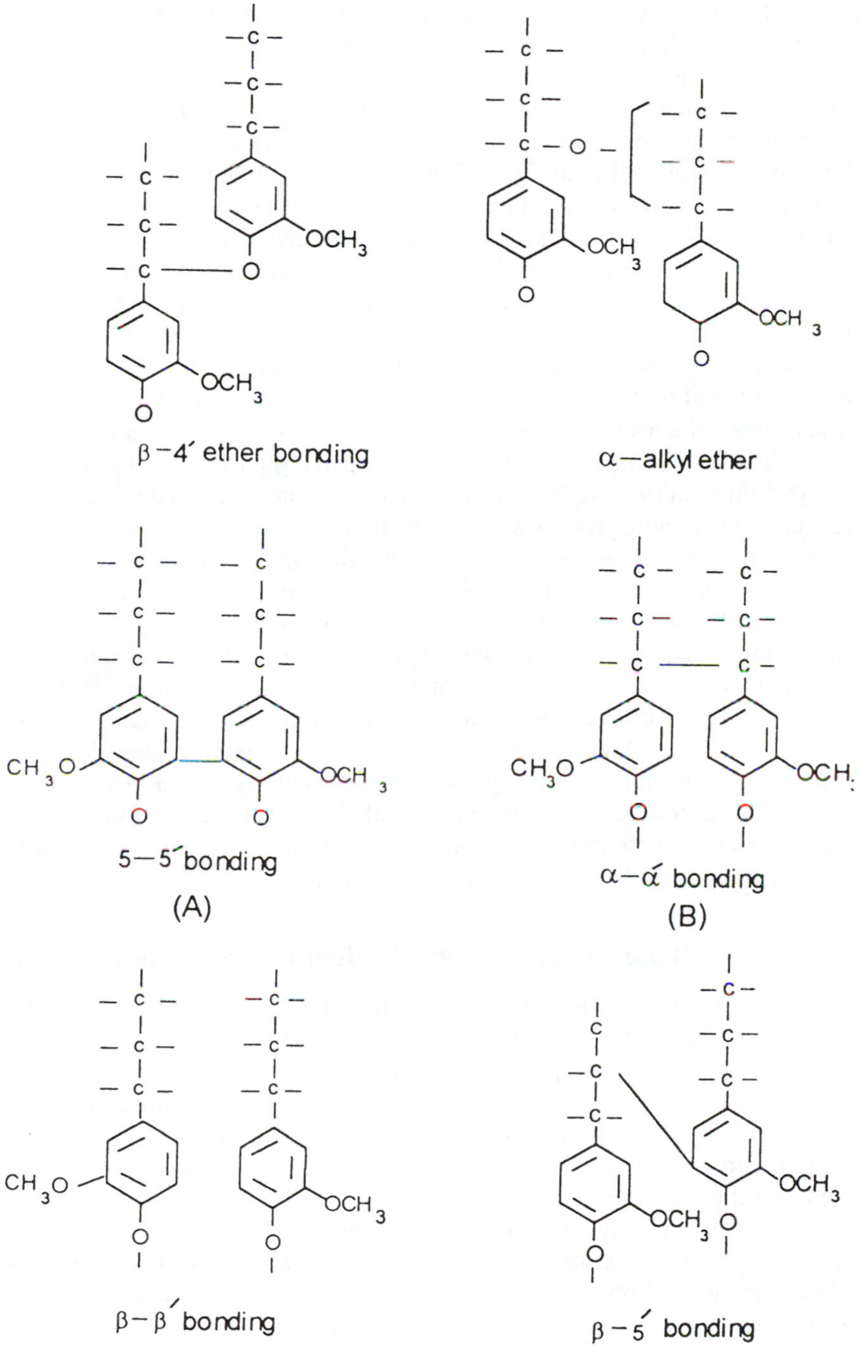

β—4′ ether bonding

α—alkyl ether

5—5′ bonding

(A)

α—α′ bonding

(B)

β—β′ bonding

β—5′ bonding

**Figure 20**  Ether and C—C bonds in lignin [53].

Lignin is highly resistant to chemical, enzymatic, or microbial hydrolysis due to extensive cross-linking. Therefore lignin is frequently removed simply to gain access to cellulose.

Lignin monomer units are similar to gasoline that has a high octane number; thus, breaking the lignin molecules into monomers and removing the oxygen makes them useful as liquid fuels. The process for lignin conversion is mild hydrotreating to produce a mixture of phenolic and hydrocarbon materials, followed by reaction with methanol to produce methyl aryl ether. The first step usually consists of two parts: hydrodeoxygenation (removal of oxygen and oxygen-containing groups from the phenol ring) and dealkylation (removal of ethyl or large side chains from the rings). The major issue is to carry out these reactions to remove the unwanted chains without carrying the reaction too far, which would lead to excessive consumption of hydrogen and produce saturated hydrocarbons, which are not as good octane enhancers as the aromatic compounds. Catalysts to carry out these reactions have dual functions. Metals such as molybdenum and molybdenum-nickel catalyze the deoxygenation, while the acidic alumina support promotes the carbon-carbon bond cleavage.

Although lignin chemicals have many applications such as in drilling muds, as binders for animal feed, and as the base for artificial vanilla, they have not been previously used as surfactants for oil recovery. According to Naae [41], lignin chemicals can be used in two ways in chemical floods for enhanced oil recovery. In one method, lignosulfonates are blended with tallow amines and conventional petroleum sulfonates to form a unique mixture that costs about 40% less to use than chemicals made solely from petroleum or petroleum-based products. In the second method, lignin is reacted with hydrogen or carbon monoxide to form a new class of chemicals called lignin phenols. These phenols, because they are soluble in organic solvents but not in water, are good candidates for further conversion to chemicals useful in enhanced oil recovery.

### 10.4.5 Energy Balance for Ethanol Production from Biomass

Biomass process development depends on the economics of the conversion process, whether it be chemical, enzymatic, or a combination of both. A number of estimates have been computed based on existing or potential technologies. One obvious factor is that, regardless of the process, transportation of the biomass material from its source to the site of conversion must be kept to an absolute minimum. Approximately 35% of the expected energy is consumed in transporting the substrate a distance of 15 miles [67]. This considerable expenditure of energy simply to transport the starting material dictates that any conversion plant be of moderate size in close proximity to the production source.

There are several objections to the production and use of ethanol as a fuel. Most important is the criticism that producing ethanol can consume more energy than the finished ethanol contains. The European analysis takes wheat as the feedstock and includes estimates of the energy involved in growing the wheat, transporting it to the distillery, making the alcohol, and transporting it to a refinery for blending with petrol. It allows credits for by-products, such as animal feed from wheat, for savings on petrol that comes from replacing 5% with alcohol, and for the energy gained from the increase of 1.25 octane points. Yet, to confine debates on biomass fuels solely to energy balance is misleading. At least 13 plants, like the one in Sweden, are working or under construction in Europe, South America, and Asia [50].

The greatest opposition to bioethanol, not surprisingly, comes from the oil industry, where the preference is to produce low-octane petrol together with octane boosters based on oil. This would reduce the investment needed at the refinery. They would also sell more petrol because the reduction of one octane point increases a car's fuel consumption by 1–1.5% [50].

Energy requirements to produce ethanol from different crops were evaluated by Da Silva et al. [74]. The industrial phase is always more energy intensive, consuming 60 to 75% of the total energy. The energy expended in crop production includes all the forms of energy used in agricultural and industrial processing, except the solar energy that plants use for growth. The industrial stage, including extraction and hydrolysis, alcohol fermentation, and distillation, requires about 6.5 kg of steam per liter of alcohol. It is possible to furnish the total industrial energy requirements from the by-products of some of the crops. Thus it is also informative to consider a simplified energy balance in which only agricultural energy is taken as the input and only ethanol is taken as the output, the bagasse supplying energy for the industrial stage, for example.

There have been several energy analyses of ethanol production from food crops; they have been characterized by conflicting results. For example, results from Brazil show that sugarcane has a favorable energy balance for ethanol production [74]. In contrast, at Iowa State University [75] it was concluded "it cannot be claimed that ethanol fermentation of ethanol produces energy, the opposite is instead true." The dichotomous nature of these analyses shows the need for more site-specific studies. It is also important to resolve this matter, as the very existence of alcohol plants in some countries could be threatened.

The energy balance results in Zimbabwe have shown that the energy ratio is 1.52 if all the major outputs are considered and 1.15 if ethanol is considered as the only output. The reported value of the net energy ratio in Brazil [74] is 2.41 and in Louisiana [59] is 1.85. The low ratio in Zimbabwe is due to the large energy input in the agricultural phase, arising from a large need for fertilizer, and the large fossil-based fuel consumption in sugarcane processing.

The cost of producing ethanol decreases with an increase in capacity of the production facility [83]. However, the minimum total cost corresponds to a point of inflection at which an increase in the production cost for every increase in the plant capacity is seen [76]. The possibility of an empirical relationship between the plant size or output and the production costs has also been examined using various production functions and the computed $F$ values at a level of significance of 5%. It was found [77] that the best function is

$$Y = 7.426 - 0.4094X + 0.0058X^2$$

This means that, for optimal plant performance, the $X$ and $Y$ should be 35.293 MI y[1], increasing the yield and product concentration and reducing the amount of enzyme required.

Xylose fermentation is being carried by bacterial, fungal, yeast, or enzyme-yeast systems. This would reduce the cost by 25% or more in the case of herbaceous-type materials. Efforts are being made to achieve a yield of 100% and an increased ethanol concentration.

Lignin, another major component of biomass, accounts for the large energy contents of biomass because it has much higher energy per pound than the carbohydrates. Because it is a phenolic polymer it cannot be fermented to sugar and converted to materials like methyl aryl ethers, which are compatible with gasoline as a high-octane enhancer. The combination of the above processes has the potential to produce fuels for under $1.0/gal.

## REFERENCES

1. Sampson, R. N., et al. Biomass management and energy. *Water, Air Soil Pollut.* 70(1–4):139–159, 1993.
2. Trebbi, G. Power-production options from biomass: The vision of a southern European utility. *Bioresource Technol.* 46:23–29, 1993.
3. MacCleery, D. W. American forests: A history of resiliency and recovery. U.S. Dept. of Agriculture Forest Service, p. 59, 1993.
4. Reese, R. A., et al. Herbaceous biomass feedstock production. *Energy Policy* 21(7):726–734, 1993.
5. Sampson, R. N. Forest management & biomass in the USA. *Water, Air Soil Pollut.* 70(1–4): 519–532, 1993.
6. Nurmi, J. Heating values of the above ground biomass of small-sized trees. *ACTA Forestalia Fennica* 236:2–30, 1993.
7. Sipilä, K. New power-production technologies: Various options for biomass & cogeneration. *Bioresource Technol.* 46:5–12, 1993.
8. Ramsay, W. Biomass energy in developing countries. *Energy Policy* August: 326–329, 1985.
9. Elliot, P. Biomass-energy overview in the context of Brazilian biomass powered demonstration. *Bioresource Technol.* 46:13–22, 1993.
10. Ellegard, A., and H. Egenéus. Urban energy: Exposure to biomass fuel pollution in Lusaka. *Energy Policy* 21(5):622–625, 1993.
11. Bylinsky, G. Biomass: The self-replacing energy source. *Fortune* 100(6):78–81, 1979.
12. Lapidus, A., et al. Synthesis of liquid fuels from products of biomass gasification. *Fuel* 73(4): 583–589, 1994.
13. Randolph, J. C., and G. L. Fowler. Energy policies & biomass tesources in the Asia-Pacific region. *Public Administration Rev.* 43(6):528–536, 1983.
14. Wright, L. L., and E. E. Hughes. U.S. carbon offset potential using biomass energy systems. *Water, Air Soil Pollut.* 70(1–4):483–497, 1993.

15. Rahmer, B. A. Alternative energy: Towards fuel farming. *Petroleum Economist*, 78:59–60, 1978.
16. Scurlock, J. M. O., D. O. Hall, J. I. House, and R. Howes. Utilizing biomass crops as an energy source: A European perspective. *Water, Air Soil Pollut.* 70(1–4):499–518, 1993.
17. Pastor, J., and L. Kristoferson. *Bioenergy and the Environment—The Challenge.* Boulder, CO: Westview Press, 1990.
18. Bain, R. L. Electricity from biomass in the United States: Status and future direction. *Bioresource Technol.* 46(1–2):86–93, 1993.
19. Freeman, H. M., ed. *Standard Handbook of Hazardous Waste Treatment & Disposal.* New York: McGraw-Hill, 1989.
20. Benson, W. R. Biomass potential from agricultural production. *Proceedings: Biomass—A Cash Crop for the Future?* Kansas City, 1977.
21. Raman, K. P., W. P. Walawender, Y. Shimizu, and L. T. Fan. Gasification of corn stover in a fluidized bed. Presented at Bio-Energy 80, Atlanta, April 1980.
22. Bodland, B., and J. Bergman. Bioenergy in Sweden: Potential, technology, and application. *Bioresource Technol.* 46(1–2):31–36, 1993.
23. Hayes, R. D. Overview of thermochemical conversion of biomass in Canada. In: *Biomass Pyrolysis Liquids: Upgrading & Utilization.* New York: Elsevier, 1991.
24. Farrell, K. Fighting OPEC with biomass: The European counteroffensive. *Europe* May-June: 18–20, 1981.
25. Bergeron, P. W., and N. D. Hinman. Fuel ethanol usage & environmental carbon dioxide production. In *Energy from Biomass & Wastes XIV*, ed. D. L. Klass, pp. 153–167. Chicago: Institute of Gas Technology, 1991.
26. Borgwardt, R. H., M. Steinberg, and E. W. Grohse. Biomass & fossil fuel to methanol and carbon via the hydrocarb proces: A potential new source of transportation and utility fuels. In *Energy from Biomass & Wastes XIV*, ed. D. L. Klass, pp. 823–853. Chicago: Institute of Gas Technology, 1991.
27. Solantausta, Y., et al. Wood-pyrolysis oil as fuel in a diesel-power plant. *Bioresource Technol.* 46:177–188, 1993.
28. Rick, F., and U. Vix. Product standards for pyrolysis products for use as a fuel. In *Biomass Pyrolysis Liquids Upgrading & Utilization*, eds. A. V. Bridgwater and G. Grassi, pp. 177–218. London: Elsevier, 1991.
29. Skelley, W. W., J. W. Chrostowki, and R. S. Davis. The energy resources fluidized bed process for converting biomass to electricity. Symposium on Energy from Biomass & Wastes VI, pp. 665–705, January 25–29, 1982.
30. Brown, R. F., et al. Economic evaluation of the coproduction of methanol & electricity with texaco gasification-combined-cycle systems, EPRI AP-2212, 1983.
31. Ismail, A., and R. Quick. Advances in biomass fuel technologies. Presented at Energy from Biomass & Wastes XV, IGT, Washington, DC, 1991.
32. Keuster, J. L. Liquid hydrocarbons fuels from biomass. In *Biomass as a Nonfossil Fuel Source*, pp. 163–184. Washington, DC: American Chemical Society, 1991.
33. Borgwardt, R. H., M. Steinberg, E. W. Grohse, and Y. Tung. Biomass & fossil fuel to methanol and carbon via the hydrocarb process: A potential new source of transportation and utility fuels. Presented at Energy from Biomass & Wastes XV, IGT, Washington, DC, 1991.
34. Schaefer, G. P. Industrial development of biomass energy sources. In *Biomass as a Nonfossil Fuel Source—I*, pp. 1–17. Washington, DC: American Chemical Society, 1981.
35. Wendt, H., V. Plzak, and B. Rohland. Conversion of biomass-obtained gases in solid oxide fuel cells. Luxembourg Report 13564, 1991.
36. Bartlett, H. D. Energy production of a 100 m$^3$ biogas generator. In *Biomass as a Nonfossil Fuel Source—I*, pp. 373–378. Washington, DC: American Chemical Society, 1981.
37. Ghosh, S., and D. L. Klass. Advanced digestion process development for methane production from biomass-waste blends. In *Biomass as a Nonfossil Fuel Source—I*, pp. 251–278. Washington, DC: American Chemical Society, 1981.

38. Anderson, E. V. Brazil's fuel ethanol program comes under fire. *C&EN* 67(March 20, 1989): 11–12, 1989.
39. Long, J. R. More energy research called for to stem oil, climate change crises. *C&EN* 68 (September 10, 1990):16–17, 1990.
40. Gaddy, J. L. Renewable energy sources and improved methods of employing fossil fuels. Proceedings, 200th National Meeting of ACS, in Symposium on Fuel Chemistry, Washington, DC, August, 1990.
41. Naee, D. G. ACS Press Conference, 200th National Meeting of the ACS, August, 1990, Washington, DC, *C&EN* 68 (September 10, 1990):17, 1990.
42. Hodzi, D. *Hydrogen Bonding.* London: Pergamon Press, 1957.
43. Ellis, C. *Chemistry of Petroleum Derivatives*, vol. 2. New York: Reinhold, 1937.
44. Judice, C. A., and L. E. Pirkle. U.S. Patent 3,095,458 (June 25, 1963) to Esso Research and Eng. Co., 1963.
45. Lewis, W. K. U.S. Patent 2,045,785 (June 30, 1936) to Standard Oil Dev. Co., 1936.
46. Miller, S. A. *Ethylene and Its Industrial Derivatives.* London: Ernest Benn, 1969.
47. Bailey, J. E., and D. F. Ollis. *Biochemical Engineering Fundamentals*, 2nd ed. New York: McGraw-Hill, 1986.
48. Park, S. C., and J. Baratti. Batch fermentation kinetics of sugar beet molasses by *Zymomonas mobilis. Biotechnol. Bioeng.* 38:304–313, 1991.
49. U.S. Congress, *Energy from Biological Processes*, 2, pp. 142–177. Washington, DC: Office of Technology Assessment, U.S. Govt. Printing Office, 1980.
50. de Groot, P., and D. Hall. Power from the farmers. *New Scientist* 112:50–55, 1986.
51. Szamant, H. Big push for a biomass bonanza. *Chem. Week* 122(14):40, 1978.
52. Goldstein, I. S. Department of Wood and Paper Science, North Carolina State University, *C&EN* 68 (September 10, 1990):20, 1990.
53. Wright, J. D. Ethanol from biomass by enzymatic hydrolysis. *Chem. Eng. Prog.* 84:62–74, 1988.
54. Diaz, L. F., G. M. Savage, and C. G. Golueke. Critical review of energy recovery from solid wastes. *CRC Crit. Rev. Environ. Control* 14(3):285–288, 1984.
55. Farina, G. E., J. W. Barrier, and M. L. Forsythe. Fuel alcohol production from agricultural lignocellulosic feedstocks. *Energy Sources* 10:231–237, 1988.
56. Vallander, L., and K. Erikkson. Enzymatic hydrolysis of lignocellulosic materials: I. Models for the hydrolysis process—a theoretical study. *Biotechnol. Bioeng.* 38:135–138, 1991.
57. Rughani, J., and G. D. McGinnis. Combined rapid-steam hydrolysis and organosolv pretreatment of mixed southern hardwoods. *Biotechnol. Bioeng.* 33:681–686, 1989.
58. Holtzapple, M. T., M. Cognata, Y. Shu, and C. Hendrickson. Inhibition of *Trichoderma reesei* cellulase by sugars and solvents. *Biotechnol. Bioeng.* 38:296–303, 1991.
59. Blotkamp, P. J., M. Takagi, M. S. Pemberton, and G. H. Emert. In *Biochemical Engineering: Renewable Sources of Energy and Chemical Feedstocks.* eds. J. M. Nystrom and S. M. Barnett, p. 74. AICHE Symposium Series No. 181. New York: AICHE, 1978.
60. Ohmine, K., H. Ooshima, and Y. Harano. Kinetic study on enzymatic hydrolysis of cellulose by cellulase from *Trichoderma viride. Biotechnol. Bioeng.* 25:2041–2053, 1983.
61. Gonzales, G., G. Caminal, C. de Mas, and J. L. Santin. *J. Chem. Tech. Biotechnol.* 44:275, 1989.
62. Ryu, D. Y., and S. B. Lee. Enzymatic hydrolysis of cellulose: Determination of kinetic parameters. *Chem. Eng. Commun.* 45:119–134, 1986.
63. Beltrame, P. L., P. Carniti, B. Focher, A. Marzetti, and V. Sarto. Enzymatic hydrolysis of cellulosic materials: A kinetic study. *Biotechnol. Bioeng.* 26:1233–1238, 1984.
64. Okazaki, M., and M. Young. Kinetics of enzymatic hydrolysis of cellulose: Analytical description of mechanistic model. *Biotechnol. Bioeng.* 20:637–663, 1978.
65. Holtzapple, M. T., H. S. Caram, and A. E. Humphrey. The HCH-1 model of enzymatic cellulose hydrolysis. *Biotechnol. Bioeng.* 26:775–780, 1984.
66. Wald, S., C. R. Wilke, and H. W. Blanch. Kinetics of the enzymatic hydrolysis of cellulose. *Biotechnol. Bioeng.* 26:221–230, 1984.

67. Batt, C. A., V. Moses, and R. E. Cape. *Biotechnology, the Science and the Business*, pp. 521–536. Harwood: Academic Press, 1991.
68. Ooshima, H., D. S. Burns, and A. O. Converse. Adsorption of cellulase from *Trichoderma reesei* on cellulose and lignacious residue in wood pretreated by dilute sulfuric acid with explosive decompression. *Biotechnol. Bioeng.* 36:446–452, 1990.
69. Vallander, L., and K. Erikkson. Enzymatic hydrolysis of lignocellulosic materials: II. Experimental investigations of theoretical hydrolysis process models for an increased enzyme recovery. *Biotechnol. Bioeng.* 38:139–144, 1991.
70. Kadam, S., and A. Demain. Addition of cloned β-glucosidase enhances the degradation of crystalline cellulose by the *Clostridium thermocellum* cellulase comolex. *Biochem. Biophys. Res. Commun.* 161(2):706–711, 1989.
71. Kyung, K. H., and P. Gerhardt. Continuous production of ethanol by yeast "immobilized" in membrane-contained fermentor. *Biotechnol. Bioeng.* 26:252, 1984.
72. Garcia, A., E. L. Lannotti, and J. L. Fischer. Butanol fermentation liquor production and separation by reverse osmosis. *Biotechnol. Bioeng.* 28:785–791, 1986.
73. Christen, P., M. Minier, and H. Renon. Ethanol extraction by supported liquid membrane during fermentation. *Biotechnol. Bioeng.* 36:116–123, 1990.
74. Da Silva, J. G., G. E. Serra, J. R. Moreira, J. C. Concalves, and J. Goldenberg. Energy balance for ethyl alcohol production from crops. *Science* 201:903–906, 1978.
75. Reilly, P. Economics and energy requirements for ethanol production. Dept. of Chemical Eng. and Nuclear Eng., Iowa State University, Ames, 1978.
76. Alam, W. and J. M. Amos. The economies of scale of fuel-grade ethanol plants. *Energy* 7(6): 477–481, 1982.
77. Gladius, L. Some aspects of the production of ethanol from sugar cane residues in Zimbabwe. *Solar Energy* 33(3/4):379–382, 1984.
78. Beall, D. S., K. Ohta, and L. O. Ingram. Parametric studies of ethanol production from xylose and other sugars by recombinant *Escherichia coli*. *Biotechnol. Bioeng.* 38:296–303, 1991.
79. Hopkinson, C. S., and J. W. Davy. Net energy analysis of alcohol production from sugarcane. *Science* 207:302–304, 1980.
80. *Kirk-Othmer Encyclopedia of Chemical Technology*, 3rd ed., vol. 9, pp. 342–351. New York: Wiley, 1978.
81. L'Italien, Y., J. Thibault, and A. Le Duy. Improvement of ethanol fermentation under hyperbaric conditions. *Biotechnol. Bioeng.* 33:471–476, 1989.
82. Sarthy, A., L. McConaughy, Z. Lobo, A. Sundstorm, E. Furlong, and B. Hall. Expression of the *Escherichia coli* xylose isomerase gene in *Saccharomyces cerevisiae*. *Appl. Environ. Microbiol.* 53:1996–2000, 1989.
83. Sperling, D. An analytical framework for siting and sizing biomass fuel plants. *Energy* 9(11–12): 1033–1040, 1984.
84. *C&EN* September 10, 1995.

# PROBLEMS

1. Define "biomass" from the viewpoint of energy source. What is included in this category?
2. List several examples of biomass utilization for energy generation.
3. Does biomass fuel contain more sulfur than conventional coal?
4. Which contributes more to the greenhouse $CO_2$: biomass fuel or coal?
5. Is the direct combustion of biomass economically viable compared to coal combustion?

6. Explain briefly the three primary conversion routes of biomass.
7. What is "biogas"? What gases are typically included?
8. Discuss the discerning characteristics of biomass gasification in comparison to coal gasification.
9. Explain briefly Sweden's VEGA gasification system.
10. What is the mixed wood feedstock comprised of?
11. Describe the pyrolysis process.
12. Explain the difference between pyrolysis and devolatilization.
13. Compare the wood pyrolysis products with the coal pyrolysis products.
14. Discuss the economics of biomass pyrolysis.
15. What are the products of the Hydrocarb process?
16. Explain briefly anaerobic digestion.
17. What makes the liquid alcohol behave as though it were largely dimerized?
18. What are the main processes to manufacture ethanol from ethylene?
19. What are the raw materials that can be used for fermentation to produce ethanol?
20. What are the technological advantages of using the black-strap molasses in alcohol fermentation?
21. Write down the stoichiometric equation for conversion of starch to ethanol.
22. Of the two steps in the alcohol production from cellulosic materials, the hydrolysis step is a rate determining one. What do we mean by a rate determining step?
23. Ethanol-water binary system can produce an azeotropic mixture. When does this happen?
24. What is cellobiose?
25. What are four major steps in enzymatic process of lignocellulose?
26. Explain briefly autohydrolysis.
27. Discuss the advantages of the Organosolv pretreatment process.
28. What is the function of cellulase?
29. Discuss the inhibition of cellulase enzyme.
30. What are the three cellulase enzymes that control the overall hydrolysis?
31. What is the role of $\beta$-glucosidase?
32. Discuss the advantages and disadvantages of the simultaneous saccharification and fermentation (SSF) process.

# ALTERNATIVE SOURCE OF ENERGY FROM SOLID WASTES

## 11.1 INTRODUCTION

Solid wastes are, by definition, any wastes other than liquids or gases that are no longer deemed valuable and therefore discarded [1]. Such wastes are generated by the residential community (i.e., municipal solid waste), as well as commercial and light industrial communities. Wastes from heavy industrial and chemical industries are typically classified as hazardous wastes. As regulations continue to increase with decreasing land availability, alternative uses for the wastes must be found. As indicated from the heating values in Table 1, the generation of waste-derived fuels appears to be promising from both the environmental and energy aspects [2, 3]. This chapter addresses the development of alternative energy sources from the various solid waste classifications.

## 11.2 ENERGY RECOVERY FROM MUNICIPAL SOLID WASTES (MSW)

### 11.2.1 Introduction

Recovery of energy from municipal solid waste (MSW) in its simplest form has existed for centuries. The burning, or incineration, of wastes such as wooden planks and miscellaneous household products was first used to produce warmth.

**Table 1 Comparison of heating values of various waste-derived fuels**

| Fuel source | BTUs per pound |
|---|---|
| Yard waste | 3,000 |
| MSW | 6,000 |
| Combustible paper products | 8,500 |
| Textiles and plastics | 8,000 |
| Bituminous coal (average) | 11,300 |
| Anthracite coal (average) | 12,000 |
| Tires | 13,000–15,000 |
| Crude oil (average) | 17,000 |
| Natural gas (425 ft$^3$) | 13,500 |

This idea was the basis for energy generation from today's MSW. For instance, each year Sweden burns 1.5 million tons of MSW to meet approximately 15% of its district heating requirements [4]. The heating value of this incinerated MSW is approximately one-third the heating value of coal. Besides incineration, gaseous "fuels" can be obtained by anaerobic digestion in conjunction with landfill gas recovery.

## 11.2.2 Gasification

One method for recovering usable energy from MSW is gasification. The Environmental Protection Agency (EPA) is currently investigating the use of the Texaco gasification process for generating a medium-Btu gas from MSW [5]. The simplified Texaco process (Figure 1) gasifies the MSW under high pressure by injection of air and steam with concurrent gas-solid flow. After separation of the noncombustible waste, water or oil is added to the combustible MSW to form a pumpable slurry. The slurry is then pumped under pressure to the gasification reactor. In the gasifier, the slurry is reacted with air at high temperatures. The resulting gaseous product is then sent to a scrubbing system to remove any impurities.

The Texaco process is well known as a coal gasification system (refer to Chapter 3). It has been modified to treat soils contaminated with hydrocarbons, as well as recover usable energy from MSW and polymeric wastes. Although the process has demonstrated 85% remediation efficiency for contaminated soils, it is still in the preliminary experimental stages for the MSW-to-energy application [5]. A demonstration of the MSW-to-energy process has been scheduled for July 1995 at Texaco's Montebello Research Laboratory in South El Monte, California [5].

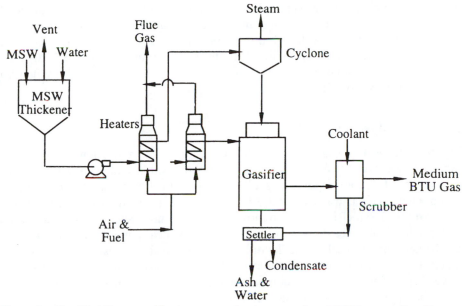

**Figure 1** Simplified Texaco gasification process for the conversion of MSW to a medium-Btu gas.

## 11.2.3 Digestion of MSW

Anaerobic digestion of solid wastes is a process very similar to that used at wastewater treatment facilities. Bacteria, in the absence of oxygen, are used to break down the organic matter of the waste. Frequently, the MSW is mixed with sewage sludge from the treatment plant to enhance the efficiency of the digestion. During the conversion, a mixture of methane and carbon dioxide gases is produced. The typical ratio of the gas mixture is 70% methane and 30% carbon dioxide. Even without further treatment, the off-gas has a heating value of 650 to 750 Btu/ft$^3$.

With rising energy costs, the use of anaerobic digestion to generate a potential fuel source from MSW is an attractive alternative. A new technology, landfill gas recovery, has been developed to aid in the collection of gases generated from the anaerobic digestion of solid wastes. In 1980, twenty-three landfills were used as a source of methane production [6]. However, most current research focuses on the generation of liquid fuels instead of gaseous fuels from anaerobic digestion because of the high capital cost associated with collecting methane.

Production of liquid fuels has several advantages. First, low-sulfur, low-ash fuels can be made for commercial use [7]. Second, liquid fuels are traditionally much easier to store, handle, and transport than their gaseous counterparts. Finally, the production of the fuel aids in the battle against pollution by municipal wastes [8, 9]. Using the waste to generate fuel means that less MSW has to be disposed of in landfills. Although this statement holds true for the production of gaseous fuel, the generation of liquid fuels utilizes more of the MSW. In fact, processes have been developed that will make over a barrel of pyrolytic fuel oil from a ton of MSW [7].

## 11.2.4 Pyrolysis of MSW

Figure 2 shows a typical material distribution for MSW generation in the United States. Since raw MSW contains both noncombustible and combustible components, the first step in producing a liquid fuel is to concentrate the combustible components. This is usually achieved with a rotating screen to remove glass and dirt. An air classifier is used to remove the "lightends" such as plastics, wood, and small metals. Heavier components, ceramics, heavy metals, and aluminum, are routed for disposal in the landfill. With removal of these noncombustible materials, the heating value of the raw MSW is approximately 7000 Btu/lb on a wet basis. The combustible components are sent to a shredder to reduce their size, then pyrolyzed to generate the fuel.

In the past, pyrolysis of MSW was used only to generate a gaseous fuel. However, current research has found that pyrolyzing a cellulose-based waste at

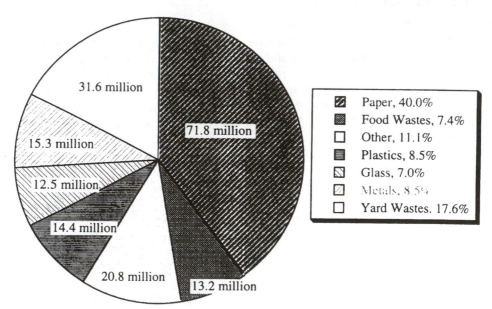

**Figure 2**  Material distribution of MSW by weight (tons).

116°C and atmospheric pressure will generate a liquid fuel. In 1988 approximately 80% of the 180 million tons of waste generated in the United States had a cellulose base [10]. Pober and Bauer [7], were able to utilize this particular type of waste to obtain a fuel having 77% of the heating value of typical petroleum fuels. The cellulosic components of the refuse comprise paper, newsprint, packing materials, wood clippings, and yard clippings [11]. A possible pyrolytic reaction for the cellulose component of MSW is:

$$C_8H_{10}O_5 \xrightarrow{\Delta} 8H_2O + 2CO + 2CO_2 + CH_4 + H_2 + 7C + C_6H_8O \quad (1)$$

where $C_6H_8O$ represents a family of liquid products. The exact composition of $C_6H_8O$ is dependent on the feedstock composition and reaction temperature. Research conducted by the U.S. Bureau of Mines has further demonstrated the successful pyrolysis of 1 ton of MSW at temperatures ranging from 500 to 900°C. At these conditions, the end product composition is similar to that given in Table 2. The light oil is primarily composed of benzene, and the liquor contains dissolved organics in water. The gaseous product resembles a typical town gas. The heating value of the gas is 447 Btu/ft$^3$, which translates to a heat recovery of 82% [13].

When the pyrolysis temperature is greater than 350°C, small quantities of polyethylene chips can be added to the cellulose feedstock. Operating temperatures ranging from 100 to 400°C decrease the amount of gaseous product and increase the formation of the liquid product ($C_6H_8O$ family).

Besides the liquid fuel, or oil, pyrolysis generates a medium-Btu value gas stream that, after purification, can be recycled as a supplemental fuel within the plant. Process water and char are also generated. All of the pyrolysis products have the potential to be useful fuels or intermediates for producing useful fuels for the petroleum industry. A simplified schematic of a pyrolysis process such as the one developed by Pober et al. and the full-scale plant in Ames, Iowa is shown in Figure 3. Figure 4 shows a schematic of the pilot plant used to demonstrate the pyrolysis of MSW that generated the results in Table 2. Although a number of MSW-derived fuel systems are in operation or starting up, they are still developmental in regard to process, equipment, and application [14].

**Table 2  Final product for the pyrolysis of 1 ton of MSW [12]**

| Component | Amount |
| --- | --- |
| Char | 154–424 lb |
| Tar | 0.5–6 gal |
| Light oil | 1–4 gal |
| Liquor | 97–133 gal |
| Gas | 7.38–18 scf |

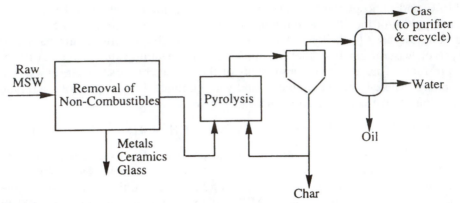

**Figure 3** Simplified process schematic for the pyrolysis of MSW.

## 11.3 ENERGY GENERATION FROM POLYMERIC WASTES

### 11.3.1 Introduction

A component of MSW is polymeric material. Although polymer waste accounts for only 8.5% of the total MSW disposed of in the United States, plastics represent over 28% by volume [15]. Polymeric wastes range from packaging materials in the food industry to various components in automobiles to high-density polyethylene (HDPE) for containers such as 2-liter soft drink bottles and laundry detergent bottles. In 1993 over 50% of all the food packaged in Europe for

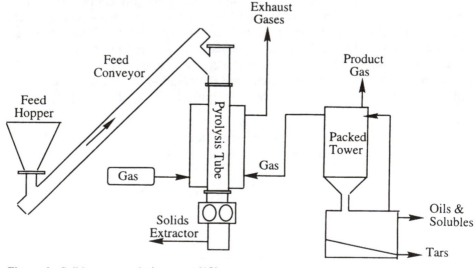

**Figure 4** Solid waste pyrolysis system [12].

distribution was packed using plastics [4]. It is estimated that over 65% of the food packaging in the United States is from plastics. As of 1978, the Ford Motor Company estimated that the plastic usage in car manufacture was expected to rise at a rate of 4% per year. At the 4% rate, each 1994 car should contain over 132 kg of plastic and rubber.

Within the next decade, the use of polymeric materials will continue to increase because of their versatility, functional values, and low energy requirements for production. Current technology enables over 200 million kg of plastic and rubber materials to be recovered from shredded automobiles [16]. The impetus to reuse the polymeric wastes instead of disposing of them in landfills is primarily due to the inability of polymers to degrade rapidly in landfills.

## 11.3.2 Mechanical Recycling

The manufacture of bottle containers from HDPE is perhaps the largest use for polymeric materials. Some of the more common items manufactured from HDPE are 2-liter soft drink bottles, juice containers, milk jugs, laundry detergent bottles, and Tupperware containers. As with other polymeric wastes, the ability to dispose of HDPE in landfills is diminishing. Austria introduced a regulation in October 1993 stating that over 90% of all HDPE containers must be recycled instead of placed in a landfill [17]. Germany has been the most aggressive in its demand for plastics recycling. In 1993 over 12% of all German municipalities were active in collecting 12,000 tons of polymeric containers [18, 19]. Officials have projected that this number will increase to collecting 80,000 tons of containers from 62% of the municipalities by 1996.

Other countries have jumped on the recycling bandwagon. The Netherlands plans to recycle 35% of all plastics and to recover energy by incinerating another 45% [20]. Italy currently recycles over 40% of its containers. Several states in the United States have also enforced mandatory recycling of HDPE. In order to ensure consumer involvement, states such as Michigan impose a deposit "tax" on all of their HDPE bottles. The consumer can recover the deposit only if the item is returned to designated stores or distributors. Other states have developed a recycling "lottery" [1]. Sanitary officials randomly pick an area and check for proper recycling of domestic waste. Homeowners who are complying with all recycling guidelines receive $200.

However, recycling alone is not the solution to polymeric waste. Research conducted at Dow Chemical Company has shown that more than 52% of all HDPE bottles would have to be recovered before mechanical recycling could save more energy than employing a waste-to-energy process such as incineration [21]. For this reason, scientists are striving to develop processes to generate fuel from polymeric wastes.

### 11.3.3 Waste-to-Energy Processes

Several technologies available are for the generation of energy, in either gaseous or liquid form, from polymeric waste. The technologies include pyrolysis, thermal cracking, catalytic cracking, and degradative extrusion followed by partial oxidation. Brief descriptions of these technologies are given below. It is important to note that these are not the only processes available. As technology advances from day to day, so do the processes used for the recovery of energy from wastes.

**Pyrolysis.** Pyrolysis of polymers is, by definition, the decomposition of plastic waste back into oil or gas using heating processes that are free of or deficient in oxygen. It is typically used for waste-to-energy generation because it has several advantages. Besides being a proven technology, pyrolysis has relatively good adaptability to fluctuations in the quality of the feedstock, is simple to operate, and is economical. In addition, refineries, which will utilize the chemically recycled polymers, already have pyrolysis units in operation. SPI's Council for Solid Waste Solutions comprises three of the leading petroleum companies: Amoco, Mobil, and Chevron. The council is already investigating the use of oil refineries equipped with pyrolysis units for the conversion of mixed plastics into hydrocarbons [22]. Ideally, these hydrocarbons would be identical to those split from petroleum oils.

Research conducted by Chambers et al. [16], implements the pyrolysis of polymeric wastes in the presence of molten salts. The molten salts are used to enhance the production of a particular desired oil-gas mix because of their excellent heat transfer properties. For instance, when a mixture of LiCl and KCl and 10% CuCl was used as the pyrolysis medium at 520°C, over 35% of the shredded polymer was converted to fuel oil [16].

Molten salts have also been successful at slightly lower operating temperatures. When the salts are used during a standard pyrolysis operation at 420°C, the gaseous fraction is minimized. This enables higher liquid and solid fractions to be recovered. Although the nature of the chemical reactions between the molten salts and polymeric waste is not yet completely understood, it appears that production of a particular product mix is optimized by the presence of the salts. Typically, the recovered fractions consist of light oils, aromatics, paraffin waxes, and monomers [23]. When pyrolysis utilizes molten salts, care must be taken to decrease the amount of corrosion and contamination to the pyrolysis chamber due to the salts.

In 1980 researchers at Germany's Federal Ministry of Research and Technology (FMRT) demonstrated the success of pyrolyzing polymeric wastes in a 10 kg/hr pilot plant. Figure 5 shows the simplified flow diagram of the laboratory test plant used in the development of the full-scale pilot plant. The main component of the FMRT pilot plant is a fluidized bed reactor with a space-time

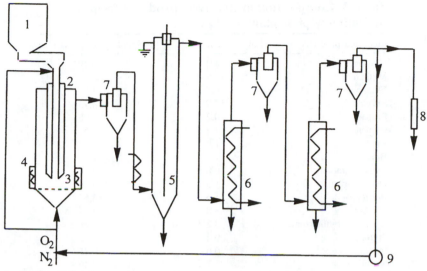

**Figure 5** Simplified flow diagram of the FMRT laboratory test plant [24]. (1) Feed hopper; (2) downpipe and cooling jacket; (3) fluidized bed reactor; (4) heater; (5) electrostatic precipitator; (6) intensive cooler; (7) cyclone; (8) gas sampler; (9) compressor.

ratio of 0.4 kg/hr/L. Preliminary experiments with the pilot plant enabled 40–60% of a polymeric feed to be recovered as a usable liquid product [24]. The major components of this liquid product were benzene, toluene, styrene, and $C_3$ and $C_4$ hydrocarbons. Table 3 shows all of the end products for two pyrolysis experiments. These experiments utilized spent polyethylene and syringes as the feed at pyrolysis temperatures of 810 and 720°C, respectively.

As of 1981, the pilot plant had accumulated over 600 hours of successful operation. The researchers at FMRT also demonstrated that the plant was self-sufficient in regard to energy needs. In other words, FMRT's facility was able to utilize one-half of the pyrolysis gas produced [24].

**Thermal Cracking.** Thermal cracking is similar to pyrolysis in that it is a high-temperature process. When polymeric wastes are cracked in an oxygen-free environment at temperatures above 480°C, a mixture of gas and liquid hydrocarbons is produced. At higher temperatures (650–760°C), more of the gaseous product is generated. Conversely, at lower temperatures, up to 85% of the product is a liquid hydrocarbon [25]. Both the gas and liquid forms of the converted mixed polymeric waste can be utilized as a feed stream by petroleum facilities.

**Catalytic Cracking.** An extrapolation of thermal cracking is catalytic cracking. The same operating principles apply; the primary difference is the addition of a catalyst to enhance the cracking process. A typical catalytic cracker consists of

**Table 3 Composition of pyrolysis products from preliminary pilot plant studies**

| Identified products | Polyethylene (wt. %) | Spent syringes (wt. %) |
|---|---|---|
| Hydrogen | 1.2 | 0.49 |
| Methane | 18.8 | 18.82 |
| Ethane | 6.2 | 7.75 |
| Ethylene | 17.9 | 13.73 |
| Propane | 0.2 | 0.08 |
| Propene | 7.2 | 10.67 |
| Butene | 1.0 | 3.32 |
| Butadiene | 1.5 | 1.39 |
| Cyclopentadiene | 0.8 | 2.79 |
| Other aliphatics | 1.3 | 3.46 |
| Benzene | 21.6 | 13.62 |
| Toluene | 3.8 | 3.84 |
| Xylene, ethylbenzene | 0.2 | Trace |
| Styrene | 0.4 | 0.43 |
| Indane, indene | 0.6 | 0.46 |
| Naphthalene | 3.7 | 2.46 |
| Methylnaphthalene | 0.6 | 0.92 |
| Diphenyl | 0.3 | 0.33 |
| Fluorene | 0.1 | 0.14 |
| Phenanthrene | 0.6 | 0.33 |
| Other aromatics | 0.7 | 1.15 |
| Carbon dioxide | 0.0 | Trace |
| Carbon monoxide | 0.0 | Trace |
| Water | 0.0 | Trace |
| Acetonitrile | 0.0 | Trace |
| Waxes, tars | 9.3 | 5.07 |
| Carbon residue, fillers | 1.8 | 5.80 |
| Balance | 99.8 | 97.05 |

two large reactor vessels, one to react the feed over the hot catalyst and the other to regenerate the spent catalyst by burning off carbon with air. Having two catalytic reactors enables the process to be run on a continuous basis. Figure 6 shows a simplified schematic of Mobil Oil's process for the generation of gasoline from polymeric wastes via catalytic cracking [26]. As the theory behind catalytic cracking is not new, the main focus of research is to determine the optimum catalyst for the cracking of the waste polymers.

Studies using organotin compounds have shown promise for generation of fuel from polyurethanes. However, the highest activity was exhibited when used for glycolysis, not cracking [27]. The use of chromium compounds has also been investigated. Scheirs and associates [28] have demonstrated the use of chromium to aid in the cracking and pyrolysis of HDPE. Although chromium compounds appear relatively successful, more research is needed on determining the opti-

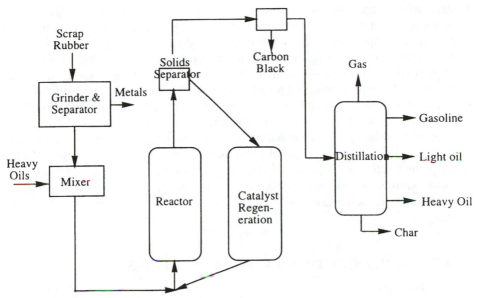

**Figure 6** Simplified schematic of Mobil Oil Corporation process for gasoline production from polymeric waste.

mum solubility of the catalyst for each polymeric compound. This is crucial because the catalyst's activity is directly proportional to the solubility of the polymer it is to be dissolved in.

The addition of platinum and iron over activated carbon has also been investigated. Specifically, the activity of the catalysts for the degradation of polypropylene waste into aromatic hydrocarbons was studied. The addition of metals increased the yield of aromatics from polypropylene. It has been speculated that the increase in activity is influenced by the methyl branching of the polypropylene. However, the exact mechanism is still not clearly understood.

Perhaps the most widely studied classification of catalysts for cracking operations is that of solid acid catalysts [29–31]. Specifically, HZSM-5, HY, and rare earth metal–exchanged Y-type (REY) zeolites and silica-alumina have been investigated [31]. Only HZSM-5 was found to be unsuitable for cracking polymeric wastes. HY, REY, and silica-alumina were all capable of producing at least 30% gasoline and 20% heavy oil. The differences between the catalysts arise in the production of the coke and gas fractions. Songip et al. [31] found that the incorporation of rare earth metals in HY zeolite increased the gasoline yield and decreased coke formation. Regardless of the catalyst used, catalytic cracking, due to lower temperature operation, appears to be more economically sound than pyrolysis.

**Degradative extrusion.** Degradative extrusion is not a new technology; however, its use as a tool for energy generation is. The basis of the technology is that at high temperatures, under the effect of shearing, it is possible to break down complex mixtures of plastics into homogeneous low-molecular-weight polymer melts [32]. These polymer melts could replace the heavy oils used in the production of synthesis gas. The polymer-derived heavy oil would be fed directly into a partial oxidation chamber (i.e., fluidized bed). The end product from the extruded waste at 800°C would be methanol, one of the primary feedstocks for the chemical industry [33]. Depending on the operating conditions, the methanol could be sent directly to a chemical plant as a feedstock without undergoing any subsequent treatment. However, if the gas undergoes low-temperature decomposition, carbon dioxide and hydrogen are the end products. Figure 7 shows a possible process schematic for the generation of methanol from polymeric wastes.

## 11.4 FUEL PRODUCTION FROM SPENT TIRES

### 11.4.1 Introduction

Scrap tire disposal has become a global problem of epidemic proportions. In 1977 the number of tires scrapped in Japan was 47 million [34]. More alarming is the fact that this number has doubled over the past 5 years. In addition, the United Kingdom currently scraps over 25 million tires per year [35]. Add the annual scrapping of 250 million passenger tires in the United States and the outlook becomes even grimmer [36].

Traditionally scrap tires have been disposed of in landfills; however, the acute shortage of viable landfills has all but eliminated this as a means of disposal. In fact, several of the midwestern states have issued laws that close landfills to tires [37]. The state-registered private collectors must dispose of the tires at approved legal dumps and recyclers. But even the number of legal landfills is dwindling. This has forced researchers to find an economical and efficient alternative for the spent tires.

Currently there are three key areas for marketing the spent tires. The first uses shredded tires as a "clean dirt" for road embankments and landfill liners [38]. Second is a rubber-modified asphalt. The asphalt, or tire crumb, can be used for playgrounds, running tracks, or as an ingredient of highway paving material. The third and most important area is that of tire-derived fuel (TDF) [39]. Although this may seem far-fetched, under proper conditions spent tires are a clean fuel with a 15% higher Btu value than coal [40]. In fact, it is estimated that by 1997 over 150 million scrap tires will be used for the generation of TDFs [41].

TDFs can be obtained by several methods. The first method is incineration [39, 42]. Britain's tire incinerator burns approximately 90,000 tons of rubber a

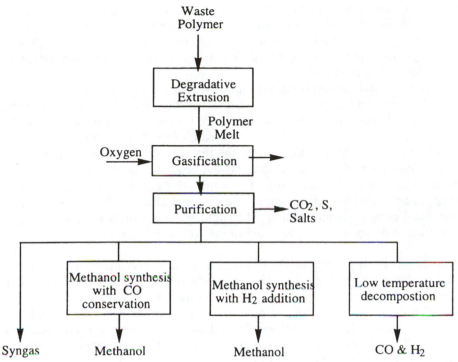

**Figure 7** Process scheme for the degradative extrusion and partial oxidation of polymeric wastes for methanol generation.

year. With this amount, the Wolverhampton facility will generate 25 MW of energy, which is enough to power a small town [43]. In the United States, Illinois Power incinerates shredded tire chips to supplement soft coal. The direct incineration of the chips will utilize approximately 15.6 million tires per year and will make up 2% of the total fuel consumed at the plant. Other processes, such as thermal cracking and depolymerization, recover the oil, char, and gases from the tires as separate product streams [44, 45]. However, the best-known method for the generation of TDFs is pyrolysis [36, 38, 46–49].

## 11.4.2 Pyrolysis

Recovery of energy from spent tires is not a new process. In 1974 the UK Department of Industry's Warren Spring Laboratory conducted the first tests to recover energy from spent passenger tires [47]. These initial experiments showed that it was possible to break the tires down into oil and gas by heating in a closed retort followed by distillation of the gaseous products. As research continued, it was found that the final bottoms product, or char, could be further treated for the manufacture of activated carbon (Figure 8) [50]. During the late-1980s, the use of pyrolysis for making TDFs became common. Several U.S. and European patents have been granted for the pyrolysis of scrap tires [51–54]. Although each patent is slightly different, the main operating principles and conditions are almost identical.

In each instance, the scrap tires are first reduced in size, either by grinding or pelletizing, and then sent to a clarifier for the removal of the scrap metal. The method presented by Williams et al. [55] represents the basic process utilized as the "starting block" by most research. The temperature is initially at 100°C when the rubber is loaded into the reactor. After 1 hour, the temperature is ramped to 300–500°C, depending on the process. The reactor is held at the final temperature for a minimum of 2 hours. The gas fraction is sent to a distillation apparatus for subsequent purification and analysis. The oil and char are separated using a second column. The slight variations on the pyrolysis process and their end products are presented in the following sections.

**Occidental flash pyrolysis.** Occidental Chemical's flash pyrolysis system was first demonstrated in late 1971. The process was able to produce a high-quality fuel oil at a moderate temperature and pressure. The advantage of the process was that the pyrolysis reaction was achieved without having to introduce hydrogen or a catalyst. The process was divided into three main sections: feed preparation, flash pyrolysis, and product collection. A simplified version of the overall process is given in Figure 9 [55, 56].

The most time-consuming aspect of Occidental's process was the feed preparation. During this stage, the tires were debeaded and shredded to approximately 3 in. A magnet was then used to remove all of the metal components. The remaining material was then shredded to 1 in. before being ground to −24 mesh. The grinding of the tires to such fine particles enabled the flash pyrolysis to occur at a higher rate. The quick vulcanization of the ground rubber enabled a shorter residence time, which in turn decreased the occurrence of product cracking [57].

After leaving the pyrolysis reactor, the gaseous stream was sent to a quench tower to separate the two end products. The product oil was collected and sent

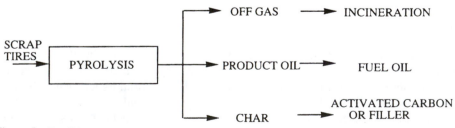

**Figure 8** Possible end products from the pyrolysis of scrap tires.

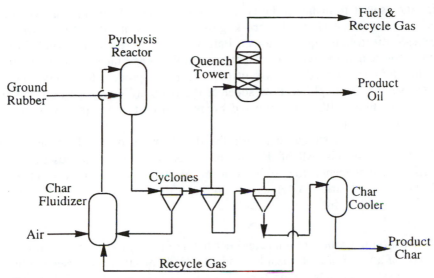

**Figure 9** Simplified schematic of Occidental's flash pyrolysis system.

to a storage facility. The recovered gas was recycled to the char fluidizer and pyrolysis reactor as a supplemental process fuel. The solid components that remained in the bottom of the pyrolysis reactor were sent through three cyclones. The cyclones were used to separate the solid particles by size, as well as to cool the material back down to room temperature. At the end of the process, a 35 wt. % carbon black was obtained. Analysis conducted on the carbon black showed that it had a high enough quality for direct reuse in the rubber industry [55].

**Fluidized thermal cracking.** The first commercial use for fluidized thermal cracking (FTC) dates back to the 1930s for coal gasification. Like other pyrolysis processes, FTC burns waste with high combustion; however, it has a few other advantages. Because it is a fluidized process, there is rapid mixing of solid particles, which enables a uniform temperature distribution; thus the operation can be simply controlled. Fluidization also enhances the heat and mass transfer rates, which in turn decreases the amount of CO emitted. Researchers have been able to adjust the FTC operating conditions to decrease the $SO_x$ and $N_x$ emissions [58].

The Nippon Zeon Company has conducted extensive studies on TDFs via thermal cracking. The precommercial process feeds the crushed tire chips to the fluidized bed using a screw feeder. Air heated by a preheating furnace is fed to the reactor bottom in order to elevate the reactor temperature near cracking (400–600°C) conditions before the chips are introduced. This enables the tire chips themselves to maintain the cracking temperature. A continuous cyclone is used to remove the char. The crack gases are brought into contact with the recovered oil from the quench tower. Part of the recovered oil is recycled to the quench tower and the remainder sent to an oil storage tank for further purification. The uncondensed gas is sent to a treatment process to remove the hydrogen sulfide. A simplified flow sheet of the Nippon Zeon plant in Tokuyama is shown in Figure 10. The estimated break-even cost of the pyrolysis plant is $0.25 per tire [59].

The preliminary studies used for the development of the Tokuyama plant had promising results. All of the end products produced could be either used directly as supplemental fuel sources at the plant or sent off-site for petroleum and chemical industries. A typical end product distribution of the FTC process is given in Table 4 [60].

**Carbonization.** Carbonization is another form of pyrolysis that can convert over one-half of a scrap tire into usable products. Carbonization processes operate at much higher temperatures than typical pyrolysis units. At these higher temperatures, the main product is char. The char is purified to carbon black, one of the main components in the manufacture of tires. In 1974 the cost of carbonization was approximately three times the cost of making carbon black from standard petroleum operations [61]. Although this number has decreased by only 15% in the past decade, the incentive of utilizing a waste material offsets the remaining cost [62].

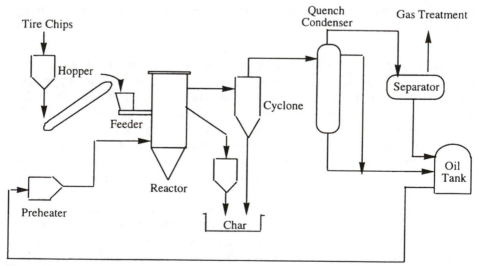

**Figure 10** Nippon Zeon scrap tire fluidized cracking process [60].

**Table 4 End product distribution
for tires cracked at 450°C**

| Product | Amount produced (kg) | Percent |
|---------|---------------------|---------|
| Oil | 257 | 52.0 |
| Char | 166 | 33.6 |
| Gas | 72 | 14.4 |
| Total | 495 | 100.0 |

## 11.4.3 Partial Oxidation via Supercritical Water Oxidation

Lee et al. [63] developed a novel process that can depolymerize spent tire ground particles into monomer quality materials. A U.S. patent for depolymerization of scrap tire materials was allowed in 1995. The process is unique in the sense that:

1. The final products are monomer quality materials that are superior to fuels, as they can be repolymerized or used as feedstocks.
2. The use of supercritical water allows the process to proceed an order of magnitude faster than pyrolysis reactions.
3. The proposed chemical mechanism of oxidative decoupling bond cleavage reactions is scientifically significant. The oxygen-deficient or partially oxidative environment created in a supercritical water system also adds a new dimension to the process technology.

They demonstrated the process feasibility using a 1-liter, semibatch (solid), continuous-flow (gas), supercritical water oxidation minipilot plant. Their reaction residence times vary from 1 to 3 minutes. Typical process conditions are quite severe: temperatures from 380 to 410°C, pressures from 220 to 250 atm. Fluids that exist beyond their critical temperatures and pressures are called supercritical fluids. Supercritical fluids normally exhibit extraordinary properties that are not easily conceivable at normal conditions. Such properties include liquid-like density, gas-like viscosity, low diffusivity, dielectric constant, and organic and inorganic solubility. In supercritical water, most polymers either dissolve into it or swell substantially. Moreover, oxygen is infinitely miscible in supercritical water, and near-homogeneous reactions take place.

Figure 11 shows a gas chromatogram of the effluent product gas at the 4-minute mark of a reaction. As can be readily seen, the isoprene monomer peak is the dominant one. The time scale on the abscissa is the GC elution time of molecules, not the process residence time. The figure also shows the fractionating nature of the process when applied to spent tires. An isoprene yield of 23.3% along with a yield of other hydrocarbons of 45.4% was obtained from

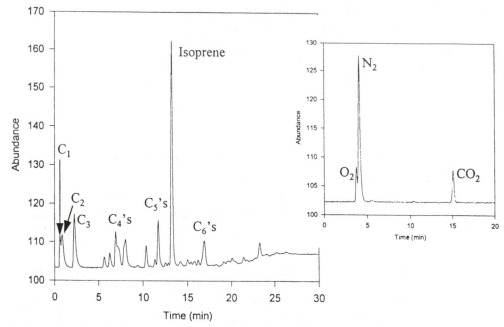

**Figure 11** Gas chromatogram of SCWO partial oxidation gaseous products at the 4-minute mark (FID on the left, TCD on the right: parallel and equal flow rates for both detectors).

their feasibility run. A schematic of the partial oxidative process is given in Figure 12.

### 11.4.4 Others

**IFP.** A relatively new approach to the decomposition of used tires is the IFP process, developed by the French Institute of Petroleum [35]. Unlike the other processes, IFP does not require pretreatment of the scrap tire. The whole tires are placed in a basket and lowered into a 600-liter reactor. Hot oil at 380°C is sprinkled onto the tire's surface. A chemical reaction between the hot oil and tires causes the depolymerization of the rubber. After the tires have been completely depolymerized, the reactor is cooled to 100°C. The off-gases are sent to a distillation column. The column separates the gas and light hydrocarbon fractions. The resultant gas is sent for further purification into a $C_4$–$C_6$ fraction and gasoline ($C_8$–$C_{10}$), and the light hydrocarbons are recycled to the reactor. The recycled hydrocarbon fraction is used to dilute the viscous fuel oil generated during the depolymerization process. Upon evacuating the reactor, the scrap metal is removed and sent off-site for salvage.

The IFP process is different in that it is a batch process [44]. Because it is operated batchwise, the concern over having enough tires for continuous feed is

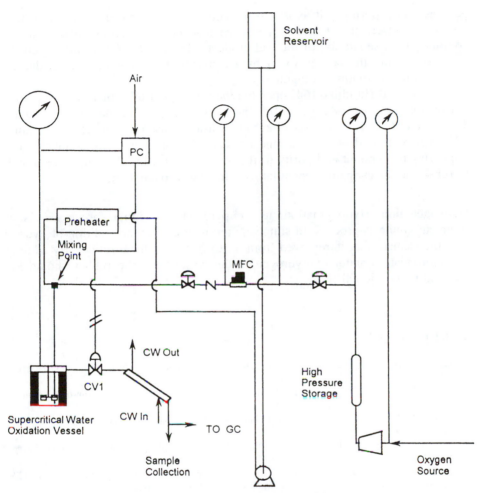

**Figure 12** Simplified schematic of supercritical water oxidation pilot plant.

eliminated. Depolymerizing 4000 ton/yr of waste tires has been estimated to generate over 15,200 tons of fuel oil, 360 tons of gasoline, 560 tons of gas, and 600 tons of metallic waste.

**Dry distillation.** Researchers have investigated the use of a specialty dry distillation apparatus to produce a gaseous fuel from spent tires. The process can be operated with either shredded or whole tires. The tires enter the top of the distillation tower via a conveyor belt. Once filled, the tower is sealed to prevent the gaseous product from escaping. Combustion is initiated by burners located at the bottom of the tower and then sustained by the introduction of process air. As combustion continues, the hot gas rises to the top of the tower. The rising gas has a dual purpose. It assists in the combustion of the tires located at the top of the tower, and it acts in the same manner as a traditional distillation column for the separation of the end products. The residue (char and tire cord) is removed from the bottom tower by a conveyor. Once cooled, the residue is separated into individual components.

Hiroshi and Haruhiko [64] operated the tower on a pilot plant scale continuously for 250 hours. At the end of that time it was found that approximately 40 wt. % of the tires was converted into a usable gaseous fuel. The remaining 60% was composed of the char and tire cord. Research is being conducted on improving the amount and purity of the recovered gas. In addition, Hiroshi and Haruhiko are investigating potential uses for the recovered char.

**Hydrogenation.** Hydrogenation, unlike pyrolysis or similar processes, is a chemical synthesis process. In simplistic terms, it entails the addition of hydrogen, the element that is removed from oil to make synthetic rubber. By adding the appropriate amount of hydrogen to the waste tires, the rubber should be returned to its original form.

## REFERENCES

1. Corbitt, R. A., ed. *Standard Handbook of Environmental Engineering.* New York: McGraw-Hill, 1990.
2. Fisher, P. M., and L. R. Evans, Jr. Whole tyre recycling: The Elm energy approach. *Rubber Europe Conference Proceedings*, No. 6.012, 1–6, June 1993.
3. Frederick, W. J., et al., Energy and materials from recycled paper sludge. *AIChE Summer National Meeting*, Denver, August 14–17, 1994.
4. Association of Plastics Manufacturers in Europe. Plastics: A vital ingredient for the food industry, RAPRA Technology, Ltd., 12–30, 1993.
5. Richards, M. K., USEPA's evaluation of a Texaco gasification technology. Proceedings: *ACS Emerging Technologies in Haz. Waste Management*, VI, Atlanta, September 19–21, 1994.
6. Stearns, R. P., Landfill methane: 23 sites are developing recovery programs. *Solid Wastes Management/Refuse Removal Journal*, 23(6):56–59, 1980.
7. Pober, K., and H. Bauer. From garbage—Oil. *Chem. Tech.*, 1(3):164–169, 1977.
8. Kagayama, M., M. Igarashi, M. Hasegawa, and J. Fukuda. Gasification of solid waste in dual fluidized-bed reactors. *ACS Symp. Ser. No. 130.* 38:525–540, 1980.
9. Henry, J. G., and G. W. Heinke. *Environmental Science & Engineering*, p. 560. Englewood Cliffs, NJ: Prentice Hall, 1989.
10. Green, A. E. S. Overview of fuel conversion. *FACT* 12:3–15, 1991.
11. Helt, J. E., and N. Mallya. Pyrolysis experiments with municipal solid waste components. *Proceedings of 23rd Intersociety Energy Conversion Engineering*, IEEE Service Society, Park Ridge, NJ, 1988.
12. Bell, P. R., and J. J. Varjavandi. Pyrolysis—resource from solid waste. *Proceedings: Waste Management, Control, Recovery and Reuse*, pp. 207–210. Ann Arbor, MI: Ann Arbor Science, 1974.
13. Schlesinger, M. D., et al. Pyrolysis of waste materials from urban and rural sources. *Proceedings: Third Mineral Waste Utilization Symposium*, pp. 423–428, Chicago, March 14–16, 1972.

14. Davis, M. L., and D. A. Cornwell. *Introduction to Environmental Engineering*, 2nd ed. New York: McGraw-Hill, 1991.
15. Scott, D. S., S. R. Czernik, J. Piskorz, J., and A. G. Radlein. Fast pyrolysis of plastic wastes. *Energy & Fuels*, 4:407–411, 1990.
16. Chambers, C., J. W. Larsen, W. Li, and B. Wiesen. Polymer waste reclamation by pyrolysis in molten salts. *Ind. Eng. Chem. Process Des. & Dev.* 23:648–654, 1984.
17. Producer responsibility for packaging waste in Austria, *ENDS Report* 224:18, 1993.
18. Baker, J. Unravelling the recycling targets. *Eur. Chem. News*, 60(1596):16–17, 1993.
19. Topfer extends plastics recycling deadlines. *Eur. Chem. News* 60(1595):25, 1993.
20. McMahon, P. Plastics reborn. *Chem. Eng.* 37–43, 1992.
21. Hunt, J. LCA endorses use of HDPE waste. *Packaging Week* 9(22):1, 1993.
22. Leaversuch, R. D. *Chem. recycling brings real versatility to solid-waste management. Mod. Plastics*, July: 40–43, 1991.
23. Bertolini, G. E., and J. Fontaine. Value recovery from plastics wastes by pyrolysis in molten salts. *Conservation and Recycling* 10(4):331–343, 1987.
24. Kaminsky, W. Pyrolysis of plastic waste & scrap tyres in a Fluid Bed Reactor. *Resource, Recovery, & Conservation* 5:205–216, 1980.
25. Romanow-Garcia, S. Plastics—Planning for the future. *Hydrocarbon Process.* 72(10):15, 1993.
26. Sittig, M. *Organic & polymer waste reclaiming encyclopedia*, p. 178. Denver, CO: Noyce Data Corp., 1981.
27. Vohwinke, F. Approach to recycling polyurethane waste. *Tin and Its Uses.* 149:7–10, 1986.
28. Scheirs, J., S. W. Bigger, and N. C. Billingham. Effect of chromium on the oxidative pyrolysis of gas-phase HDPE as determined by dynamic thermogravimetry. *Polym. Degradation and Stability* 38(2):139–145, 1992.
29. Venuto, P. B., and E. T. Habib, Jr. *Fluid Catalytic Cracking with Zeolite Catalysts.* New York: Marcel Dekker, 1979.
30. Hashimoto, K., et al. *New Developments in Zeolite Science and Technology*, pp. 505–510. Amsterdam: Elsevier, 1986.
31. Songip, A. R., T. Masuda, H. Kuwahara, and K. Hashimoto. Test to screen catalysts for reforming heavy oil from plastic wastes. *App. Catal B: Environmental* 2:153–164, 1993.
32. Menges, G. Basis and technology for plastics recycling. *Int. Poly. Sci. Technol.* 20(8):10–15, 1993.
33. Semel, G. Study of gasification of plastic waste, *UAPG Symposium*, Ser. No. 19, 9, 1991.
34. Kawakami, S., K. Inoue, H. Tanaka, and T. Sakai. Pyrolysis process for scrap tires. *ACS Symp. Ser. No. 130*, 40:557–572, 1980.
35. Used tyres: A crumb of comfort. *Recycling and Resource Management* 25–26, 1992.
36. English, D. Scrap tire problem could be gone by 2003. *Tire Business* 11(7):34–35, 1993.
37. Greenhut, S. Dealers, retreaders confront growing scrap tire problem. *NTDRA Dealer News* 49(1):36–40, 1986.
38. Clark, T. Scrap tyres: Energy for the asking, *British Plastics & Rubber* 35–37, March 1985.
39. McCarron, K. Maker of scrap-tire boilers sees bright future. *Tire Business* 11(17):14, 1993.
40. Sikora M. C., ed. Whitewall Cement fires kiln with scrap tires: Tires-to-fuel process becomes a reality at LaFarge Cement. *Scrap Tire News* 7(11):1–10, 1993.
41. Kokish, B. Organization seeks scrap tire solutions. *Rubber Plastics News* 23(6):44–46, 1993.
42. SSI: Lucas Furnaces—tyres disposal by incineration. *Tyres & Accessories* 11:68–69, 1992.
43. Pearce, F. Scrap tyres: A burning issue. *New Sci.* 140(1900):13–14, 1993.
44. Audibert, F., and J. P. Beaufils. Thermal depolymerization of waste tires by heavy oils: Conversion into fuels. Final Report EUR 8907 EN, Commission of European Communities: Energy, 1984.
45. Saeki, Y., and G. Suzuki. Fluidized thermal cracking processes for waste tires. *Rubber Age.* 108(2):33–40, 1976.

46. Braslaw, J., R. L. Gealar, and R. C. Wingfield, Jr. Hydrocarbon generation during the inert gas pyrolysis of automobile shredder waste. *Polym. Prep.* 24(2):434–435, 1983.
47. Reed, D. Tyre pyrolysis comes on stream. *Euro. Rubber J.* 166(6):29–33, 1984.
48. Earle, B. A. Dallas investors purchase tire pyrolysis plant. *Rubber Plastic News* 23(2):3, 1993.
49. Wyman, V. Turning a profit from old tyres. *The Engineer* 273(7066/7):36, 1991.
50. Jackson, D. V. Resource recovery. *Warren Spring Laboratory Report* No. C95/85, 1985.
51. Apffel, F. Recovery process. *U.S. Patent* 4,647,443, 1987.
52. Breu, R. A. Pyrolytic conversion system. *European Patent* 446,930,A1, 1991.
53. Grispin, C. W., Jr. Pyrolytic process and apparatus. *European Patent, No.* 162,802, 1984.
54. Roy, C., Vacuum pyrolysis of scrap tires. *U.S. Patent* 4,740,270, 1988.
55. Williams, P. T., S. Besler, and D. T. Taylor. The pyrolysis of scrap automotive tires: Influence of temperature and heating rate on product composition. *Fuel* 69:1474–1481, 1990.
56. Che, S. C., W. D. Deslate, and K. Duraiswamy. The Occidental flash pyrolysis process for recovering carbon black and oil from scrap rubber tires. *ASME*, 76-ENAs-42, 1976.
57. Nag, D. P., K. C. Nath, D. C. Mitra, and K. Raja. A laboratory study on the utilization of waste tire for the production of fuel oil and gas of high calorific value. *J. Mines, Metals, & Fuels*, 473–476, 1983.
58. Chang, Y. M., and M. Y. Chen. Industrial waste to energy by circulating fluidized bed combustion. *Resouces, Conservation, and Recycling* 9(4):281–294, 1993.
59. Kroschwitz, J. I., ed. *Encyclopedia of Polymer Science and Engineering*, vol. 14, pp. 787–794. New York: Wiley, 1988.
60. Saeki, Y., and G. Suzuki. Fluidized thermal cracking process for waste tires. *Rubber Age* 108(2): 33–40, 1976.
61. Kiefer, I. *U.S. EPA Report* No. SW-32c.1, 1974.
62. Jarrell, J. *International Patent* WO 93/12198, 1993.
63. Lee, S., F. O. Azzam, and B. S. Kocher. *U.S. Patent Appl. 08/105,881, allowed October 1995*.
64. Hiroshi, K., and A. Haruhiko. Process and apparatus for dry distillation of discarded rubber tires. *European Patent 0,072,387*, 1982.

## PROBLEMS

1. As for the MSW gasification, what are gasifiable?
2. Does the Texaco gasification of MSW require injection of air and steam?
3. What are the general chemical reactions taking place in MSW gasification?
4. What is contaminated soil? Where do we find such soils? What are the typical contaminants?
5. What is PAH? What is PNA?
6. What is involved in anaerobic digestion of solid wastes?
7. What are the advantages of liquid fuels over gaseous fuels?
8. List the following in order of its average heating value in terms of Btu/lb. Methane, Premium gasoline, Bituminous coal, MSW.
9. What are, in general, so-called recyclable polymers?
10. What are the typical products from pyrolysis of HDPE?
11. What are the advantages of using molten salts in pyrolysis?
12. What chemicals are typically used in molten salt process?
13. What is the difference between thermal cracking and pyrolysis?

14. Is pyrolysis an endothermic reaction? How about the partial oxidation reaction?
15. Why is catalytic cracking used instead of thermal cracking? What does the catalyst do in general senses?
16. What can be used as catalyst for catalytic cracking of polymeric materials?
17. Why is the catalyst regeneration important?
18. Explain briefly how the degradative extrusion works. What are the other unconventional uses of extruders?
19. What is the approximate heating value of spent tire? Is this higher or lower than that of bituminous coal?
20. Why does spent tires and their disposal cause environmental concerns?
21. How are the spent tire materials currently being used other than landfilling?
22. What are the concerns involved in incineration of spent tires? Why is incineration unpopular in the United States?
23. Why is char formed in various pyrolysis processes, in general?
24. What are the differences between carbonization and pyrolysis, in treatment of spent tires?
25. Define supercritical water. Discuss the properties of supercritical water.
26. Define supercritical water oxidation (SCWO).
27. What is Hastelloy C-276? What does this alloy possess as unique properties that are not generally expected from stainless steel?
28. What was the predominant hydrocarbon species, when spent tire samples were subjected to a controlled, supercritical partial oxidation?
29. Discuss briefly the IFP process.
30. What are the common difficulties faced with various tire treatment processes?

# INDEX